Cybersecurity Issues in Smart Grids and Future Power Systems

Cybersecurity Issues in Smart Grids and Future Power Systems

Editor

Arshad Arshad

MDPI

Basel • Beijing • Wuhan • Barcelona • Belgrade • Novi Sad • Cluj • Manchester

Editor
Arshad Arshad
Glasgow Caledonian University
Glasgow, UK

Editorial Office
MDPI
St. Alban-Anlage 66
4052 Basel, Switzerland

This is a reprint of articles from the Special Issue published online in the open access journal *Sensors* (ISSN 1424-8220) (available at: https://www.mdpi.com/journal/sensors/special_issues/Smart_Grids_Power_Systems).

For citation purposes, cite each article independently as indicated on the article page online and as indicated below:

Lastname, A.A.; Lastname, B.B. Article Title. *Journal Name* **Year**, *Volume Number*, Page Range.

ISBN 978-3-0365-9410-1 (Hbk)
ISBN 978-3-0365-9411-8 (PDF)
doi.org/10.3390/books978-3-0365-9411-8

Contents

About the Editor

Arshad Arshad

Dr. Arshad has over 11 years' experience of working in industry and academia in the UK, Pakistan and Canada. He is currently a Lecturer in Electrical Power Engineering at Glasgow Caledonian University where he led various courses on power system design, renewable power integration and high voltage engineering.

Arshad has a PhD in high voltage engineering from Glasgow Caledonian University and MSc in electrical power system from University of Strathclyde with distinction. Since completion of his PhD in 2017, he has been involved in various projects with Whittaker Engineering, University of Strathclyde, and University of Quebec Chicoutimi. While working for University of Strathclyde, Arshad was involved in the design, development and deployment of a novel data acquisition and instrumentation system for HVDC substations. Arshad has over 50 publications to his name in prestigious conferences and peer-reviewed journals.

In 2020, Arshad was recognized as Global Talent in Electrical Power Engineering by UK Research and Innovation (UKRI) for his critical contribution to the EPSRC project on DC Networks, Power Quality and Plant Reliability. He has been awarded grants by National ICT R&D Pakistan, Quebec Research Council Canada, Santander Universities Europe and IEEE USA.

sensors

MDPI

Review

Augmented Reality (AR) and Cyber-Security for Smart Cities—A Systematic Literature Review

Nouf M. Alzahrani [1] and Faisal Abdulaziz Alfouzan [2,*]

[1] Information Technology Department, Collage of Computer Science and Information Technology, Al Baha University, Al Bahah 65731, Saudi Arabia; noufalzahrani@bu.edu.sa

[2] Department of Forensic Sciences, College of Criminal Justice, Naif Arab University for Security Sciences (NAUSS), Riyadh 14812, Saudi Arabia

* Correspondence: falfouzan@nauss.edu.sa; Tel.: +966-(0)-11-246-3444 (ext. 1515)

Abstract: Augmented Reality (AR) and cyber-security technologies have existed for several decades, but their growth and progress in recent years have increased exponentially. The areas of application for these technologies are clearly heterogeneous, most especially in purchase and sales, production, tourism, education, as well as social interaction (games, entertainment, communication). Essentially, these technologies are recognized worldwide as some of the pillars of the new industrial revolution envisaged by the industry 4.0 international program, and are some of the leading technologies of the 21st century. The ability to provide users with required information about processes or procedures directly into the virtual environment is archetypally the fundamental factor in considering AR as an effective tool for different fields. However, the advancement in ICT has also brought about a variety of cybersecurity challenges, with a depth of evidence anticipating policy, architectural, design, and technical solutions in this very domain. The specific applications of AR and cybersecurity technologies have been described in detail in a variety of papers, which demonstrate their potential in diverse fields. In the context of smart cities, however, there is a dearth of sources describing their varied uses. Notably, a scholarly paper that consolidates research on AR and cybersecurity application in this context is markedly lacking. Therefore, this systematic review was designed to identify, describe, and synthesize research findings on the application of AR and cybersecurity for smart cities. The review study involves filtering information of their application in this setting from three key databases to answer the predefined research question. The keynote part of this paper provides an in-depth review of some of the most recent AR and cybersecurity applications for smart cities, emphasizing potential benefits, limitations, as well as open issues which could represent new challenges for the future. The main finding that we found is that there are five main categories of these applications for smart cities, which can be classified according to the main articles, such as tourism, monitoring, system management, education, and mobility. Compared with the general literature on smart cities, tourism, monitoring, and maintenance AR applications appear to attract more scholarly attention.

Keywords: augmented reality; cybersecurity; smart city; systematic literature review

Citation: Alzahrani, N.M.; Alfouzan, F.A. Augmented Reality (AR) and Cyber-Security for Smart Cities—A Systematic Literature Review. *Sensors* **2022**, *22*, 2792. https://doi.org/10.3390/s22072792

Academic Editor: Hyoungshick Kim

Received: 8 February 2022
Accepted: 30 March 2022
Published: 6 April 2022

1. Introduction

Rapid technological advancement has taken place in the last decade, propelled by developments and advances in information and communication technologies (ICT). This has revolutionized the way people communicate, live, work, and travel, among other things. As a result, smart cities have emerged, evolving towards intelligent, dynamic infrastructures that serve civilizations while accomplishing the criteria of sustainability and energy efficiency [1]. According to [2,3], the meaning of smart cities implies the integration of existing substructures with novel ICTs to produce an all-inclusive system of efficient urban services. Correspondingly, Ref. [4] note that a smart city is a conurbation that connects physical infrastructure, business infrastructure, social infrastructure, and information technology (IT) infrastructure to strengthen the city's collective intelligence. Although there exists no general consensus on the definition of a smart city, there is a unanimity within literature that one of the main goals of smart cities is to improve the quality and efficiency of city services, while at the same time making better use of public resources and reducing operational costs [5,6]

One of the new technologies that are widely deployed and exploited within smart cities is Augmented Reality (AR), which, as noted by [7], can enhance human-machine interaction. Immersive technologies, such as AR and virtual reality (VR), which either superimpose digital content into a physical world or immerse users into an altogether different, interactive, and digital environment, are shown to have exceptional and convincing uses in smart cities. As discussed by scholars such as [8–10], the availability of high-speed and consistent network connectivity has enabled AR technology to mature enough to impact cities in becoming smart, digital, and connected in various ways, including disaster response, enabling medical services, and navigation management. Related technologies such as VR have also enabled services and programs such as police training, education, and urban planning. In particular, Ref. [11] mention that AR can be described as a coincidental combination between virtual objects and the real world, which enables real-time interaction as well as three-dimensional virtual registration.

With human perception regarding the environment increasingly changing with modern technology, AR becomes progressively prominent. Ref. [11] further note that AR adds virtual information to a real environment, ultimately impacting user cognition. This augmentation of the virtual intangible information into the tangible world impacts how people live and interact in smart cities. The application of AR into smart cities offers a unique immersion into the Internet of Things (IoT) applications. Recent research suggests that this can guide an interactive demonstration of how public services such as street control [12] video surveillance [13] solid waste collection [14], and parking management [15] can be accomplished and controlled from a single platform, making cities safer, cleaner, and more livable. As suggested previously herein, the key objective of smart cities is to connect everything together and to people. Consequently, AR technology enables smart city inhabitants to have an instant and immersive connection with everything around them.

The use of smart technologies such as AR, however, raises new issues and challenges. In smart cities, the vulnerable action of users and organizations can put the entire city at risk of cybercrimes. Ref. [2] notes that due to the reliance of various components of smart cities on ICT, cyber-security challenges, including malicious cyber-attacks and leakage of sensitive information, may affect a smart city's behavior. When smart cities lose control of their technologically developed systems, it can negatively impact quality of life, people's security and privacy, the city's economy, technological axis, and even more, can put people at risk [16]. Given this, in order to respond to the increasing fervent acceptance of global smart city technologies such as AR, cyber-security programs must be developed in the same direction. Congruently, Ref. [17] argue that it is clear that cyber-threats for smart cities need to be taken extremely seriously. The researchers identify key areas for possible solutions, including the development of procedures and action plans for responding to cyber-attacks; the implementation of manual overrides and failsafes on all smart city systems; and the creation and use of security checks for encryption, authorization, authentication, as well as software updates while implementing new smart city systems.

In this work, an overview of AR applications in smart cities around the world is provided. The paper also investigates the topic of cyber-security for smart cities, demonstrating how specific aspects of smart cities give rise to cyber-security challenges in an augmented reality world. In other words, this article focuses on cyber-security risks that may affect data privacy and the outputs generated by the AR applications through the device, and how this could impact the implementation of AR technology in smart cities. The paper adopts a systematic literature review approach in order to achieve a comprehensive research synthesis on the basis of evaluation and analysis of literature under investigation in this study. In the end, it is believed that the findings of this study will have implications for academicians, practitioners, and societies at large. It also adds to the academic literature on the use of AR technology in the building of smart city infrastructure, where government officials and urban planners can use this technology to enhance cities' environments, infrastructure, services, and the quality of life of urban dwellers.

The study consists of six parts: Introduction, Smart City Concept, Methods, Results, Discussion, Conclusion and Future Work. The second chapter describes the concept, elements, and general structure of a smart city. The third chapter presents the methodology adopted in conducting the systematic literature review, including the search strategy and inclusion and exclusion criteria. The fourth chapter presents the results of the review, while the fifth presents the discussion. In the final chapter, the conclusion and future research are stated.

2. Smart City Concept

As noted by researchers such as [7], smart cities describe cities that use ICT systems to disseminate information to citizens, enhance operational efficiency, and ultimately improve the quality of public services. In their view, some key aspects of smart cities include smart education, smart governance, smart environment, smart homes, smart mobility, smart energy, and smart health. In terms of energy, Ref. [18] agree that smart cities can have the ability to control and monitor the amount of energy consumed and distributed using ICT technologies, which can enable cities to improve reliability and provide greater power quality and profit growth. When it comes to health, Ref. [19] note that the monitoring of people's health through IT technologies such as sensor devices help cities provide real-time information on patient health indicators (breathing, temperature, heartbeat), enabling faster and better decisions. Ref. [18] further note that environmental parameters such as air quality, humidity, and temperature are vital for smart cities. As such, ICT technologies such as AR can aid in the management of such parameters, with applications already successful in areas such as water and air quality and garbage management. In terms of traffic and mobility, Ref. [19] argue that efficient exploitation of IoT technologies can help in solving the problem of traffic congestion, as well as issues in existing transport infrastructure.

The infrastructure of a smart city can create a unique collaborative system where citizens, educational institutions, industries, prosumers, and researchers can develop innovative products, services, and solutions. Contrary to traditional double-sided marketplaces, Ref. [20] argue that the ecosystem that results from a smart city allows a multitude of actors to be engaged in private and public consumption, production, professional activities, entertainment, research, and education. As suggested herein and noted by [17], a smart city can be divided into six fundamental aspects: smart people, smart living, smart mobility, smart government, smart environment, and smart economy (see Table 1), with each element having its key indicators and benefits.

Table 1. Key Smart City Components (Alibasic et al., 2016).

Component	Indicators & Benefits
Smart People	Inclusive, creative, educationally excellent
Smart Living	Safety and health, happiness, culturally vibrant
Smart Mobility	Mixed modalities, clean and non-motorized options, ICT, connected
Smart Government	Open data, transparency, e-government application, ICT, supply and demand-side policies
Smart Economy	Local and global business interconnectedness, productivity, innovation, entrepreneurship

Presently, there are many existing urban projects that have been transformed to meet the aforementioned smart city criteria. Some projects have been developed from scratch, while others consist of the transition and modernization of existing cities into smart cities.

As noted by [1], some of the worldwide smart cities that have been developed from scratch include Skolkovo in Russia, PlaIT Valley in Portugal, Lavasa in India, Masdar City in Abu Dhabi, Meixi Lake in China, and Songdo in IBD. Some of the elements and features of such cities include the development of large tech-driven business centers, innovative education, cultural and medical services to citizens, sustainable urban infrastructure, green buildings, power supply through renewable energy, and collective broadband connectivity. Smart cities that have been developed from existing urban centers include London, Santander, and Portland. According to [1], some actions that have been undertaken by urban developers to make such cities smart include the installation of intelligent systems for waste collection and management, implementation of payment systems through smartphones, control and monitoring of public security through video monitoring technologies, implementation of smart devices for tracking and monitoring people's health, improving tourism through interactive mobile applications and immersive technologies such as AR and VR, and management and control of traffic through ICT technologies.

3. Methods

In order to achieve this study's objectives, a Systematic Literature Review (SLR) methodology was followed. As pointed out by [21], the SLR approach aims at searching, appraising, synthesizing, and analyzing all studies relevant to a specific field of research. Given the nature of this study, the SLR research approach was adopted to aid in planning, searching, screening, extracting data, and synthesizing and reporting findings.

3.1. Search Strategy

One of the fundamental goals of this study was to be as inclusive as practically possible. However, the idea was also to obtain scholarly papers within the last decade to ensure that materials gathered were the latest, and coincided with the digital age and era of smart cities. As such, all papers published in journals and conferences between 2010 and 2021, which included the phrases 'Augmented Reality for Smart Cities' and 'Cyber-security for Smart Cities', and related user studies were considered. In the planning phase, some of the most utilized online scientific databases were used to search for peer-reviewed literature. Three relevant literature databases were selected, including Emerald Insight (EI), Science Direct (SD), and IEEE Xplore (IX). As research sources, these multidisciplinary databases were selected and recognized for their indexing and coverage. They were consulted, and subsequently, the results obtained were cross-checked. Furthermore, the databases used in these academic studies have met the protocol requirements and used protocol-specific parameters.

3.2. Inclusion and Exclusion Criteria

The results from the search process were screened against pre-set inclusion and exclusion criteria, as mentioned in Table 2. In terms of inclusion criteria, only studies published between 2010 and 2021 were considered to ensure that interventions and applications related to AR and cyber-security for smart cities were relevant and up-to-date. Secondly, studies were only considered if they reported the application of AR and/or cyber-security for smart cities. As such, studies that reported the advantage, limitations, uses, challenges, effectiveness, and scope of AR and cyber-security in smart cities were considered for inclusion. Only studies that were published in the English language were considered for this review's inclusion for practical reasons.

Among the exclusion criteria, this study excluded studies that only reported on AR and cyber-security initiatives in other contexts. Studies that did not mention AR or cyber-security in smart cities were also excluded. The same applied to studies that claim to report AR but referred to VR or mixed reality (MR) instead. Studies not considered as peer-reviewed journals, book chapters, or conference papers in the context of AR and cyber-security in smart cities were excluded.

Table 2. Inclusion and Exclusion Criteria.

Inclusion Criteria	Exclusion Criteria
Addressed the application of AR and cyber-security for smart cites	Addressed related concepts that were not AR and cyber-security in smart cities
Published between 2010 and 2021	Published before 2010
Written in English	Non-English publications
Peer-reviewed journal, conference papers and related publications such as book chapters and seminal works	Unpublished studies (e.g., proposals, theses, and ongoing projects)
Original publications	Duplicates (by title or content)
Publications available online	Publications not available online

3.3. Search Results

After performing the search based on the identified keywords, as shows in Figure 1, 421 documents were initially identified from the three selected scientific databases. The initial search was done on 6 September 2021 and the last was done on 14 September the same year. An initial filter was conducted on the identified 421 studies among scientific articles, book chapters, and conference articles. This included evaluating the inclusion and exclusion criteria, considering the titles and abstracts of studies, and cross-checking the results of the three databases to remove duplicates. After this initial filter, 176 studies remained.

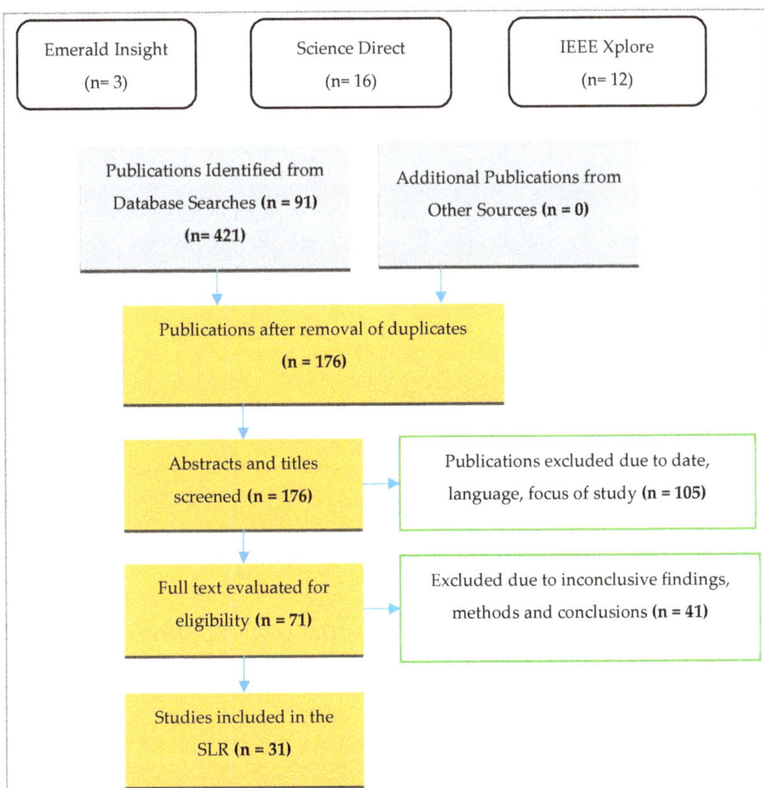

Figure 1. The Preferred Reporting items for SLR flow diagram.

After further screening titles, abstracts, introductions, and conclusions, a further 105 studies were removed. Finally, following full-text reading and examination of each of 71 articles, a total of 31 studies met the criteria proposed and definition for inclusion in this review.

3.4. Data Extraction

This phase was a two-part process in which each individual article was analyzed to determine final inclusion into the SLR. To minimize bias, this process was conducted by two individuals, each screening articles independently. The initial part involved skimming through the title and abstracts, as well as introductions and conclusions, to determine whether the obtained articles were relevant in addressing this study's research objective. The second stage involved skimming through and subsequently reading through full texts to obtain relevant data for SLR. In this stage, five QA criteria were developed to assess each study's quality.

- Q1: what was the application area (education, construction, tourism, design, or healthcare)?
- Q2: what was the type of data collected?
- Q3: what was the experimental design (methodology)?
- Q4: where was the research conducted (based)?
- Q5: what type of study was conducted (case study, field, experiment, formal, or pilot)?

In order to systematically and accurately record data based on these QAs, an excel spreadsheet was developed. During the review of the paper and data extraction, the researcher also flagged certain publications (especially in references) for additional discussion and cross-referencing. The data extraction form only contained information relevant to this study for analysis and descriptive purposes (in the discussion chapter) later in the SLR.

3.5. Synthesis

Since this SLR spans across a number of academic disciplines and fields, a thematic synthesis approach was adopted as the modulus of analysis. As discussed by [22], this approach in data synthesis is particularly suitable for analyzing and synthesizing multi-disciplinary datasets. To address the key aims of this study, common themes across the included studies were identified and analyzed in detail. The identified concepts of AR and cyber-security applications were seen as the starting point from which to introduce a common language to compare and contrast identified perspectives and findings in included studies. As the common concepts in this study, AR and cyber-security applications and interventions in smart cities were used to provide a common denominator for developing new themes from the included studies.

4. Results

The search process produced 31 studies that were included in the SLR. The ensuing sections in this chapter provide a summary of results according to the pre-stated data extraction criteria intended to fulfill this study's research objectives. One of the findings is that most of the studies related to AR and cyber-security for smart cities contexts are mostly published in Science Direct (16) and IEEE Xplore (12). The third database for this investigation, Emerald Insight, only produced three results; two relating to AR and one to cybersecurity, under the researched contexts. The majority of included studies were published in 2019 (14). Of the rest, three were published in 2014, three in 2016, five in 2017, seven in 2018, and nine in 2020. Based on this SLR, the number of publications increased from 2014 to 2020. This drastic increase may have been caused by technological advances and the development of smart cities. Between 2018 and 2020, there is particularly a consistent number of publications related to AR and cybersecurity for smart cities. As highlighted in the subsequent sub-sections, AR and cybersecurity papers for smart cities are divided into various applications and relating aspects.

4.1. AR for Smart Cities

Results under this segment indicate that the most discussed potential AR applications for smart cities are in various categories including, tourism, information dissemination, mobility, risk management, construction, education, energy management, and traffic monitoring. These application categories can be further grouped into five distinct classifications, including healthcare, robotics, public sector, tourism, (system) monitoring, (system) management, education, and mobility. Through this SLR, these categories have been critically compared based on aim, methodology, and chronological perspective. The results indicate that a number of AR application studies for smart cities focus on tourism and management/monitoring aspects comparing to other aspects.

4.1.1. Tourism

There is considerable consensus that AR allows for the simultaneous perception of the real environment and a virtual audio overlay. This is particularly vital in mobile applications, where users are continuously aware of their surroundings, such as in the case of urban smart and urban tourism, where travelers explore foreign sights and cities. Recent studies report that AR continues to become increasingly popular within the travel industry, especially because it enables attraction sites, hotels, and businesses operating in this industry to enhance physical environments while encouraging both local and international tourists to visit. As seen subsequently in the Discussion Chapter, previous studies in this area have shown that AR can be used in a variety of ways, including augmented reality gasification, beacon technology and push notifications, augmented reality destinations, and interactive hotel and attraction elements. However, AR application in tourism for smart cities is not well explored, and as such, publications relating to the same are limited. From database search, this study only obtained three studies: [23–25].

Ref. [23], in his paper on smart tourism, discusses issues, challenges, and opportunities presented to the industry by smart technologies such as AR. In this paper, the scholar clarifies that the smart concept signifies the integration of organizational networks and smart features such as AI, IoT, big data VR, and AR to enable automation, facilitate daily activities for users, and enrich the ecosystem and way of life. These latest technologies have given rise to the concepts of 'smart destination,' 'smart city,' and 'smart planet' that have become increasingly popular in recent times. Thanks to the visualization feature, the researcher argues that AR application, in particular, has enabled smart cities to increase the quality of tourist experience while also enhancing interaction with the physical world. During on-site travel experience, it is noted in this paper that AR can provide information about the destination, including image recognition platforms, multiple viewpoints of the destination, and landscape information that can be viewed in adaptive screens. Given the power of this technology, Ref. [23] notes that it can be applied in various areas, including accommodation enterprises, F&B businesses, and museums, among others.

Correspondingly, Ref. [25] believe that advances in AR are expected to further push the boundaries of what data can be collected and how they can be utilized to improve touristic experiences for smart cities. They argue that, in connection with the physical infrastructure (e.g., smart tourism and smart city), the focus is on blurring the lines between the physical and the digital, and fostering digital integration. In the context of tourism, smart technologies such as AR are essential in altering consumer experience and are fostering creative tourism business models. In connection with other technologies such as Social Networking Services (SNSs), VR, beacon technology, geo-tag services, location-based technologies, mobile apps, big data, and cloud computing, AR is enabling businesses in smart cities new ways of advertising, novel collaborative ventures, better tourist services, and better ways of managing tourist flows to innovate beyond traditional industry boundaries. Ref. [25] further note that smart tourism powered by these technologies allows travelers to better interact and communicate with and in cities, and to establish closer relationships, not only with residents but also with city attraction highlights, local government, and local businesses.

In the third paper for AR application in tourism for smart cities in this study, Ref. [24] explore an AR-based prototype in the pilot region of Gökova Mugla, Turkey. In their study, a mobile application prototype was developed using AR aimed at introducing sightseeing places, hotels, restaurants, touristic destinations, and other important centers for both domestic and international tourists. Key information such as price, social media data, comments, ratings, and intensity about these areas was simultaneously provided on the mobile application, while location data and image processing techniques were also used for implementing AR technology. The researchers suggest that a successful implementation of such an application can be important for smart tourism in a number of ways. First, social media integration can increase an attraction's recognition and popularity. Second, AR-based applications can remove the need to carry maps and brochures, allowing tourists to obtain real-time information conveniently. Third, such an application can provide current and simultaneous instant price, intensity, position, and evaluation information, while also allowing extra virtual images with actual images simultaneously.

4.1.2. System Monitoring

Performing monitoring and maintenance tasks in complex environments has long been shown to be challenging and difficult due to complexity, and possibly due to the use of multifaceted processes, heavy machinery, human factors, uneasy access, and underground facilities, among others. The underlying technology is often shown to be inefficient because of simultaneous supervision of people working together under extreme settings, missing multi-input interfaces, and significant delays in communication and data transmission. Recently, however, AR as a tool for information visualization is said to provide solutions to these issues, where real-time reports are essential for monitoring systems and aiding in decision making. This SLR identified two applications where this technology can be applied for smart cities. The first is the study of [26] that analyses the application of AR service for efficient information dissemination based on a deep learning algorithm for the smart city of Jeddah in Saudi Arabia. Their proposed framework uses the system architecture of the iMARS system that consists of user alerts, deep learning, databases, and municipality services. These entire components are correlated with each other to enable the application to provide real-time and prioritized information to users, while also aiding governmental departments and other organizations to prioritize services based on user needs.

In the second paper under this category, Ref. [8] explore the application of AR for traffic monitoring in smart cities. By integrating a number of modern technologies, including AR, cloud computing, machine learning, Internet of Vehicles, and IoT, the researchers intended to develop a smart traffic control system that can produce traffic updates as well as road status based on the strength of vehicles in smart cities. They used two types of sensors: roadside sensors to provide information about the condition of the road, and vehicle sensors to keep track of the entire information of the vehicle. VEINS (Vehicle in Network Simulator), which is an amalgam of network simulator OMNET++ and road traffic simulator SUMO, were used for analyzing traffic density according to different simulations. Their proposed Internet of Vehicle with AR for smart traffic monitoring model indicates that AR-based systems can effectively be used for smart transportation and real-time traffic monitoring.

4.1.3. System Management

This study identified four studies that examine how AR can be used in management systems in various areas, including rent management, emergency management, lighting system management, and energy management. In the first study under this set, Ref. [27] analyze how visualization by AR can be used for smart rent portals in smart cities. Their study proposes a recommender system that is managed and visualized through AR and Vuforia to provide a platform that allows users to hold out a preference-based cooperative filtering search on rental properties. This recommender system is shown to enable rental service seekers to refine their preferences based on shallow learning. This study is grounded on the premise that locating services or products online that meet every users' preference is more and more troublesome, owing to the massive pool of decisions to think about before inbound at the required choices. With a management system powered by technologies such as AR, the researchers conclude that users in smart cities can now be able to visualize products and services (such as rental properties) to enable more informed decisions than traditional frameworks and algorithms.

Similarly, Ref. [28] illustrate how the same can be applied for emergency management in smart cities. In their review of relevant literature through the Web of Science Core Collection and snowballing, the scholars note that immersive technologies such as AR and VR can be applied to emergency management systems to allow for better response to emergency situations that may arise in cities. They provide an instance where AR can be used combined with databases to explore how to efficiently maintain fire safety equipment and cope with related fire hazards. They argue that AR can serve as an effective approach through which disaster scenes can be reproduced in a way that information can be extracted from the virtual setting for safety assessment, and research results can be applied in real settings. Given the potential this technology has, Ref. [28] further note that it can be used in emergency simulation systems to study potential dangers in specific situations such as potential coastal flooding, evacuation safety, and productivity in manufacturing plants, oxygen deficiency in the steel industry, as well as accidental perception in construction sites.

A wireless, AR-powered LED streetlight system with centralized and remote-control technology has also emerged as an innovative application for smart cities. In their study, Ref. [29] illustrate how such a lighting management system with remote control capability can provide numerous benefits for smart cities, including monitoring and real-time control capabilities, as well as reduced energy consumption and operational costs. In their study, they present two forms of applications powered by modern technologies such as AR and IoT to aid in lighting management for a learning institution. The first is a mobile application that can generate the safest walking paths on campus by integrating streetlights with various pedestrian-counting video sensors. The second is an emergency response aid application that integrates streetlights with on-campus 911 emergency buttons. Both applications were designed specifically with the goal of public safety improvement. By integrating visualization and intelligence into such systems, the scholars suggest that this can enable controlling of light brightness in real-time based on environmental dynamics to lower power consumption, while also managing the schedule of light systems. Maintenance of the streetlights can also be done efficiently by relying on a network system with sensors that provide usage statistics and operational states.

The search results also indicated that AR could be used in energy management for human-computer interaction in a smart city. Traditional energy management systems are often based on graphical and numerical values through monitor screens. Ref. [30] suggest that such systems are prone to errors and are often very difficult for consumers to interpret. As such, the researchers propose a novel, AR-enabled interface that can easily be accessed and interpreted by users in smart cities. Through a diorama that can visualize energy data intuitively, Ref. [30] illustrate how this new EMS system can enable users to check operation, renewable energy production, energy consumption, as well as environmental information of the zone where the systems are installed through Augmented Reality Interface (ARI). Such information can then be used for improving the indoor environment (maintenance) and reducing unnecessary energy consumption.

4.1.4. Education and Instruction

The combination of educational content with AR technology to create a new type of automated applications and enhance the effectiveness and attractiveness of teaching, learning, and instruction has been explored widely in recent times. However, this exploration is yet to attract considerable attention under the smart city context. In fact, this study's SLR only retrieved two studies from the selected databases that met the inclusion criteria. The first was that of [9] who attempted to explore an AR application for smart campus urbanization using a Mugla Sitki Kocman University (MSKU) campus prototype.

Although their study falls under the education and instruction category in this study, their application was not designed to aid in learning or instruction. Rather, they suggested a platform that can offer more technical services to users in the university, improve campus technological infrastructure, increase the interaction between the campus and students, improve training services and presentation of technological learning environment for the learner, and enable easier identification of buildings and other locations in the campus.

In the second paper under this category, Ref. [31] examine how AR can be used to present work instructions for factory employees in order to lighten the workload of employees (particularly those in assembly, maintenance, and set up) and boost factory efficiency. The researchers developed a new smart solution for designing and presenting work instructions through AR, including virtual instructions on the screen (with or without special controllers, or with or without in-situ projections), video instructions, and traditional 'paper' instructions. These are designed as a software system for developing and working with virtual assembly instructions. Ref. [31] designed and tested this software capable of creating 3D animated assembly instructions. The results suggest that this software can provide easy and efficient ways for both technical and non-technical employees to receive instruction to improve their work efficiency and productivity.

4.1.5. Mobility

Mobility, although examined limitedly, is another area of application of AR in smart cities. This study only produced one study that explores this application in the smart city concept. Ref. [32] investigated how AR and IoT can be used to improve the accessibility of people with motor disabilities in smart cities. The researchers designed a system that enabled wheelchair users to interact with items placed beyond their arm's length with the help of Radio Frequency Identification (RFID) and AR. This application was an interactive AR that ran on different interfaces to enable users to digitally interact with physical items thanks to the updated inventory by an RFID system. The result of the study suggests that an AR-powered system can not only enable disabled people with wheelchairs to interact with the physical world, it can also enable them to visit and experience various site activities in an autonomous way.

4.2. Cyber Security for Smart Cities

The implementation of technologies such as AR and IoT in smart cities is normally hailed as the solution to numerous urban problems such as environmental protection, waste and energy management, and transportation. However, the implementation of these technologies is seen to be the source of numerous cybersecurity issues. Scholars agree that as smart cities become increasingly interconnected and the level of digital infrastructure becomes more complex, such cities will become more vulnerable to cyber-attacks. This SLR produces 17 publications that address the topic of cybersecurity for smart cities, with the majority of articles published in the Science Direct database. After reviewing the articles, the researcher classified the results under two main categories: challenges (issues) and opportunities (solutions).

4.2.1. Challenges/Issues

One of the key papers under this category is from the Oxford Analytica Daily Brief that analyses industrial [33], business, social, economic, and geopolitical developments on a global and regional basis, providing various sectors with a timely and authoritative analysis of various issues and solutions. According to this brief, the cyber security of smart cities will be critical in harnessing the cost and efficiency gains brought about by increased connectivity. It is believed that systems that analyze and interpret consumer behavior in these cities will continue to present unique challenges of security and privacy, as they create attractive targets for malicious actors, intelligence agencies, and repressive regimes. This expert briefing further states that cybersecurity protections will be a point of differentiation among technology vendors; citizens will continue to bear ultimate responsibility for what data they share and how they are used, and innovation among technology vendors will be inhibited so long as they see a risk of liability when collaborating with cities. Similar challenges have also been reported in [34–37].

In a systematic literature review, Ref. [2] also identified a number of security threats and problems that may affect smart cities—first, eavesdropping on information sent from sensors and equipment, giving cyber-security attackers sufficient knowledge to undertake malicious acts. Second, distortion of messages sent to subsystems. As a result, incorrect messages are sent to various devices, and the normal operation of systems is disturbed. Third, attackers are also able to delay the messaging and communication between systems to have the effect of denial of service. Some of the messages exchanged in the transmission and distribution systems may be time-critical and must be transmitted within a short period of time. When this does not happen, major negative repercussions may follow.

Reference [38] provide a thorough insight into the smart city threat landscape for components related to acquisition and storage of smart city data emerging from components such as smart vehicles, unnamed aerial vehicles, building automation systems, and smart grids, along with enabling technologies such as cloud computing and IoT sensors. From the cloud, they identify a variety of cybersecurity issues, including system and application vulnerabilities, malware injection attacks, denial of service (DoS), malicious insider threats, and data leakage. From IoT sensors, a number of issues can also materialize, such as remote exploitation, sensor failure, data storage, and management problems, insecure communication, and confidentiality leaks. With smart grids, problems such as the attack on internet-connected devices, rogue/infected devices, eavesdropping, privacy, and protocol vulnerability can also occur.

Reference [38] also identify a variety of physical threats from such systems, such as introducing data glitches to gain unauthorized access to debug interfaces, side-channel attacks to leak information, or fault-injection into the ECU to defeat central locking systems. Other scholars including [39,40] also report similar cybersecurity challenges for smart cities, but further address other issues such as man-in-the-middle attack, phishing, and spoofing. The man-in-the-middle attack is when cyber criminals intercept communication channels to manipulate transmitted data and falsified operators' actions. Spooning is when they duplicate data by a third malicious party and send it to the reader after revealing the security protocol. Lastly, phishing is when criminals impersonate trusted and reputable parties to gain critical information such as credit card numbers and passwords.

4.2.2. Opportunities/Solutions

Although the exact shape that smart cities will finally take remains to be indeterminate, scholars agree that there needs to be a variety of precautions, interventions, and solutions to cyber threats to guarantee a smoother implementation process and, ultimately, more secure infrastructure. Some studies included in this SLR provide some solutions and applications that need to be implemented for smart cities to provide some form of security against cyber-criminal activities. For instance, Ref. [41] believe that the current cybersecurity developments for smart cities cannot keep up with the eager adoption of advanced technologies, so there need to be corresponding measures that avert associated cyber threats. In their study, they propose designing correct preventive measures based on deep learning methods to ensure that technological systems in smart cities are robust and well protected. Their paper provides a summary of the knowledge and interpretation of deep learning, cyber security, and smart city concepts, as well as discusses existing related work on IoT security in this setting. Specifically, they review a number of deep learning models that can enhance the security of these cities, including Boltzmann machines, generative adversarial networks, convolutional neural networks, recurrent neural networks, and deep belief networks.

In their paper, Ref. [42] propose investigating the security concern of the smart city's infrastructure and taking into account the views of both technological and business operations before building a preventive framework. They state that it is vital to analyze the threats before developing safe data. The researchers put forth a Hybrid Smart City Cyber

Security Architecture (HSCCA) framework to enhance cybersecurity for smart cities. In this model, they consider key factors such as vulnerable data collection, recovery, memory storage, and well-organized network source supply. To address the key security challenges, their model first highlights and analyzes these problems along with aspects of risks associated with them, and then provides recommendations for solutions and prevention. A similar approach is suggested by ([43,44], and in particular [45]), who design a threat 'hunting' model based on Sparse Representation based Classifier (SRC). Their framework identifies cyber threats by Opcode, Bytecode, and system call views to provide suggestions on ways of addressing and preventing such threats.

Reference [46] suggests that, other than focusing on the technical part of the solution, governments should strengthen policies relating to cybersecurity in smart cities. In this paper, the researcher analyzes the impact of China's 2016 cybersecurity on foreign technology organizations and China's big data and smart city dreams. He suggests that to reduce threats, cybersecurity and informatization should be seen as 'two wings, one body' and must be planned together, arranged together, and moved forward together. Implementation of technologies in smart cities must therefore be integrated with security measures that are built for the long-term. Similar strategies have also been suggested by a number of other scholars, including [47,48] who agree that there should be set standards and avert strategies that govern the implementation of technologies in smart cities to prevent the stated security threats. Such studies also highlight the role of third-party risk management and security ownership in such contexts.

5. Discussion

In the context of smart city initiatives in different parts of the world, this SLR reveals that upcoming digital technologies such as AR play a fundamental role in the development of various systems and infrastructures. Indeed, several studies have already been conducted in this area with a keen focus on the application of AR in different areas, including tourism, management, monitoring, education, shopping, transportation, marketing, interior design, and smart parking. Although this exploration remains limited for smart cities, the study identified five main areas where recent scholars have investigated the application of AR for smart cities, including tourism, system monitoring, system management, education and instruction, as well as mobility (aid in movement). Under the tourism category, scholars such as [23–25] highlighted some of the areas of AR application for smart cities. Some of these include: smart tourism, smart experience, smart destination, smart trade, geographical information systems, destination managers and marketers, destination image formation, image recognition platforms, multiple viewpoints of the environment, landscape information, scene discovery by image detection, GPS and radar implementation, and prototype screens. Their analyses of these applications are consistent with smart tourism literature in other contexts.

For example, in exploring the value of AR for tourism, Ref. [49] found five value dimensions for AR in tourism, including marketing, organizational, epistemic, touristic, and economical. Ref. [50] also took an internal stakeholder perspective to examine the role of AR in tourism, and found that this technology adds value in this area by modernizing the existing offerings, making it more attractive for new markets. Ref. [51] also suggested the use of AR as robotic tour guides to augment multimedia elements such as 3D objects, sound clips, and video clips to real artifacts in museums. Similarly, Ref. [52] recommended the use of AR games for pre-historical places to enhance tourist interaction. This suggests that AR can be a vital addition for smart cities intending to revamp their tourism sector to attract both domestic and international tourists.

The SLR also discovered that AR could be effectively used for system management and monitoring. Studies included in this research, including [8,26–30], which illustrated how AR could be applied in areas such as real-time information dissemination, smart traffic monitoring, rent management, emergency management, lighting system management, and energy management to eliminate bottlenecks associated in such systems, and in turn, improve efficiency and productivity while reducing costs. Similar studies have also been conducted in other contexts to complement studies included in this SLR. For instance, Ref. [53] found that AR can successfully be used in system management and monitoring in construction, particularly in quality and defects management, time and cost management, safety monitoring and management, worker training, process tracking, and project scheduling. Other researchers that have explored the successful implementation of AR in these areas include [54] (traffic monitoring), Ref. [55] (facility management), and [56] (emergency management), among others.

Education and training are other areas where AR can maximize the potential for smart cities. In this study's SLR, Refs. [9,31] highlighted how this technology could effectively be used for campus urbanization and assembly-line instruction. Although this area is investigated sparingly in the smart city context, it is one of the most explored in other settings. Researchers such as [57–59] have all explored opportunities, challenges, and provided recommendations for the use of AR in education and training. Most of these academic sources agree that the capacity to overlay rich media through AR onto the real world for viewing via web-enabled devices such as smartphones and tablets means that instruction can be made available to learners at the exact time and place of need. This has the benefit of reducing cognitive overload, improving concentration, and making education interesting.

The last category, mobility, is also examined and reported by researchers studying the application of AR in differing frameworks. In the included study for this SLR, Ref. [32] revealed that this digital tech could be used by smart cities to enhance mobility for people with motor disabilities. In line with this, other scholars have also suggested that AR can be used to improve both indoor and outdoor mobility for different people. For example, Ref. [60] recommended an AR indoor positioning system that can be used to track user location and their angles of vision, thereby creating a high adaptability space. Ref. [61] also proposed an indoor navigation system to be used in libraries to help users navigate to shelves and find books. Consistently, Ref. [62] suggested a new AR indoor navigation system for a wheeled robot that can be used in shopping malls and museums. On the other hand, Ref. [63] presented a novel tracking system that can be used for outdoor spaces using beacons. Ref. [64] also proposed a navigation system that can caution drivers about unseen obstacles, and suggested a visual/audio information model powered by AR. For smart cities, such suggestions provide groundwork from which AR-based systems can be implemented to guide navigation and mobility for people.

While technologies such as AR hold great promise for smart cities, the SLR also revealed great concern relating to cybersecurity. Included studies suggest that the interconnectedness of smart cities through advanced technologies gives room for cyber threats. It is suggested that lack of protection against these threats, poor understanding of social engineering, weaponized machine learning technologies by cyber-attackers, non-existent secure device onboarding services, poor encryption key management, as well as lack of cryptographic measures are some of the key issues that contribute to the intensification of cyber threats in smart city ecosystems. Studies included in the SLR, including [34–38], suggest that some of the key challenges in these systems include system and application vulnerabilities, privacy invasion, malware injection attacks, denial of service (DoS), malicious insider threats, and data leakage. Away from smart cities, similar challenges have also been reported by researchers such as [65–68]. Although solutions to such challenges have been suggested, both technical and non-technical, literature suggests that there is no unanimously agreed solution, and cyber threats will continue to emerge as technologies advance.

6. Conclusions and Future Work

This SLR intended to address the study's main topic: AR and cybersecurity for smart cities. In other words, it intended to investigate the application of AR in this context, and the cybersecurity issues that may arise as a result of the adoption of digital technologies. A review of the final 31 articles was then provided. Based on this review, the application of AR for smart cities can be categorized into five main classifications, including tourism, system monitoring, system management, education and instruction, and mobility (navigation). Tourism, monitoring, and maintenance AR application appear to attract more scholarly attention than education and mobility for smart cities, compared to the general literature on the topic. Regardless of attention, it is believed that a close connection between industry and academic fields is going to be connected. The adoption of these technologies and continued interconnectedness in smart cities is shown to give rise to a host of cybersecurity threats, with among them loss of privacy and confidentiality, physical threats, systems and applications vulnerability, malware injection attacks, denial of service (DoS), malicious insider threats, and data leakage. Solutions to these problems have been suggested by various researchers, but they remain inconsistent and widely unproven. Until now, the majority of studies have attempted to prove concepts rather than describe well-established analytical approaches. In the future, the need for more practical- and analytically based studies is emphasized in order to evaluate discussed hypotheses from this SLR in real smart city contexts.

Author Contributions: Conceptualization, N.M.A.; methodology, N.M.A. and F.A.A.; software, N.M.A.; validation, N.M.A. and F.A.A.; formal analysis, N.M.A.; investigation, N.M.A. and F.A.A.; resources, N.M.A. and F.A.A.; data curation, N.M.A. and F.A.A.; writing—original draft preparation, N.M.A.; writing—review and editing, F.A.A.; visualization, N.M.A. and F.A.A. All authors have read and agreed to the published version of the manuscript.

Funding: This research received no external funding.

Data Availability Statement: Not applicable.

Conflicts of Interest: The authors declare no conflict of interest.

References

1. Pellicer, S.; Santa, G.; Bleda, A.L.; Maestre, R.; Jara, A.; Skarmeta, A.G. A Global Perspective of Smart Cities: A Survey. In Proceedings of the 2013 Seventh International Conference on Innovative Mobile and Internet Services in Ubiquitous Computing, Taichung, Taiwan, 3–5 July 2013; pp. 439–444.
2. Ma, C. Smart city and cyber-security; technologies used, leading challenges and future recommendations. *Energy Rep.* **2021**, *7*, 7999–8012. [CrossRef]
3. Maruf, M.H.; ul Haq, M.A.; Dey, S.K.; Al Mansur, A.; Shihavuddin, A.S.M. Adaptation for sustainable implementation of Smart Grid in developing countries like Bangladesh. *Energy Rep.* **2020**, *6*, 2520–2530. [CrossRef]
4. Kumar, S.; Prasad, J.; Samikannu, R. Barriers to implementation of smart grids and virtual power plant in subsaharan region—focus Botswana. *Energy Rep.* **2018**, *4*, 119–128. [CrossRef]
5. Chaves-Diéguez, D.; Pellitero-Rivero, A.; García-Coego, D.; González-Castaño, F.J.; Rodríguez-Hernández, P.S.; Piñeiro-Gómez, Ó.; Gil-Castiñeira, F.; Costa-Montenegro, E. Providing IoT Services in Smart Cities through Dynamic Augmented Reality Markers. *Sensors* **2015**, *15*, 16083–16104. [CrossRef] [PubMed]
6. Zanella, A.; Bui, N.; Castellani, A.; Vangelista, L.; Zorzi, M. Internet of things for smart cities. *IEEE Int. Things J.* **2014**, *1*, 22–32. [CrossRef]
7. Badouch, A.; Krit, S.D.; Kabrane, M.; Karimi, K. Augmented Reality services implemented within Smart Cities, based on an Internet of Things Infrastructure, Concepts and Challenges: An overview. In Proceedings of the Fourth International Conference on Engineering & MIS, Istanbul, Turkey, 19–20 June 2018; pp. 1–4.
8. Dey, M.R.; Sharma, S.; Shit, R.C.; Meher, C.P.; Pati, H.K. Iov based real-time smart traffic monitoring system for smart cities using augmented reality. In Proceedings of the 2019 International Conference on Vision Towards Emerging Trends in Communication and Networking (ViTECoN), Tamilnadu, India, 30–31 March 2019; pp. 1–6.
9. Ozcan, U.; Arslan, A.; Ilkyaz, M.; Karaarslan, E. An augmented reality application for smart campus urbanization: MSKU campus prototype. In Proceedings of the IEEE 2017 5th International Istanbul Smart Grid and Cities Congress and Fair (ICSG), Istanbul, Turkey, 19–21 April 2017; pp. 100–104.
10. Wang, X.; Kim, M.J.; Love, P.; Kang, S.-C. Augmented Reality in built environment: Classification and implications for future research. *Autom. Constr.* **2013**, *32*, 1–13. [CrossRef]
11. Kaji, S.; Kolivand, H.; Madani, R.; Salehinia, M.; Shafaie, M. Augmented reality in smart cities: Applications and limitations. *J. Eng. Technol.* **2018**, *6*, 28–45.

12. Medenica, Z.; Kun, A.L.; Paek, T.; Palinko, O. Augmented reality vs. street views: A driving simulator study comparing two emerging navigation aids. In Proceedings of the 13th International Conference on Human Computer Interaction with Mobile Devices and Services, Stockholm, Sweden, 30 August – 2 September 2011; pp. 265–274.
13. Milosavljević, A.; Rančić, D.; Dimitrijević, A.; Predić, B.; Mihajlović, V. Integration of GIS and video surveillance. *Int. J. Geogr. Inf. Sci.* **2016**, *30*, 2089–2107. [CrossRef]
14. Somayaji, S.R.K.; Kaliyaperumal, S.; Velayutham, V. Managing and Monitoring E-Waste Using Augmented Reality in India. In Proceedings of the International Conference on Sustainable Communication Networks and Application, Erode, India, 30–31 July 2019; Springer: Cham, Switzerland, 2019; pp. 32–37.
15. Moguel, E.; Preciado, M.Á.; Preciado, J.C. A smart parking campus: An example of integrating different parking sensing solutions into a single scalable system. *Smart Cities* **2014**, *98*, 29–30.
16. Andrade, R.O.; Yoo, S.G.; Tello-Oquendo, L.; Ortiz-Garces, I. A Comprehensive Study of the IoT Cybersecurity in Smart Cities. *IEEE Access* **2020**, *8*, 228922–228941. [CrossRef]
17. Alibasic, A.; Al Junaibi, R.; Aung, Z.; Woon, W.L.; Omar, M.A. Cybersecurity for smart cities: A brief review. In Proceedings of the International Workshop on Data Analytics for Renewable Energy Integration, Riva del Garda, Italy, 23 September 2016; Springer: Cham, Switzerland, 2016; pp. 22–30.
18. Dhumale, R.B.; Thombare, N.D.; Bangare, P.M.; Gawali, C. *Internet of Things for Smart City*; Sinhgad College of Engineering: Pune, India, 2017; Volume 4.
19. Talari, S.; Shafie-Khah, M.; Siano, P.; Loia, V.; Tommasetti, A.; Catalão, J.P.S. A Review of Smart Cities Based on the Internet of Things Concept. *Energies* **2017**, *10*, 421. [CrossRef]
20. Appio, F.P.; Lima, M.; Paroutis, S. Understanding Smart Cities: Innovation ecosystems, technological advancements, and societal challenges. *Technol. Forecast. Soc. Chang.* **2019**, *142*, 1–14. [CrossRef]
21. Yusof, A.; Ahmad, N.S.; Hamid, A.; Khong, T.K. Effects of honey on exercise performance and health components: A systematic review. *Sci. Sports* **2018**, *33*, 267–281. [CrossRef]
22. Gough, D.; Oliver, S.; Thomas, J. (Eds.) *An Introduction to Systematic Reviews*; Sage publications Ltd.: London, UK, 2017.
23. Akdu, U. Smart Tourism: Issues, Challenges and Opportunities. In *The Emerald Handbook of ICT in Tourism and Hospitality*; Emerald Publishing Limited: Bingley, UK, 2020; pp. 291–308.
24. Demir, O.F.; Karaarslan, E. Augmented reality application for smart tourism: GökovAR. In Proceedings of the 2018 6th International Istanbul Smart Grids and Cities Congress and Fair (ICSG), Istanbul, Turkey, 25–26 April 2018; pp. 164–167.
25. Gretzel, U.; Zhong, L.; Koo, C. Application of smart tourism to cities. *Int. J. Tour. Cities* **2016**, 2. [CrossRef]
26. Basori, A.H.; bin Abdul Hamid, A.L.; Mansur, A.B.F.; Yusof, N. iMars: Intelligent Municipality Augmented Reality Service for Efficient Information Dissemination based on Deep Learning algorithm in Smart City of Jeddah. *Procedia Comput. Sci.* **2019**, *163*, 93–108. [CrossRef]
27. Satapathy, S.M.; Jhaveri, R.; Khanna, U.; Dwivedi, A.K. Smart Rent Portal using Recommendation System Visualized by Augmented Reality. *Procedia Comput. Sci.* **2020**, *171*, 197–206. [CrossRef]
28. Zhu, Y.; Li, N. Virtual and augmented reality technologies for emergency management in the built environments: A state-of-the-art review. *J. Saf. Sci. Resil.* **2021**, *2*, 1–10. [CrossRef]
29. Jin, D.; Hannon, C.; Li, Z.; Cortes, P.; Ramaraju, S.; Burgess, P.; Buch, N.; Shahidehpour, M. Smart street lighting system: A platform for innovative smart city applications and a new frontier for cybersecurity. *Electr. J.* **2016**, *29*, 28–35. [CrossRef]
30. Cho, K.; Jang, H.; Park, L.W.; Kim, S.; Park, S. Energy Management System Based on Augmented Reality for Human-Computer Interaction in a Smart City. In Proceedings of the 2019 IEEE International Conference on Consumer Electronics (ICCE), Las Vegas, NV, USA, 11–13 January 2019; pp. 1–3.
31. Horejsi, P.; Novikov, K.; Simon, M. A Smart Factory in a Smart City: Virtual and Augmented Reality in a Smart Assembly Line. *IEEE Access* **2020**, *8*, 94330–94340. [CrossRef]
32. Rashid, Z.; Melià-Seguí, J.; Pous, R.; Peig, E. Using Augmented Reality and Internet of Things to improve accessibility of people with motor disabilities in the context of Smart Cities. *Future Gener. Comput. Syst.* **2017**, *76*, 248–261. [CrossRef]
33. Oxford Analytica. *Cybersecurity Will Lag Development of Smart Cities*; Oxford Analytica: Oxford, UK, 2019.
34. Elmaghraby, A.S.; Losavio, M.M. Cyber security challenges in Smart Cities: Safety, security and privacy. *J. Adv. Res.* **2014**, *5*, 491–497. [CrossRef] [PubMed]
35. Habibzadeh, H.; Nussbaum, B.H.; Anjomshoa, F.; Kantarci, B.; Soyata, T. A survey on cybersecurity, data privacy, and policy issues in cyber-physical system deployments in smart cities. *Sustain. Cities Soc.* **2019**, *50*, 101660. [CrossRef]
36. Hamid, B.; Jhanjhi, N.Z.; Humayun, M.; Khan, A.; Alsayat, A. Cyber Security Issues and Challenges for Smart Cities: A survey. In Proceedings of the 2019 13th International Conference on Mathematics, Actuarial Science, Computer Science and Statistics (MACS), Karachi, Pakistan, 14–15 December 2019; pp. 1–7.
37. Mohammadpourfard, M.; Khalili, A.; Genc, I.; Konstantinou, C. Cyber-Resilient Smart Cities: Detection of Malicious Attacks in Smart Grids. *Sustain. Cities Soc.* **2021**, *75*, 103116. [CrossRef]
38. Baig, Z.A.; Szewczyk, P.; Valli, C.; Rabadia, P.; Hannay, P.; Chernyshev, M.; Johnstone, M.; Kerai, P.; Ibrahim, A.; Sansurooah, K.; et al. Future challenges for smart cities: Cyber-security and digital forensics. *Digit. Investig.* **2017**, *22*, 3–13. [CrossRef]

39. AlDairi, A.; Tawalbeh, L. Cyber Security Attacks on Smart Cities and Associated Mobile Technologies. *Procedia Comput. Sci.* **2017**, *109*, 1086–1091. [CrossRef]
40. Lee, J.; Kim, J.; Seo, J. Cyber attack scenarios on smart city and their ripple effects. In Proceedings of the IEEE 2019 International Conference on Platform Technology and Service (PlatCon), Jeju, Korea (South), 28–30 January 2019; pp. 1–5.
41. Chen, D.; Wawrzynski, P.; Lv, Z. Cyber security in smart cities: A review of deep learning-based applications and case studies. *Sustain. Cities Soc.* **2021**, *66*, 102655. [CrossRef]
42. Sengan, S.; Subramaniyaswamy, V.; Nair, S.K.; Indragandhi, V.; Manikandan, J.; Ravi, L. Enhancing cyber–physical systems with hybrid smart city cyber security architecture for secure public data-smart network. *Futur. Gener. Comput. Syst.* **2020**, *112*, 724–737. [CrossRef]
43. Belgaum, M.; Alansari, Z.; Jain, R.; Alshaer, J. A Framework for Evaluation of Cyber Security Challenges in Smart Cities. In Proceedings of the Smart Cities Symposium. Institution of Engineering and Technology (IET), Bahrain, 22–23 April 2018; pp. 1–6.
44. Mohammad, R.M.A.; Abdulqader, M.M. Exploring Cyber Security Measures in Smart Cities. In Proceedings of the IEEE 2020 21st International Arab Conference on Information Technology (ACIT), Giza, Egypt, 28–30 November 2020; pp. 1–7.
45. Fard, S.M.H.; Karimipour, H.; Dehghantanha, A.; Jahromi, A.N.; Srivastava, G. Ensemble sparse representation-based cyber threat hunting for security of smart cities. *Comput. Electr. Eng.* **2020**, *88*, 106825. [CrossRef]
46. Parasol, M. The impact of China's 2016 Cyber Security Law on foreign technology firms, and on China's big data and Smart City dreams. *Comput. Law Secur. Rev.* **2018**, *34*, 67–98. [CrossRef]
47. Vitunskaite, M.; He, Y.; Brandstetter, T.; Janicke, H. Smart cities and cyber security: Are we there yet? A comparative study on the role of standards, third party risk management and security ownership. *Comput. Secur.* **2019**, *83*, 313–331. [CrossRef]
48. Wibowo, S. Enriching Digital Government Readiness Indicators of RKCI Assessment with Advance Https Assessment Method to Promote Cyber Security Awareness Among Smart Cities in Indonesia. In Proceedings of the IEEE 2018 International Conference on ICT for Smart Society (ICISS), Semarang, Indonesia, 10–11 October 2018; pp. 1–4.
49. Cranmer, E.E.; tom Dieck, M.C.; Fountoulaki, P. Exploring the value of augmented reality for tourism. *Tour. Manag. Perspect.* **2020**, *35*, 100672. [CrossRef]
50. Cranmer, E.; Jung, T.; tom Dieck, M.C.; Miller, A. Understanding the acceptance of augmented reality at an organisational level: The case of Geevor Tin Mine Museum. In *Information and Communication Technologies in Tourism*; Springer: Cham, Switzerland, 2016; pp. 637–650.
51. Han, B.-O.; Kim, Y.-H.; Cho, K.; Yang, H.S. Museum tour guide robot with augmented reality. In Proceedings of the 2010 16th International Conference on Virtual Systems and Multimedia, Institute of Electrical and Electronics Engineers, Seoul, South Korea, 20–23 October 2010; pp. 223–229.
52. Thon, S.; Serena-Allier, D.; Salvetat, C.; Lacotte, F. Flying a drone in a museum: An augmented-reality cultural serious game in Provence. In Proceedings of the IEEE 2013 Digital Heritage International Congress (DigitalHeritage), Marseille, France, 28 October–1 November 2013; Volume 2, pp. 669–676.
53. Ahmed, S. A Review on Using Opportunities of Augmented Reality and Virtual Reality in Construction Project Management. *Organ. Technol. Manag. Constr. Int. J.* **2019**, *11*, 1839–1852. [CrossRef]
54. Rohacs, J.; Rohacs, D.; Jankovics, I. Conceptual development of an advanced air traffic controller workstation based on objective workload monitoring and augmented reality. *Proc. Inst. Mech. Eng. Part G J. Aerosp. Eng.* **2016**, *230*, 1747–1761. [CrossRef]
55. Gheisari, M.; Irizarry, J. Investigating human and technological requirements for successful implementation of a BIM-based mobile augmented reality environment in facility management practices. *Facilities* **2016**, *34*, 69–84. [CrossRef]
56. Luchetti, G.; Mancini, A.; Sturari, M.; Frontoni, E.; Zingaretti, P. Whistland: An Augmented Reality Crowd-Mapping System for Civil Protection and Emergency Management. *ISPRS Int. J. Geo-Inf.* **2017**, *6*, 41. [CrossRef]
57. Lee, K. Augmented Reality in Education and Training. *TechTrends* **2012**, *56*, 13–21. [CrossRef]
58. Chen, P.; Liu, X.; Cheng, W.; Huang, R. A review of using Augmented Reality in Education from 2011 to 2016. In *Innovations in Smart Learning*; Springer: Berlin/Heidelberg, Germany, 2017; pp. 13–18.
59. Wu, H.-K.; Lee, S.W.-Y.; Chang, H.-Y.; Liang, J.-C. Current status, opportunities and challenges of augmented reality in education. *Comput. Educ.* **2013**, *62*, 41–49. [CrossRef]
60. Kuo, C.; Jeng, T.; Yang, I. An invisible head marker tracking system for indoor mobile augmented reality. *Autom. Constr.* **2013**, *33*, 104–115. [CrossRef]
61. Mahadik, A.; Katta, Y.; Naik, R.; Naikwade, N.; Shaikh, N.F. A review of Augmented Reality and its application in context aware library system. In Proceedings of the IEEE 2016 International Conference on ICT in Business Industry & Government (ICTBIG), Indore, India, 18–19 November 2017; pp. 1–6.
62. Kundu, A.S.; Mazumder, O.; Dhar, A.; Lenka, P.K.; Bhaumik, S. Scanning Camera and Augmented Reality Based Localization of Omnidirectional Robot for Indoor Application. *Procedia Comput. Sci.* **2017**, *105*, 27–33. [CrossRef]
63. Azuma, R.; Neely, H.; Daily, M.; Leonard, J. Performance analysis of an outdoor augmented reality tracking system that relies upon a few mobile beacons. In Proceedings of the 2006 IEEE/ACM International Symposium on Mixed and Augmented Reality, Santa Barbara, CA, USA, 22–25 October 2006; pp. 101–104.
64. Schwarz, F.; Fastenmeier, W. Augmented reality warnings in vehicles: Effects of modality and specificity on effectiveness. *Accid. Anal. Prev.* **2017**, *101*, 55–66. [CrossRef] [PubMed]

65. Ilves, L.K.; Evans, T.J.; Cilluffo, F.J.; Nadeau, A.A. European union and nato global cybersecurity challenges. *Prism* **2016**, *6*, 126–141.
66. Pan, J.; Yang, Z. Cybersecurity Challenges and Opportunities in the New "Edge Computing + IoT" World. In Proceedings of the 2018 ACM International Workshop on Security in Software Defined Networks & Network Function Virtualization, Association for Computing Machinery (ACM), Tempe, AZ, USA, 21 March 2018; pp. 29–32.
67. Tisdale, S.M. Cybersecurity: Challenges from a Systems, Complexity, Knowledge Management and Business Intelligence Perspective. *Issues Inf. Syst.* **2015**, *16*, 191–198.
68. Alfouzan, F. Energy-Efficient Collision Avoidance MAC Protocols for Underwater Sensor Networks: Survey and Challenges. *J. Mar. Sci. Eng.* **2021**, *9*, 741. [CrossRef]

sensors

MDPI

Review

Impact, Vulnerabilities, and Mitigation Strategies for Cyber-Secure Critical Infrastructure

Hugo Riggs [1], Shahid Tufail [1], Imtiaz Parvez [2], Mohd Tariq [1,*], Mohammed Aquib Khan [1], Asham Amir [1], Kedari Vineetha Vuda [1] and Arif I. Sarwat [1,*]

1 Department of Electrical and Computer Engineering, Florida International University, Miami, FL 33174, USA
2 Department of Computer Science, Utah Valley University, Orem, UT 84058, USA
* Correspondence: tmohd@fiu.edu (M.T.); asarwat@fiu.edu (A.I.S.)

Abstract: Several critical infrastructures are integrating information technology into their operations, and as a result, the cyber attack surface extends over a broad range of these infrastructures. Cyber attacks have been a serious problem for industries since the early 2000s, causing significant interruptions to their ability to produce goods or offer services to their clients. The thriving cybercrime economy encompasses money laundering, black markets, and attacks on cyber-physical systems that result in service disruptions. Furthermore, extensive data breaches have compromised the personally identifiable information of millions of people. This paper aims to summarize some of the major cyber attacks that have occurred in the past 20 years against critical infrastructures. These data are gathered in order to analyze the types of cyber attacks, their consequences, vulnerabilities, as well as the victims and attackers. Cybersecurity standards and tools are tabulated in this paper in order to address this issue. This paper also provides an estimate of the number of major cyber attacks that will occur on critical infrastructure in the future. This estimate predicts a significant increase in such incidents worldwide over the next five years. Based on the study's findings, it is estimated that over the next 5 years, 1100 major cyber attacks will occur on critical infrastructures worldwide, each causing more than USD 1 million in damages.

Keywords: computer networks; cyber attack; signal detection; machine learning; smart grid

Citation: Riggs. H.; Tufail, S.; Parvez, I.; Tariq, M.; Khan, M.A.; Amir, A.; Vuda, K.V.; Sarwat, A.I. Impact, Vulnerabilities, and Mitigation Strategies for Cyber-Secure Critical Infrastructure. *Sensors* **2023**, *23*, 4060. https://doi.org/10.3390/s23084060

Academic Editors: Hyoungshick Kim and Hai Dong

Received: 22 February 2023
Revised: 4 April 2023
Accepted: 9 April 2023
Published: 17 April 2023

1. Introduction

With many examples of cyber attacks affecting critical infrastructure (CI) in recent years, it has become evident that these incidents are a major threat to the existing critical infrastructure and, thus, society as a whole [1–10]. In this paper, we define CI based on the definition from the Cyber & Infrastructure Security Agency. Critical infrastructure includes cyber and physical assets, systems, and networks of chemical and commercial facilities, communications, critical manufacturing, dams, defense industrial base, emergency services, energy, financial, food and agriculture, government facilities, healthcare, and public health, information technology (IT), nuclear reactors, materials and waste, transportation systems, and water and waste management. Not only are these sectors highly significant to modern countries but they also have strong interdependencies. A disruptive effect on one CI sector can have a cascading failure effect on other CIs, specifically, outages in electrical CIs affect most other CIs [11–13]. CIs are sets of physical and virtual assets, systems, and networks that provide a nation with economic security, public health, and safety [12]. This review focuses on electrical grid CIs due to the dependence that other CIs have on the electrical grid. The electrical grid has numerous remote terminal units for controlling physical systems and is a balanced system between load and generation. Cyber attacks against the electrical grid can lead to a cascading failure affecting most other CIs.

CI was listed in [13] as agriculture and food, water, public health and safety, emergency services, government, defense industrial base, information and telecommunications, energy, transportation, banking and finance, industry/manufacturing, postal, and shipping. CI is not immune to cyber attacks; for example, parts of the power grid in Ukraine have been taken down by malware named BlackEnergy3. Other examples show that industry-scale food processing plants have closed due to ransomware, and multiple businesses and their retail functions have been shuttered by malware embedded in trusted third-party updating services [14]. In March 2019, a denial of service (DoS) attack was launched against part of the Supervisory Control and Data Acquisition (SCADA) infrastructures of electric utilities in Utah, which shut down some of their observation capabilities. The exploitative programs that are in use today on the internet have enabled cybercriminals to acquire and update malicious programs with ease. The avenue for attack, once opened, allows the attacker to explore further attacks over potential vulnerabilities in the victim's system. Communications in a network are often assumed to be private; however, an attacker may exploit the network in a man-in-the-middle attack to steal confidential information, sabotage a cyber-physical system, or maliciously alter information. Cyber attacks have malicious intents; they were identified in [15] as obstruction of information, undermining cybersecurity measures, retardation of the decision-making process, denial in providing public services, abatement of public confidence, lowering the reputation of the victim country, and destroying a legal interest.

1.1. Evolution of Cyber Attacks

Cybercrime has existed since the early days of computer networks, with ransomware attacks seen as early as 1989. The digitization of control systems in CI, which previously operated from electromechanical systems, embeds the vulnerabilities of the digital system. The opening for cyber attackers grows as CIs have evolved their operational technologies [16]. More advanced malware has been developed over the past three decades, posing a constant threat to CIs. Many types of malware are being developed by professional software development organizations and purchased by cyber attackers. This division of malware development and deployment depends on the growing cybercrime economy [16]. Over time, the complexity of malware has increased, and it is used for the ransom of computer systems and CI system sabotage. A modern attacker can source customized malware tools from third-party providers.

Ransomware is expected to be more commonly experienced in CI through the Internet of Things (IoT) and CPS. While the technical specifics of a cyber attack can vary, the general flow of such attacks follows a trend. The trend in industrial control system (ICS) cyber attacks involves initiating a phishing attack to obtain access or insider access to facility computers. With access, a download or local pen drive can deliver spying and control malware. This malware carries out the primary sabotage actions, and then the exfiltration of the computer system is done, often preceded by a kill disk operation. The kill disk operation writes a binary zero value for all bits in the computer system storage, temporarily rendering it useless [16].

An emerging type of attack is a false data injection attack (FDIA) that targets the data stream of state estimation measurement outputs to cause the system operator to take incorrect control actions, which can have a detrimental physical and economic impact on the power system. The FDIA depends on three assumptions. The first is that the attacker has experience with power system operations and the capabilities of the targeted system. Secondly, the attacker is capable of manipulating meter measurements. Thirdly, the attacker has knowledge of the network topology, system electrical parameters, an understanding of the SCADA system, and existing cybersecurity mechanisms [1,8]. Furthermore, as the power grid digitizes and transitions to a smart grid, implementing neural networks for prediction has been shown to be highly sensitive to even small manipulations of data [7]. A set of case studies involving FDIA attacks against voltage and current sensors of power converters in a photovoltaic (PV)-based microgrid concluded that malfunction due to FDIA in these sensors can damage the PV modules.

The falsified signal was removed from the control process using sensor malfunction detection and ride-through operations [17].

Furthermore, the FDIA vulnerability of the power grid being researched is automatic generator control vulnerability. Such an attack affects the frequency of the power grid by interrupting the ability of the load control center to calculate control values. This could be very damaging and potentially cause a blackout [2]. In a specific use-case presented in [9], considering distributed energy resources, FDIA on PV production meter data used in 15-minute ahead forecasting is simulated and studied. The FDIA causes disruption in the control center communication with distributed energy resources (DER) assets simulated on an IEEE 34 bus system with three PVs, one synchronous generator, and one energy storage. The results showed that the FDIA can potentially lead to cascading failures by creating an overcurrent and voltage collapse.

1.2. Contributions in This Paper

This paper has the following key contributions: (a) provides a comprehensive set of major cyber attack categories for a holistic understanding of the threats and damages that can be expected from a cyber attack; (b) identifies standards and organizations that address cybersecurity in IT; (c) summarizes some of the historical major cyber attacks against critical infrastructure; and (d) identifies strategies and tools that cybersecurity teams will use as they build their defenses through both passive and active methods. The paper provides a description of seven major categories of cyber attacks, presents a 20-year history of significant (more than USD 1 million in damages) cyber attacks, and summarizes the way the systems were compromised.

1.3. Organization of the Paper

This paper focuses on the primary driving factors in cyber attacks and the types of cyber attacks. It also enumerates methods that are used by cyber security teams to counter these threats. The rest of the paper is structured as follows. Section 3 discusses the cybercrime economy and the types of cyber attacks on the CI. Section 4 provides a systematic process for building cyber defenses, it also discusses attribution techniques and the role of attribution; Section 5 covers existing standards that detail the frameworks and best practices that address cyber attacks; the section also shows a process for developing cyber-secure infrastructure. In Section 6 a discussion of the findings of the review is conducted. Finally, our conclusions are presented in Section 7.

2. Methodology for Review

The aim of our paper is to review and understand the reported information on cybersecurity research and incidents targeted at critical infrastructures, with a focus on the energy critical infrastructure, in order to identify areas that require future research. Research questions were devised to motivate the review and evaluate the identified publications. The contents of the publications and reports in the review are contextualized with the adversary techniques matrix published by the MITRE organization.

2.1. Research Questions

To evaluate the existing works on the impacts on critical infrastructure from cyber-incidents, research questions were identified.

- Question 1: *What are the motivations of cyber attackers?* To understand the motivations of the adversary, we searched for publications that detailed the modern economic structure of cybercrime. The cybercrime economy has been growing parallelly with the internet, enabling a robust network of adversarial actors and advanced persistent threats (APTs). Answering this question provides insight into the degree of sophistication of the adversary and the level of development.
- Question 2: *What are the types of cyber attacks on critical infrastructures?* To understand the technical aspects of a cyber attack, this question aims to list and describe the cyber

attacks that can be used against critical infrastructures. Answering this question will help develop knowledge of cyber attacks and the means and mediums through which they can occur.

- Question 3: *How does a cyber attack impact critical infrastructure and how does it affect citizens?* To identify how critical infrastructure is impacted by cyber attacks, this question aims to create an understanding of the expected impacts that a cyber attack will have on critical infrastructure and gain some understanding of the effect it will have on citizens who depend on critical infrastructure services.
- Question 4: *How many significant cyber attacks on critical infrastructure have occurred, and which critical infrastructures are targeted?* By analyzing the records of significant cyber attacks to account for all of the attacks on the various critical infrastructures, this provides a perspective on the trend in cyber attacks and which CIs can be targeted.
- Question 5: *What mitigation strategies are in use to mitigate the effects of a cyber attack?* Answering this question will inform security operators about the various solutions that exist to enhance infrastructure protection against cyber attacks. If the mitigation strategies do not address all cyber attack techniques, they could help to identify areas that require future research.

The research questions require a systematic approach to be answered in-depth and such an approach was taken in developing the answers to these questions.

To answer these research questions a review and synthesis of the literature was conducted and followed the sequential process shown in Figure 1.

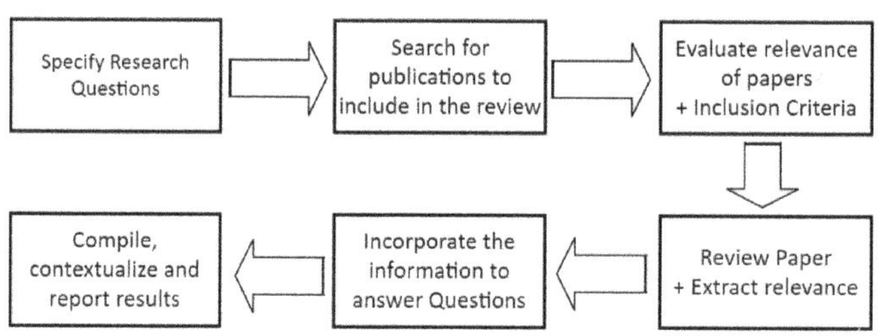

Figure 1. Method of approach to the review.

2.2. Selection of Papers and Reports

The selection criteria for identifying publications for inclusion in the review are as follows.

- Directly reports the events of a historical cyber attack.
- Directly implements a study of attack detection and prevention.
- Describes the organizational aspects of cybercrime.
- Describes the implementable network technology practices for mitigation of cyber attacks.

Peer-reviewed publications in international journals and conferences, along with theses, are all considered for review. The search for publications was primarily conducted through online databases, such as IEEE Xplore, ACM, Science Direct, and Springer Link. Additionally, white papers from high authority publishers and organizations in the cybersecurity industry were included, including the SANS Institute.

The sourcing of publications and articles along with the evaluation of the selected materials is done in a systematic way, shown in Figure 2. The various publishing entities, the numbers of selected works, and the process of evaluation, inclusion, and reporting on the selected works are listed.

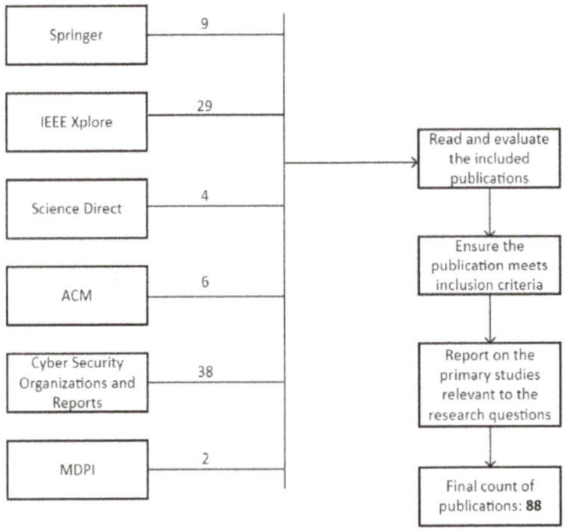

Figure 2. Publication selection and inclusion.

3. Types of Cyber Attacks on Critical Infrastructure

The projection shown in Figure 3 was conducted from the data collected by the Center for Strategic and International Studies (CSIS) in Washington, D.C. The CSIS provides a significant cyber attack list [18]. The CSIS defines a significant cyber attack as one that results in at least USD 1 million in damage. Significant cyber attacks are defined as cyber attacks on government agencies, defense, and high-tech companies, or attacks on other CIs that cause losses of more than USD 1 million Figure 3 shows the total number of significant cyber attacks measured and includes a projection of expected attacks through 2025. The projection, using polynomial regression, shows that there will be more significant cyber attacks in the next five years than the combined significant cyber attacks since 2005. The list from CSIS was further analyzed based on a keyword search to relate the cyber attack to a specific critical infrastructure. For example, if the cyber attack targeted a military base, it was attributed to the military CI, and if an attack contained the words financial or banking, it was included in the financial CI. The significant attacks per-CI are shown in Figure 4. The rest of this section expands on the discussion of the disruptive cybercrime economy. The section also enumerates the various top-level cyber attack types with some of their sub-variants.

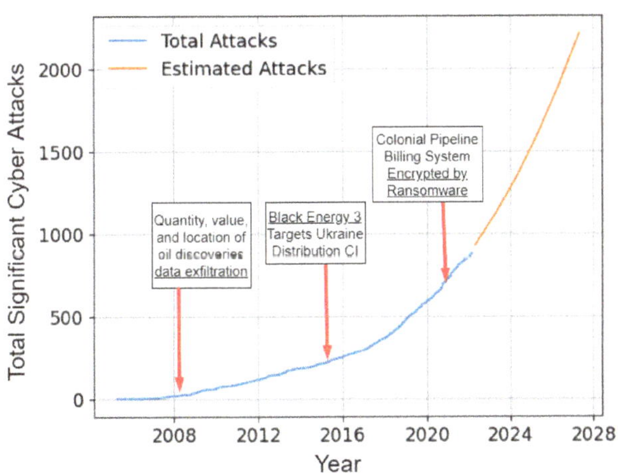

Figure 3. Estimate: cyber attacks will increase exponentially.

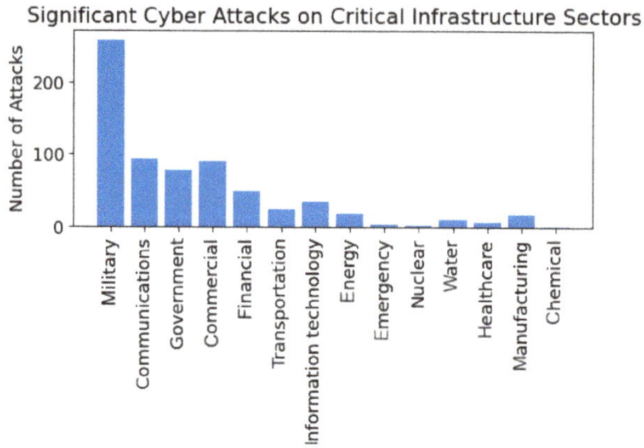

Figure 4. Significant cyber attacks by the CI sector since 2006, analyzed from the CSIS incidents list.

3.1. Cybercrime Economy

The cybercriminal economy has emerged worldwide, enabling many types of cyber attack functions as a service. However, while the focus is on cyber attacks in these sections, the cybercrime economy enables many other types of criminal activity. In [5], a literature review yields an extensive and consistent survey of the services used by the cybercrime business, organized using the value chain perspective, to understand cyber attacks systematically. Further, an understanding of the specialization, commercialization, and cooperation in coordinating a cyber attack is developed. They identify 24 value-added activities and their relations in the cybercrime market. These can be offered "as a service" for use in a cyber attack. The framework in [5] of cyber attacks "as a service" helps us understand the modern cybercriminal ecosystem and hacking innovations. Some services that facilitate cyber attacks include training and recruiting, development of exploitative software, scanning networks, denying service, phishing, target ranking, and money laundering. These services are provided as subscriptions, licenses, pay-per-records, or commission-based services [5,19]. The prominent concern for CIs is APTs. APTs are groups that are supported by their host nations and perform long-term targeting of the victim's CI. The general goal of APTs is to steal data from the victim. However, they also target the control management systems and components [19] of CI. The critical importance of the power infrastructure to the socioeconomic stability and the effect of blackouts make the smart grid a primary target [6]. APTs represent a subset of the cybercrime economy, and an APT is often a benefit to the host nation's economy, as they are compensated for their actions. This is due to the subterfuge of critical infrastructures slowing the economies of competitors to the host nation. An emergent factor for the electrical infrastructure is electricity theft, which is a major contributor to nontechnical losses in the distribution systems of the smart grid [4].

Money Laundering, Theft, Black Markets, and Ransom

One role of the cybercrime economy is in money laundering. This activity is evident in the use of cryptocurrencies for financial exchange from the victims to the attacker. A cryptocurrency transaction occurs, such as a ransom payment, and it is exchanged into another currency by the attacker. Cryptocurrencies lend themselves to this practice as they are functional currencies for communication networks that operate outside of traditional banks [20]. Trojan malware can facilitate information theft. If an enterprise system is compromised and the database is accessed to steal personally identifiable information, this information can be sold online. Online black markets exist, and they are frequently pursued by law enforcement and shut down. However, popular and well-known digital black markets commonly re-emerge at a new location, as moving software frameworks throughout different IT infrastructures is easily facilitated [21–23]. Another example of the function of the cybercriminal economy involves the ransom of critical computer systems. These ransomware-based attacks are targeted against critical services, such as utilities and hospitals [24]. The reasons for targeting these services are clear. They are critical for the public, and victims are willing to pay significant amounts of money to free their computer systems from ransomware. This is simply because it is less expensive for them to pay the ransom and recover their systems than remain out of operation [25].

3.2. The Ransomware Cyber Attack

A ransomware's malicious action is to either encrypt, lock, or exfiltrate data, and the ransomware will be specialized for the target platform. The variety of operating systems means that system-specific libraries and functions will be used by the ransomware to perform malicious actions. Mostly, they will target PC/workstations with a Windows operating system [16]. Within the cybercrime economy, some groups operate as Ransomware-as-a-Corporation (RAAC). Attackers operating as RAAC frequently issue press releases and use corporate language in their communications. If the ransom is not paid, then the victims' operational systems will remain inaccessible, and any critical personal information that has been exfiltrated will be posted on a dark web leak site to damage the company's reputation and business processes [21].

Although current ransomware campaigns do not target CPS, the installation of more intelligent electronic devices in the field by CI makes the CI and its CPS a more likely target for ransomware. As smart technologies continue to expand and integrate into homes, transportation, buildings, and throughout cities, they will become a growing target in the future development of ransomware that targets this new environment. Thus, ransomware that targets industrial CPS intelligent electronic devices will become more prevalent [16,26–28].

Most commonly, e-mails are the delivery method of ransomware. Malicious e-mails carry ransomware as an attachment, which contains the malware. These messages are often sent as spam broadcast to as many e-mail addresses as possible or can be directed and tailored to specific individuals or organizations. More details on targeted e-mailing are given in Section 3.5. The attachment can provide a link or file that initiates the installation of ransomware [16].

Encryption ransomware prevents victims from accessing their files by encrypting them with a secret key. The key and decryption software are then used for ransom. With advances in ransomware design, more targeted algorithms are used in encryption to specifically target file types of higher value to the victim. This reduces the time needed to perform the malicious encryption action after infecting the victim's computer. Locking ransomware has a similar goal to encryption-based malware, but it targets locking mechanisms designed to lock a system, such as a master boot record lock, screen lock, or computer desktop lock. The malware uses built-in security systems to lock the victim out of their computer system [29]. Finally, an information theft ransomware exfiltrates personally identifiable information (PII) from a victim's computer. The stolen PII is advertised to the victim as blackmail, and ransom is paid to prevent the publishing of the PII.

Supply Chain Ransomware

This type of ransomware is distributed through a trusted software distribution mechanism, particularly through a software updater provided by an IT service company. The attack was worldwide and affected businesses such as pharmacies, railways, and storefronts. The attack exploited a vulnerability in the IT service company's software updating system, which compromised the businesses that relied on it for updates [30].

3.3. Denial of Service

In the DoS attack the attacker prevents the intended user from accessing a resource. The attacker can reduce the intended user's access to the server by flooding the network i.e., increasing the traffic to disrupt access to a service. The attacker also attempts to break the connectivity between two systems [31]. Flooding services make the system receive too much traffic for the servers to handle. Flooding a system slows the system down and can ultimately halt the system.

The implication of DoS-based electricity theft against the energy CI is shown in the experimental results. The growing installation of intelligent electronic devices in CPS and the Internet of Things (IoT) domestic devices, such as connected homes and smart appliances, also increases the potential damage to CI from DoS attacks. The proliferation of more internet-connected grid technologies creates an increased vulnerability to such attacks [3].

3.3.1. Flooding in Mesh Networks

A utility can implement advanced meter infrastructure (AMI) using large wireless mesh networks. However, delays in wireless sensor networks can be caused by network flooding attacks. A malicious node in a wireless mesh network can tamper with messages that are sensitive to flooding attacks, resulting in a saturation of the AMI network. The DoS attack will come from a malicious node or nodes in the mesh network, sending excessive unnecessary data packets throughout the network and issuing excessive requests for communication. This traffic congests the mesh network and forms the basis of the flooding attack, which is identified as a DoS and impacts the network by increasing the latency of the communications [32,33].

IEC 62351 assigns digital signatures as a requirement for low-latency critical communication in ICS. However, digitally signed messages in wireless mesh networks are vulnerable to flooding DoS attacks, as demonstrated in [34], in which a model of phasor measurement data collection and transmission was subjected to flooding DoS. The flooding blocked the phasor measurement unit from transmitting data to the load flow control center. This type of interruption can affect the decision-making processes of the control center and generation control centers. In [3], an experiment with a consumer meter was performed, in which the meter was subjected to a flooding cyber attack. The flooding attack caused the meter to under-report the average watt-hour consumed at a rate of 1.77% less reported power consumption after four days. Other intelligent electronic devices may also be targeted. In [35], experimental signal jamming is performed on wireless networks against IEC 62351-based technologies. The GOOSE substation protocol is evaluated on a WiFi-based wireless power network, and the reactive jamming resulted in an 88% degraded throughput. Time-critical messaging is affected, resulting in latency overshooting the maximum message delay constraints.

3.3.2. Incidents of Denial of Service Attack

In 2000, a DoS attack on Yahoo rendered the site non-operational for more than 3 h. The attack was based on a Smurf attack and a Tribe Flood Network Technique. Through this attack, Yahoo received data requests of around or greater than one gigabyte per second [31]. Another DoS attack on the electric grid operations of Los Angeles County in California and Salt Lake County in Utah interrupted the electrical system operations for more than 10 h. It affected the computer systems used within the electrical utilities responsible for running the office functions. The attack had little impact on power delivery, but it raises concerns about the future if proper steps are not taken to mitigate such attacks [36].

3.4. *Man-in-the-Middle*

This Man-in-the-Middle (MITM) cyber attack is a kind of cyber attack where an outsider enters between two communication nodes and tries to remain undetected. The MITM can change the routed information before the information reaches the other node. This cyber attack accesses, reads, changes, or modifies the secret information without the victim's detecting manipulation. One capability involves injecting new messages and another involves the capacity to intercept all messages. Despite cryptography, a successful MITM attacker compromises exchanges between two systems. The MITM is either a passive listener or imitates one of the parties and manipulates the data sent. There may be many objectives for an attack either using the data overheard for a subsequent action or changing the data before it reaches the other party. The attacker extracts information to be used in many ways: fraud, unapproved support exchanges, blackmail, credential theft, and spying [37].

A MITM attack intercepts the victim's activity through the attacker's system before it is routed to its intended destination. The attacker gains access to an unsecured network, often targeting networks in public areas such as Wi-Fi access points [37]. This provides the attacker with an avenue to deploy tools that intercept information between the victims, often targeting personal computers where their connection to websites is monitored. This can result in credentials, financial details, and personally identifiable information being captured [37]. There are several types of MITM attacks, and the man-in-the-browser variant injects malicious software into the victim's computer or mobile device through phishing. Upon clicking on a phishing e-mail link or opening the attachment, the user loads the malware, and the malware installs itself on the browser without the user's knowledge. The malware enables the attacker to capture the information between the victim and specific websites. Exploits that are used to enter a MITM include internet protocol (IP) spoofing, address resolution protocol (ARP) spoofing, global navigation satellite system (GNSS) spoofing, and domain name system (DNS) spoofing [38].

3.4.1. IP Spoofing

In IP spoofing, the attacker modifies the source address in the IP packet header to make the receiver believe that the packet was received from a trusted site. From the victim's side, the packets will be received as though they were sent from a trusted source. However, the IP source reported in the packet is modified and does not represent the actual source [39].

3.4.2. ARP Spoofing

ARP spoofing involves sending a false ARP reply message to the default network gateway, claiming to associate the MAC address with the target's IP address. This ARP protocol translates IP addresses to MAC addresses. MITM ARP packets transmit over LAN by sending malicious ARP packets to a default gateway on the local area network [40]. The re-association request from the attacker can enable them to appear as the default gateway for traffic; thus, all other hosts in the network will transmit their data through the MITM.

3.4.3. DNS Spoofing

In DNS spoofing, the IP address in a DNS record is replaced by an IP address in the control of the attacker. This redirects internet traffic to fraudulent websites that resemble intended destinations [41–43].

3.4.4. HTTPS/SSL Hijacking

Stolen data can be decrypted using several methods, including HTTPS spoofing, SSL hijacking, SSL stripping, and others. In HTTPS spoofing, the attacker uses a domain that appears identical to the target website's domain. In SSL hijacking, the attacker passes the produced authentication keys to both the client and application during a TCP handshake [44]. This seems, by all accounts, to be a safe association when the MITM controls the whole session. In SSL stripping, the attacker sends a decoded form of the application's site to the client by maintaining the anchored session with the application. Meanwhile, the client's whole session is noticeable to the attacker.

3.5. Phishing and Remote Execution

Phishing and remote attacks rely on social engineering methods designed to have the victim reveal sensitive information or use malicious software. Phishing is highly prevalent in cyber attacks on CIs and it is identified in many of the significant cyber attacks in Table 1. Attackers send fraudulent communication to coerce a victim into sharing classified credentials or other information. Credentials obtained can be used to perform other attacks, such as the installation of malware, remote access, or the theft of information. Attackers may ransom credentials through the threat of publication [45–47].

Table 1. Abridged list of significant cyber attacks in recent years.

Adversary Technique	Types of Cyber Attacks Used	CI	Impact on CI Operations
Initial Access, Execution, Lateral Movement, Impact	Phishing, Remote Desktop, BlackEnergy3 (FDIA)	Energy (2015–2016)	Opened breakers in substations in Ukraine, causing 230,000 customers to lose power.
Initial Access, Execution, Impact	Ransomware	Energy (2021)	Fuel shortages for Southeast US with gas prices rising (9–16 cents per gallon) and 10,600 stations without gas. The Colonial Pipeline billing system shutdown for six days.
Initial Access, Execution, Impact	Stuxnet Worm, Zero-day Vulnerabilities	IT (2009–2011)	Shutdown of uranium enrichment facilities in Natanz, Iran.
Initial Access, Execution, Persistence, Collection	Trojan Laziok, reconnaissance malware	Energy (2014)	Gathered information from devices on the network that has vulnerabilities.
Initial Access, Execution, Collection, Impact	Ransomware	Food (2021)	Data of customers, suppliers, and employees were stolen. Productivity was reduced, access to some systems was blocked. A USD 11 million ransom was paid. Operation servers were shutdown and operations halted.
Initial Access, Execution, Impact	Ransomware	Healthcare (2021)	All Hospital appointments and radiology services were impacted, the ransomware affected Windows operating systems. The failure was experienced across national networks.
Initial Access, Execution, Impact	Phishing and Ransomware (Roobinhood)	Financial (2019)	Trading services of exchange halted and maintained offline, as computer systems were maliciously encrypted.
Initial Access, Execution, Collection	Phishing	Financial (Disclosed 2017)	The Equifax data breach resulted in the theft of personal data belonging to 140 million Americans and caused the company's share price to drop by 13%.
Initial Access, Execution, Persistence, Collection	Trojan Malware	Financial (2016)	The malware recorded debit cards and their pins from compromised ATM machines. Approximately USD 194,000 was stolen.
Initial Access, Execution, Collection, Impact	WannaCry Ransomware Cryptoworm	Energy (2017)	Worldwide Microsoft Windows operating systems were ransomed using an older Windows systems vulnerability EternalBlue.
Initial Access, Execution, Collection	Phishing, Ransomware	IT (2021)	The Accellion data breach and ransomware attack led to the theft of data in the Accellion data management service.
Initial Access, Execution, Lateral Movement, Impair Process Control	Supply Chain Ransomware	IT (2021)	Over 1500 businesses and organizations halted operations. A software updater released by an IT company, operating as a managed service provider
Initial Access, Execution, Inhibit Response Function, Impact	Ransomware	Municipal Services (2018)	Required over 5000 government computers to be shut down for 5 days to resolve the attack. Affected servers that were used to issue police warrants and employ new hiring processes, as well as official city complaints could not be submitted.

Phishing methods are also used to introduce ransomware infections on the victim's network infrastructure [48]. The 2020 Federal Bureau of Information's Internet Crime Report lists phishing as the most common cyber attack performed against US citizens by a wide margin, likely due to the increasingly sophisticated methods that cybercriminals use. The report lists 241,342 complaints of phishing in 2020; the next highest reported crime was non-payment or non-delivery of goods through online transactions, with a total of 108,869 complaints [49].

Phishing was used to target the Ukrainian Power grid. In the lead-up to Christmas in 2015, attackers took full control of remote terminal units in the Ukrainian power distribution grid and used them to change the set points on breakers. This action triggered the opening of critical breakers, de-energizing around 225,000 customers for an extended duration [14]. Phishing was the initial means by which attackers gained access to perform remote connection sabotage. Furthermore, in December 2016, attackers disabled energy delivery from a Kiev transmission station by using phishing to initiate remote sabotage, which caused a one-hour outage [50]. The flow of the phishing and remote execution cyber attack against the energy distribution system CI is shown in Figure 5. The attack sequence is framed as a separation of the cyber and physical planes, highlighting the sequential process of the attack by the APT. The process starts with reconnaissance, followed by a phishing campaign, gaining access, tunneling into OT, installing malware in OT, and finally using human-machine interfaces to sabotage physical systems in the field. The flow is captured in Figure 5.

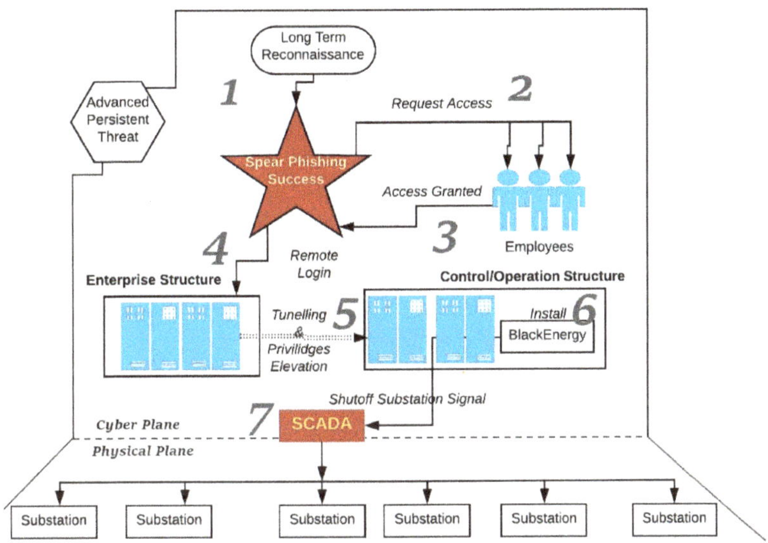

Figure 5. Targeting employees with socially engineered phishing campaigns, leading to remote sabotage.

3.5.1. Bulk Phishing

The most common form of phishing (bulk phishing) involves broadcasting messages through emails that are not personalized or targeted towards a specific individual or company. Attackers typically impersonate banking services, email/cloud providers, and streaming services to obtain credentials from potential victims.

3.5.2. Spear Phishing

In contrast to bulk phishing, 'spear phishing' includes methods of attack intended to target a specific organization or person with tailored communication. To increase the chances of deceit, attackers gather and use personal information about their target. Spear phishing targeted Hilary Clinton's 2016 presidential campaign by Threat Group-4127 [51].

3.5.3. CEO Phishing and Whaling

Whaling and chief executive officer (CEO) fraud represent two specific types of spear phishing tactics. Whaling involves phishing targeting CEOs or senior executives. CEO fraud is a reciprocal tactic in which the phishing attempt is made to impersonate the CEO [52].

3.5.4. Clone Phishing

Clone phishing is another phishing attack; in this tactic, attackers manipulate the link/attachment files included in an otherwise legitimate email. Using a previously delivered email, attackers will attempt to clone an email and include malicious attachments in place of original files and links. This form of phishing typically requires that one of the parties, either the sender or the recipient of the email, has previously had their account compromised [45].

3.5.5. Additional Phishing Tactics

Phishing is practiced in attacks outside of email communication, as well. Voice phishing involves attackers spoofing a phone number to resemble a trusted institution. Attackers will dial large quantities of phone numbers and play automated recordings that try to coerce sensitive information to help resolve an issue on the victim's account [53]. Finally, page hijacking is another form of phishing in which attackers will compromise or mimic legitimate web pages and redirect users to malware or an exploit kit utilizing cross-site scripting [54].

3.6. False Data Attack (Parameter/Command Injection)

False data injection is an attack that attempts to corrupt the control data. FDIA is presented in three types [7]:

- Targeted constrained FDIA: In this type of attack, data are injected after clear analysis, with a known amount of data inserted to appear realistic.
- Targeted unconstrained FDIA: In this type of attack, the attacker attempts to corrupt the values of some variables, and those variables in turn corrupt the remaining dependent variables.
- Random FDIA: In this type of attack, data packets are randomly distributed without consideration of the real values.

3.6.1. Protocols without Encryption in the CI

Digitization and ubiquitous computing have found their way into areas once solely operated by electromechanical controls. False data injection in CI control systems of the energy sector can damage the power electronics hardware. Protocol-level challenges in securing cyber-physical systems within the energy distribution grid are apparent in Distributed Network Protocol 3 (DNP3), GOOSE, and Modbus, as these protocols transmit data without encryption [55]. These systems should operate on physically isolated networks. An additional method to enhance their security is through the use of *bump in the wire*, an encryption hardware that encrypts the transmitted data before they travel the wider network. A methodology for layer-by-layer analysis of protocols to identify vulnerabilities is provided in [55]. Understanding protocol-level weaknesses is key to a secure network. Cyber-physical systems that utilize data generated from sensors in their processing and interact with information are prime targets for FDIA attacks. The cyber-physical system uses sensor data to implement the network and control adjustments of power electronics. In certain cases, these systems also require low latency in communication, which can make encryption of communications impossible, such as in the IEC 61850 GOOSE standard. If voltage is incorrectly controlled, it can cause damage to the power electronics.

3.6.2. Automatic Generator Control

FDIA on automatic generator control is a vulnerability that enables the manipulation of data in closed-loop control of generator control signals. This type of attack can cause significant damage to the generation and transmission equipment of the power grid, potentially leading to blackouts. This control system—if attacked by FDIA—will lead to overloading transmission lines by excessive power generation [56].

3.6.3. Parameter Modification in Inverters

As the power grid becomes increasingly dependent on renewable energy sources, new grid services will emerge based on smart inverters (SI) connected to these sources. The settings of these smart inverters are critical for these grid services to operate optimally. The settings of these inverters represent a point where FDIA can be particularly damaging to the smart grid [10]. A SI attack can affect the SI functions for volt–var, volt–watt, and a constant power factor. Such attacks potentially impact voltage profiles, system losses, and the operation of voltage control legacy devices. In such cases of FDIA, the severity depends on the prevailing SI functions [10].

3.7. *Worm and Trojan Malware*

A computer worm is a computer virus that is characterized as a self-replicating malware that spreads across networks executing disruptive payloads [57]. A worm targets hosts by following these scan types:

- An active selective random scan or sequential scan, in which the worm scans for vulnerable hosts.
- A hit-list scan, where the worm creates a target list and then searches for susceptible hosts.
- A routable scan, which utilizes information about a network to select and scan the IP address space [57].

Using a routable IP address allows the worm to propagate quickly and effectively, avoiding some detection methods. Another characteristic of a worm is the target space or medium through which it propagates. This includes the internet, email, P2P, USB local, and more. The worm propagates either as self-carried or through a second channel. In the second channel method, the main malware payload is remotely downloaded by the base installer. The activation of a worm on a system uses a vulnerability in the host, and the worm may protect itself by modifying its binary code with encryption [58].

The Stuxnet Worm

Stuxnet is a computer worm that was initially found in Iran but has since spread worldwide. This worm targets the control systems of a nation's critical infrastructure, and a successful attack by Stuxnet can result in the manipulation of the control system, causing disruption and damage to critical infrastructure and posing a threat to modern society. In 2010, Iran identified over 30,000 infected industrial computer systems, with Stuxnet specifically targeting nuclear power plant operational technology (OT) computers. The initial infections were at reactor core sites with flash memory used to introduce the worm locally. The worm targets an industrial control system that runs on Windows from Siemens [59].

3.8. *Trojan*

A Trojan can be installed on a computer through phishing or a local device. The purposes of a Trojan can vary, but often this malware hides its files under well-known directories, such as the user's documents, under the name of a trusted program, such as a web browser. Trojans are commonly used as a backdoor device to collect information from the infected computer. A keylogger is a type of data collection Trojan that can operate over a network or locally through a universal serial bus (USB) as an insider threat attack vector [60–62].

Additionally, hardware Trojan attacks refer to malicious modifications of electronic hardware at various stages of its operation. These attacks are a serious security concern for the electronics industry as they can lead to control interference and the leaking of secret data. The growing global demand for electronics makes it a larger point of vulnerability. It requires the adversary to have physical access to the integrated circuits [63,64].

4. Processes For Building Cyber Defenses

A successful cyber attack resulting in the unavailability of a critical infrastructure (CI), such as power delivery, can have an economic impact that extends beyond the systems sustaining direct and physical damage. The effects can impact regional and global economies. To identify security risks, analysis is based on what assets are valuable, who wants to attack them, and how they can be compromised. Security decisions are based on understanding the potential damage that can be done to these assets. Recommended cybersecurity for enterprise systems is provided by NIST. Recommended cybersecurity practices for control systems are provided by various organizations, including the Department of Homeland Security (DHS), the North American Electric Reliability Corporation Critical Infrastructure Protection (NERC CIP) standards, and the National Institute of Standards and Technology (NIST) [55]. Two of the common IT technologies to mitigate cyber attacks against networks are illustrated in Figures 6 and 7. The protection scheme for web services and the distribution of content uses load-balancing proxy servers with regionally specific deployments. These proxies are used to absorb DDoS attacks while secondary proxies continue to serve legitimate user requests, as shown in Figure 6. In remotely accessed CI devices, a virtual private network (VPN) can isolate a CI network from the greater internet while leveraging the internet for communication routing, as depicted in Figure 7.

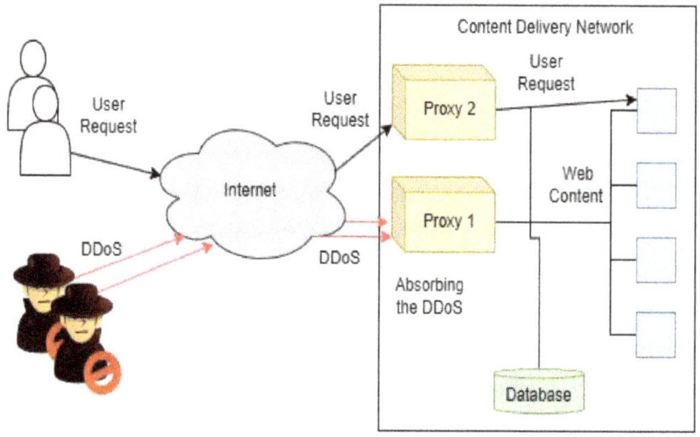

Figure 6. A proxy for protection against network and transport layer DDoS attacks.

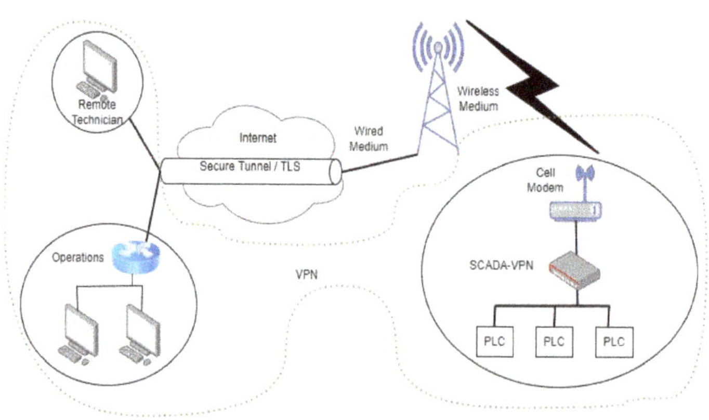

Figure 7. Securing communications (against MITM and FDIA) for SCADA with a VPN.

4.1. Reason to Train Personnel

Surveys of CI sectors have shown an increased vulnerability to cyber attacks as advances in information technology (IT) are implemented in these sectors. Furthermore, results show that the lack of common knowledge of cybersecurity among personnel is prevalent. Finding personnel lacking knowledge emphasizes the need to increase training in cybersecurity practices for CI personnel. With cyber threat-aware personnel, CIs can be hardened against cyber attacks. The emergence of the IoT in ICS has led to new security challenges, which will require newly developed expertise to prepare for a wide variety of attacks that may emerge from the integration of these systems [65–69].

4.2. Threat Matrix and Protection Development Process

Historical cyber attacks have been cataloged and studied through the efforts of Mitre.org, where they have generated an attack matrix for enterprise systems. This matrix provides the sequential stages of a cyber attack from reconnaissance, resource development, initial access, execution, persistence, privilege escalation, defense evasion, credential access, discovery, lateral movement, collection, command and control, exfiltration, and impact. The matrix is organized by techniques and the procedures to execute them. The matrix for ICS is shown in Table 2. For example, one technique is spear phishing, which can involve attaching an (xlsx) file to an email, which side-loads malware into the computer once opened. For each tactic, many procedures may be used depending on the attacker. Mitre and other organizations collaborate in the gathering of procedures under each tactic, and cybersecurity teams around the world can use this matrix to map out their vulnerabilities and areas in which they should develop defenses [70]. The approach to covering an organization's cyber vulnerabilities can involve an iterative defense development process, as summarized in Figure 8. In this process, the list of vulnerabilities is ranked according to the security operating center (SOC). The time to live of a particular vulnerability is the maximum amount of time that a vulnerability can be ignored. The values for L, D, and T are threshold values for the decisions, also set by the SOC. Such a process can leverage the MITRE matrix to have a comprehensive set of knowledge on attack techniques, with a security team analyzing the priority of vulnerabilities for their organization and iteratively developing defenses based on priority.

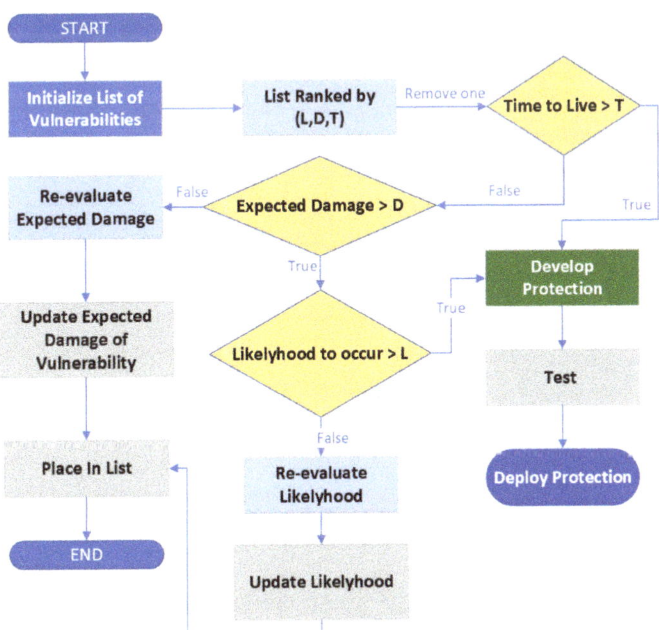

Figure 8. Cybersecurity development process for protecting IT (iterated at minimum acceptable cadence).

Table 2. Complete MITRE ATTACK Matrix for ICS—top level techniques as of November 2022 [70].

Initial Access	Execution	Persistence	Privilege Escalation	Evasion	Discovery	Lateral Movement	Collection	Command and Control	Inhibit Response Function	Impair Process Control	Impact
Drive-by Compromise	Change Operating Mode	Hardcoded Credentials	Exploitation for Privilege Escalation	Change Operating Mode	Network Connection Enumeration	Default Credentials	Adversary-in-the-Middle	Commonly Used Port	Activate Firmware Update Mode	Brute Force I/O	Damage to Property
Exploit Public-Facing Application	Command-Line Interface	Modify Program	Hooking	Exploitation for Evasion	Network Sniffing	Exploitation of Remote Services	Automated Collection	Connection Proxy	Alarm Suppression	Modify Parameter	Denial of Control
Exploitation of Remote Services	Execution through API	Module Firmware		Indicator Removal on Host	Remote System Discovery	Hardcoded Credentials	Data from Information Repositories	Standard Application Layer Protocol	Block Command Message	Module Firmware	Denial of View
External Remote Services	Graphical User Interface	Project File Infection		Masquerading	Remote System Information Discovery	Lateral Tool Transfer	Detect Operating Mode		Block Reporting Message	Spoof Reporting Message	Loss of Availability
Internet Accessible Device	Hooking	System Firmware		Rootkit	Wireless Sniffing	Program Download	I/O Image		Block Serial COM	Unauthorized Command Message	Loss of Control
Remote Services	Modify Controller Tasking	Valid Accounts		Spoof Reporting Message		Remote Services	Monitor Process State		Data Destruction		Loss of Productivity and Revenue
Replication Through Removable Media	Native API					Valid Accounts	Point & Tag Identification		Denial of Service		Loss of Protection
Rogue Master	Scripting						Program Upload		Device Restart/ Shutdown		Loss of Safety
Spearphishing Attachment	User Execution						Screen Capture		Manipulate I/O Image		Loss of View
Supply Chain Compromise							Wireless Sniffing		Modify Alarm Settings		Manipulation of Control
Transient Cyber Asset									Rootkit		Manipulation of View
Wireless Compromise									Service Stop		Theft of Operational Information
									System Firmware		

4.3. Actions of a Defending Organization during a Cyber Attack

Cyber defenders at an organization run through several phases to protect their systems. Initially, a risk assessment phase identifies vulnerabilities. In a subsequent protection phase, the organization develops hardware and software measures to achieve its security goals. In an attack on normal operations, a detection phase will begin in which monitoring mechanisms along with intrusion detection system(s) (IDS) classify abnormal and legitimate network behaviors. Normal and malicious network traffic inside the system are detected. Certain systems will be isolated as they are identified as the root of the cyber attack. The attack will be survived, and in the case of incapacitated OT and IT computer systems, those systems will be brought back into operation as quickly as possible with new security precautions.

Numerous CI sectors are facing challenges in identifying the highest-risk new threats and vulnerabilities. Due to the increased volume and sophistication of cyber attacks, it is crucial to allocate resources strategically and prioritize stopping the most dangerous and likely attacks first [71]. The risk to CI is escalating as the shift from isolated environments to "systems-of-systems" that integrate large information and communications infrastructures continues. The SOC has the responsibility of short-term and long-term planning for the IT/OT future. Guidelines for a cybersecure smart grid system are outlined in [19], emphasizing the enforcement of access control and authentication for all communication throughout the system.

- Every node in the network must have lightweight cryptographic functions.
- Attack detection and mitigating actions are necessary and must be used throughout the smart grid.
- Cyber-security testbeds must be created for the purpose of investigating vulnerabilities in the infrastructure [72].
- The security of network protocols must be designed from the application layer to the MAC layer [19].

4.4. IT/OT Practices to Mitigate Malware

A defense strategy against computer malware employs detection and removal. Either signature-based or anomaly-based detection can be used. Furthermore, patching systems to the latest security needs and anti-virus software are other methods used to prevent worms [57,58,73]. A specialized framework for handling computer log-generated data from honeypots, IDS, etc., is proposed for data ingestion, contextualization, and decision-making in formulating an effective and timely response to cyber attacks [74,75]. Approaches to detecting malware are listed below [76]:

- Statistics: An algorithm that utilizes statistical analysis of sample characteristics to determine if the sample is malware.
- Blacklist: The system uses a list of malicious domain names or IP addresses known to be used by ransomware families to identify malware.
- Rule Driven System: A rule-based decision model that finds malware. Rules may include scores such as perceived threats, various threshold values, or rules compatible with malware detection engines.
- Machine Learning-based: Through the use of machine learning (ML) models created with a variety of analysis features, the system can identify malware. ML approaches may analyze instruction opcodes, application programming interface (API) calls, and dynamically linked libraries (DLLs) to build ML classifiers. Systems for detecting malware can identify patterns in the behavior of the malware program.

Active network monitoring using computer network traffic collection tools is an active approach to the detection of security incidents. The network monitoring tools include functionalities for network data collection, parsing of data, the combination of sources into a single data stream, detection of anomalous events in the stream, further exploration of data, and automatic action on the network. Traffic statistics of packet transmission can be analyzed using the Wireshark tool. Collections of packets can be exported by the tool for future analysis. The Elastic Stack distributed database technology with data analytics tools is another option for parsing network activity [77].

IDSs analyze packets or packet flows to detect intrusion into the network by an adversary. The detection method can be signature-based, anomaly-based, or a hybrid detection approach. Intrusion detection systems can also be deployed as a centralized architecture, decentralized, or distributed [78]. It has been found that many existing IDS signatures are based on obsolete attack classes that do not map to modern attacks. Antonia et al. [78] identify and describe the behavior of modern attacks that are not mapped in the IDS attack classes.

4.5. The Role of Attribution in Holding Attackers Accountable and Methods to Attribute Attack Network Traffic

The Department of Defense of the U.S. has created techniques for tracing the origins of a cyber attack through intermediaries to the source. The work is presented as a set of techniques for network analyses to attribute a cyber incident to an original perpetrator [79]. Attribution is also discussed in [80]; the authors explain the legal problem that states often avoid penalties for being hosts of cybercrime due to the limited abilities of victims to attribute the attack to the perpetrating state. The paper also suggests that a legal system, specifically an international tribunal, is a more suitable approach to handling attribution, as compared to a technological approach to this aspect of cybercrime. Attribution is a principal aspect of research and spans from the research domains of computer science to international law [81–83]. The attribution techniques that can be used are:

- Logging and trace-back queries: Routers may log messages passing through their networks. Backward requests can go up a chain of routers and check if they have seen a previously seen message. As a result, messages that had not previously been classified as harmful can now be attributed. This requires that logging routers be placed in advance, which can lead to cost overruns, poor performance, and numerous other issues. Implementations can also give rise to privacy concerns.
- Input debugging: Defenders can provide an attack pattern as a query to nearby routers, and the router can then report any instances of noticing the pattern. Currently, some distributed denial-of-service (DDoS) attacks are defended against using this strategy. However, it is mostly reactive and only effective against attacks that continually stream data.
- Transmitted message modification: As communications are transferred, routers label them so that their path may be traced. This might affect network performance, increase bandwidth, or interfere with various authentication methods.
- Transmit separate messages: When routing a message, routers also transmit a different message to help with attribution.
- Reconfigure and observe network: Reconfigure the network and go back to a previous phase using the knowledge of what (if anything) changed. Large networks may find it challenging to implement this, and it could lead to new security flaws. On networks owned by others, "controlled flooding" is permitted, but it should only be utilized in specific situations because it could be seen as an attack on third parties.
- Host monitor functions: If a host does not already offer this information, querying functionality can be added (similar to "Query Hosts"). Without the owner's consent, this is known as a "hack back," and it calls for strong legal oversight. The information may become substantially less reliable if the host is under the control of an attacker, alerting the attacker.
- Match streams (e.g., via headers, content, timing): Determining which input streams correspond to which output streams involves keeping track of the data streams that are entering and leaving a network or host. This can aid in attribution without requiring knowledge of the network's or host's internal state. However, matching can be a

challenging technical problem, particularly when dealing with delayed attacks and internal encryption.

- Exploit/force attacker self-identification: To identify the attacker, any information they may have sent, whether on purpose or accidentally, can be used. In some cases, the defender may be able to force the attacker to submit this information. However, many of these techniques rely on extremely technical and specialized approaches (such as beacons, web bugs, cookies, and watermarking) that are easily defeated once an attacker becomes aware of them. When this technique succeeds, it can directly reveal the attacker regardless of how well they have concealed themselves.
- Honeypot/Honeynet: As decoy systems, honeypots and honeynets are used by defenders to bait attackers. Zombie traps (compromised and maliciously controlled computers) and honeynets can instantly reveal any zombies attempting to access the network. However, honeypots and honeynets can only attribute attacks that pass through them, necessitating extensive experience in monitoring and analysis.
- Intrusion detection systems: These systems should be positioned as close as possible to potential attackers (instead of near the defended assets). The placements of the IDSs (which should be close to the attackers) will determine how effective this strategy is. IDSs are notorious for producing many false positives and false negatives, so this strategy frequently necessitates intensive monitoring.
- Filtering of Ingress: Messages can be filtered so that specific links only allow them through if they fulfill particular criteria that make attribution easier. The information for attribution is contained in the message itself, which has the advantage of being transparent to users and requiring no extra storage. The technique's main limitation is that it can only be used to attribute the locations of internal attacks and often only provides a range of potential attribution values, not a specific location or identity. Frequently, there must be several possible routes for a message to take, leading to uncertainties that reduce the technique's potency. Network ingress filtering mandates that every message entering a network has a source address in an acceptable range for that network entry point. Using the existing transmission control protocol (TCP)/IP infrastructure, network ingress filtering for IP can be developed and scaled incrementally (one network at a time). The implementation of network ingress filtering by virtually all of the network's entrance points is necessary for a given network to be successful.
- Spoof prevention: Improving the resistance of protocols or their implementation against fabricated information can significantly reduce the need for examining intermediate systems. However, frequently protocols and/or implementations are difficult to modify to achieve this.
- Secure hosts/routers: The aim is to limit the number of trustworthy intermediate systems that an attacker can access. Although perfect security is unrealistic and this does not accomplish attribution, it simply makes the problem easier to address. This is nevertheless necessary for computer security.
- Surveil Attacker: Direct surveillance of potential or known attackers can prevent advanced attacker strategies. However, this requires prior knowledge of the identity of the expected attacker, and some attackers are very challenging to surveil.
- Employ reverse flow: Data being sent back to the attacker should be marked specifically, and intermediate systems should be able to identify these markings. This can be tracked by stepping stones but requires reverse flow detectors and may be prevented by encryption.
- Combine techniques: Combine multiple approaches. Although it will typically cost more to accomplish, this has a higher chance of success than any other strategy. Special attention must be paid when merging strategies because there is limited expertise in doing so.

5. Standards That Address Cyber Attacks

The ISO, NIST, and other high-authority organizations create standards for codifying best practices in establishing cyber-secure environments. Some standards apply to data management in specific sectors, such as health records management. By following these standards, the critical infrastructure industry will be able to mitigate the risks of cyber attacks. Table 3 shows the standards organizations and primary standards that involve cybersecurity. The standards serve as frameworks to develop secure networks and can be used as guides and best practice definitions [84]. The standard NISTIR 7628 provides guidelines for smart grid cybersecurity, including wide-area measurement systems. Although both are considered in the engineering of communications systems for CPS [85].

Table 3. Standards that govern the frameworks, best practices, and specifications for cybersecure information systems.

Body	Standard	Core Contribution	Adversary Technique Mitigated
ISO/IEC	27018:2019	Provides security techniques and a code of practice for the protection of personally identifiable information in public clouds with guidelines based on ISO/IEC 27002.	Initial access, discovery, collection
ISO/IEC	27037:2012	Secure techniques for identifying, collecting, and preservation of digital evidence, and will assist organizations in attributing blame based on digital evidence.	All Techniques
ISO/IEC	27040:2015	Guidelines on the creation of a low-risk data management security system. This includes security for devices and media, applications, and services, and security relevant to end-users.	Initial access, lateral movement, inhibit response function, privilege escalation
ISO	22301	A framework for organizations to be resilient and continue business operations during and after a cyber attack. Develops business continuation plans in the event of a disruption.	Inhibit response function, impair process control, impact
ISO/IEC	27001	A framework for the implementation of secure corporate enterprise computer systems. Details of the implementation of security controls to manage risks.	All Techniques
ISO/IEC	27002	Extends ISO/IEC 27001 standard with guidelines on best practices. Provides organizations with generic information security controls, including implementation guidance. Defined use for an information security management system based on ISO/IEC 27001.	All techniques
ISO/IEC	27031	Information and communication technology guidelines for business continuity. Covering concepts and best practices.	-
ISO/IEC	27032	Guidance on cybersecurity management system, and best practices for information security.	Lateral movement, inhibit response function, collection
ISO/IEC	27701	Specifications for building a privacy information management system that is based on ISO 27001. Published to address a growing need for a framework for global data privacy.	Collection
NIST	Cyber-security framework	Guideline for managing cybersecurity risks based on the existing best practices, guidelines, and standards provided in three components: core, implementation tiers, and profiles.	All techniques

The ISO 27000 family of standards is an exhaustive and evolving set of standards for traditional and new environments, such as cloud computing. Certifications provided by the ISO for the various standards it has put forth create a guarantee for service users that the system they interact with will be secured. The standard ISO 27018 governs PII security in cloud computing. By applying this standard with existing ISO 27000 family standards, an organization can have a layered approach to managing its data in the context of software as a license on-premise, extending to the cloud context of infrastructure as a service, platform as a service, and software as a service. Agreements between ISO certified service providers and their users are guaranteed by contract. The technology services these standards protect include data, applications, runtime, middleware, operating system, virtualization, servers, storage, and networking. When an organization complies with the standards, it protects its IT and gives confidence to its clients that their information will be managed securely [86].

NIST Guidelines for Smart Grid Cybersecurity

The National Institute of Standards and Technology (NIST) Guidelines for Smart Grid Cybersecurity [87] provide a framework for developing smart grid cybersecurity. The guide includes use cases to identify high-risk assets, threats, and impacts, as well as high-level security requirements, security architecture, privacy assessment, smart grid standards assessment, and conformity assessment. The guide identifies several adversaries to information systems, including nation-states, organized crime, industrial competitors, disgruntled employees, careless or poorly trained employees, hackers, cyber terrorists, and other criminal elements. The guide also distinguishes various forms of critical infrastructure (CI) on the energy CI, including definitions of (1) physical attacks informed by cyber; (2) cyber-attacks enhancing physical attacks; and (3) the use of a cyber system to cause physical harm.

The use of IDS, antivirus software, and cryptography, are combined in a defense-in-depth approach that focuses on securing PII, power systems assets, IT infrastructures, and communications through layered defenses. Many defenses should be combined to cover the many types of cyber attack threats. The defense-in-depth approach places a focus on people, processes, and technologies. The defense-in-depth strategy aims to place barriers at multiple levels for any cyber attack against the CI. The attacker should be delayed, thus helping the CI to make timely corrective actions. Some of the specific infrastructure mentioned in [87] includes cryptography supporting key, privilege, and certificate management deployed on IT communication technologies, as well as intrusion detection and prevention systems. The cyber attacks experienced are DoS, unauthorized vulnerability probes, botnet command and control, data exfiltration, data destruction, potential physical destruction via alteration of critical software/data. The attacker will combine social engineering and malware to continue their access. The largest threat comes from APTs that select a target and plan and execute a cyber attack against that target over a long time period, with the most damaging attacks being very difficult to detect initially.

6. Discussion of Results

This paper has reviewed cyber attack techniques and mitigation strategies; however, it is limited by the vast extent of vulnerabilities that cannot be covered within the paper but which are documented by the community in the Common Weakness Enumeration database. The review provides detailed descriptions of top-level categories of cyber attacks to develop an understanding of the scope of the threat and potential damages. It serves as an introductory point for researchers and industry professionals to enhance their knowledge of existing cyber attacks and mitigation strategies. The review identifies standards that certify an organization's IT as cyber-secured and offers a retrospective of major cyber attacks launched against critical infrastructures globally in the past 20 years. The paper lists tools and strategies for cyber security teams to defend their infrastructure. Phishing techniques are identified as the initial access point in many cyber attacks on CI and phishing detection and prevention should be further researched.

The reports of major cyber attacks on critical infrastructures have been compiled to understand the types of cyber attacks that are executed, the vulnerabilities that exist, and the typical victims and attackers. The standards that guide the development of cyber-secure infrastructure for organizations, along with practical approaches, are listed in one location to help mitigate cyber attacks. Moreover, this review projects that over the next five years, there will be over 1100 significant cyber attacks on global critical infrastructures. The projection shows the rapid growth of significant cyber attacks globally. By reviewing several papers, we developed a framework for phishing and remote sabotage in Figure 5; such an approach utilizes phishing for initial access and lateral movement for access to OT, deploying remote attacks through the OT to impact the critical infrastructure. By further analysis of the adversary techniques, we created a flow chart (Figure 8), which is followed by the SOC in order to develop the necessary defenses for the CI information systems by securing their vulnerabilities via the highest priority ranking.

7. Conclusions

The increasing damages caused by cyber attacks, along with their estimated rapid increase in the coming years, make it critical to study them and document their origins, effects, the APTs perpetrating them, and the greater cybercrime economy. The use of ransomware is incredibly profitable when successful, and society pays the price for these interruptions across critical CI sectors such as food, energy, water, etc. [16,21,28,29]. With sophisticated phishing attacks targeting the weakest links in an organization, it is difficult for security teams to secure their networks. Emails, text messages, phone calls, and web pages can all be vectors for a phishing attack. After gaining initial access, escalation of a cyber attack against CI can lead to actions of remote sabotage. FDIA is one potential cyber attack against control systems that is being researched. The need to train all personnel within CIs to be vigilant of such attacks is a valuable investment for a utility [43–46].

The projection provided by this paper is that the number of significant cyber attacks on critical infrastructures will continue to grow exponentially in the next five years, highlighting the need for increased research in attack detection and prevention. The projection estimates over 1,100 significant cyber attacks on critical infrastructure worldwide in the next five years. While best efforts are made to secure systems, those affected by cyber attacks will mostly be due to selective targeting and efforts by state-sponsored cyber attacks. Certain zero-day exploits and socially engineered credential theft are expected to remain perpetual weak points in computer networks for the foreseeable future [5]. Cyber defenses should include training all personnel to be aware of cyber threats. Further, a process for developing protection mechanisms for the IT/OT should be used at the SOC. The ability to attribute a cyber attack to the original attacker is possible through many approaches. A combined approach to attribution is likely to be the most successful in identifying the original attacker and allows the CI to cooperate with law enforcement. With the dynamic and computerized nature of the smart grid, there are now more pressing cybersecurity requirements on the energy CI than ever before.

Author Contributions: The research was solely developed and conducted at the Florida International University under the guidance of A.I.S. at FIU, the authors H.R., S.T., I.P. and M.T. performed the survey, compiled the information, and wrote the paper. The authors M.A.K., A.A. and K.V.V. contributed to the writing of the paper. All authors have read and agreed to the published version of the manuscript.

Funding: This research received no external funding.

Institutional Review Board Statement: Not applicable.

Informed Consent Statement: Not applicable.

Data Availability Statement: Not applicable.

Acknowledgments: The authors would like to acknowledge the Department of Electrical and Computer Engineering (ECE) at the FIU College of Engineering for supporting this work.

Conflicts of Interest: The authors declare no conflict of interest.

Abbreviations

SCADA	Supervisory Control and Data Acquisition
DoS	Denial of Service
CIA	Confidentiality, Integrity, and Authenticity
DNP3	Distributed Network Protocol 3
CI	Critical Infrastructure
IT	Information Technology

OT	Operational Technology
RAAC	Ransomware-as-a-Corporation
CSIS	Center for Strategic and International Studies
VPN	Virtual Private Network
MITM	Man-in-the-Middle
DNS	Domain Name System
TCP	Transmission Control Protocol
SSL	Secure Socket Layer
FDIA	False Data Injection Attack
ICMP	Internet Control Message Protocol
CWE	Common Weakness Enumeration
ARP	Access Resolution Protocol
SOC	Security Operating Center
APT	Advanced Persistent Threat

References

1. Liang, G.; Weller, S.R.; Zhao, J.; Luo, F.; Dong, Z.Y. The 2015 Ukraine Blackout: Implications for False Data Injection Attacks. *IEEE Trans. Power Syst.* **2017**, *32*, 3317–3318. [CrossRef]
2. Li, Y.; Zhang, P.; Ma, L. Denial of service attack and defense method on load frequency control system. *J. Frankl. Inst.* **2019**, *356*, 8625–8645. [CrossRef]
3. Kumar, S.; Kumar, H.; Gunnam, G.R. Security Integrity of Data Collection from Smart Electric Meter under a Cyber Attack. In Proceedings of the 2019 2nd International Conference on Data Intelligence and Security (ICDIS), Island, TX, USA, 28–30 June 2019; pp. 9–13. [CrossRef]
4. Wei, L.; Sundararajan, A.; Sarwat, A.I.; Biswas, S.; Ibrahim, E. A distributed intelligent framework for electricity theft detection using benford's law and stackelberg game. In Proceedings of the 2017 Resilience Week (RWS), Wilmington, DE, USA, 18–22 September 2017; pp. 5–11. [CrossRef]
5. Huang, K.; Siegel, M.; Madnick, S. Systematically Understanding the Cyber Attack Business: A Survey. *ACM Comput. Surv.* **2018**, *51*, 70. [CrossRef]
6. Tufail, S.; Parvez, I.; Batool, S.; Sarwat, A. A Survey on Cybersecurity Challenges, Detection, and Mitigation Techniques for the Smart Grid. *Energies* **2021**, *14*, 5894. [CrossRef]
7. Tufail, S.; Batool, S.; Sarwat, A.I. False data injection impact analysis in ai-based smart grid. In Proceedings of the SoutheastCon 2021, Atlanta, GA, USA, 10–13 March 2021; pp. 1–7.
8. Riggs, H.; Tufail, S.; Khan, M.; Parvez, I.; Sarwat, A.I. Detection of False Data Injection of PV Production. In Proceedings of the 2021 IEEE Green Technologies Conference (GreenTech), Virtual Conference, 7–9 April 2021; pp. 7–12. [CrossRef]
9. Olowu, T.O.; Dharmasena, S.; Hernandez, A.; Sarwat, A. Impact Analysis of Cyber Attacks on Smart Grid: A Review and Case Study. In *New Research Directions in Solar Energy Technologies*; Tyagi, H., Chakraborty, P.R., Powar, S., Agarwal, A.K., Eds.; Springer: Singapore, 2021; pp. 31–51. [CrossRef]
10. Olowu, T.O.; Dharmasena, S.; Jafari, H.; Sarwat, A. Investigation of False Data Injection Attacks on Smart Inverter Settings. In Proceedings of the 2020 IEEE CyberPELS (CyberPELS), Miami, FL, USA, 13 October 2020; pp. 1–6. [CrossRef]
11. Sarwat, A.I.; Sundararajan, A.; Parvez, I.; Moghaddami, M.; Moghadasi, A. Toward a Smart City of Interdependent Critical Infrastructure Networks. In *Sustainable Interdependent Networks: From Theory to Application*; Springer International Publishing: Cham, Switzerland, 2018; pp. 21–45. [CrossRef]
12. Cyber Security & Infrastructure Security Agency; Critical Infrastructure Sectors. Available online: https://www.cisa.gov/topics/critical-infrastructure-security-and-resilience/critical-infrastructure-sectors (accessed on 9 January 2023).
13. Kovacevic, A. Cyber Attacks on Critical Infrastructure: Review and Challenges. In *Handbook of Research on Digital Crime*; IGI Global: Hershey, PA, USA, 2015, pp. 1–18.
14. Robert, M.; Lee, M.J.; Assante, T.C. Analysis of the Cyber Attack on the Ukrainian Power Grid. Available online: https://www.eisac.com/s/ (accessed on 2 February 2022).
15. Uma, M.; Padmavathi, G. A Survey on Various Cyber Attacks and their Classification. *Int. J. Netw. Secur.* **2013**, *15*, 390–396.
16. Oz, H.; Aris, A.; Levi, A.; Uluagac, A.S. A Survey on Ransomware: Evolution, Taxonomy, and Defense Solutions. *ACM Comput. Surv.* **2022**. *54*, 238. [CrossRef]
17. Mohammadhassani, A.; Teymouri, A.; Mehrizi-Sani, A.; Tehrani, K. Performance evaluation of an inverter-based microgrid under cyberattacks. In Proceedings of the 2020 IEEE 15th International Conference of System of Systems Engineering (SoSE), Budapest, Hungary, 2–4 June 2020; pp. 211–216.
18. Significant Cyber Incidents. Center for Strategic & International Studies. Available online: https://www.csis.org/ (accessed on 4 December 2022).
19. Gunduz, M.Z.; Das, R. Cyber-security on smart grid: Threats and potential solutions. *Comput. Netw.* **2020**, *169*, 107094. [CrossRef]

20. Worlds Largest Meat Processing Company Hit by Cyber Attack (JBS). *BBC*, 2 June 2021. Available online: https://www.bbc.com/news/world-us-canada-57318965 (accessed on 7 July 2021).
21. Ransomware on the Rise in Critical Infrastructure Sector. *JD Supra*, 13 May 2021. Available online: https://www.jdsupra.com/legalnews/ransomware-on-the-rise-in-critical-1687319/ (accessed on 4 December 2022).
22. The Curious Case of the Baltimore Ransomware Attack: What You Need to Know. *Heimdal Security Blog*, 8 September 2020. Available online: https://heimdalsecurity.com/blog/baltimore-ransomware (accessed on 14 November 2022).
23. WannaCry Ransomware Attack Summary. *Data Protection Report*, 17 May 2017. Available online: https://www.dataprotectionreport.com/2017/05/wannacry-ransomware-attack-summary/ (accessed on 5 December 2022).
24. Chokshi, N. Hackers Are Holding Baltimore Hostage: How They Struck and What's Next. *The New York Times*, 22 May 2019. Available online https://www.nytimes.com/2019/05/22/us/baltimore-ransomware.html (accessed on 5 December 2022).
25. 'Number of Days' before Systems back Working—HSE 2021. Section: News. *ProteusCyber*, 17 May 2021. Available online: https://proteuscyber.com/it/privacy-database/news/4482-number-of-days-before-systems-back-working-hse (accessed on 22 July 2022).
26. Hanna, Y.; Cebe, M.; Mercan, S.; Akkaya, K. Efficient Group-Key Management for Low-bandwidth Smart Grid Networks. In Proceedings of the 2021 IEEE International Conference on Communications, Control, and Computing Technologies for Smart Grids (SmartGridComm), Aachen, Germany, 25–28 October 2021; pp. 188–193.
27. Zhi, Y.; Fu, Z.; Sun, X.; Yu, J. Security and privacy issues of UAV: A survey. *Mob. Netw. Appl.* **2020**, *25*, 95–101. [CrossRef]
28. Newaz, A.I.; Sikder, A.K.; Rahman, M.A.; Uluagac, A.S. A survey on security and privacy issues in modern healthcare systems: Attacks and defenses. *ACM Trans. Comput. Healthc.* **2021**, *2*, 1–44. [CrossRef]
29. Al-rimy, B.A.S.; Maarof, M.A.; Shaid, S.Z.M. Ransomware threat success factors, taxonomy, and countermeasures: A survey and research directions. *Comput. Secur.* **2018**, *74*, 144–166. [CrossRef]
30. Menn, J. Kaseya ransomware attack sets off race to hack service providers-researchers. *Reuters*, 4 August 2021.
31. Lau, F.; Rubin, S.; Smith, M.; Trajkovic, L. Distributed denial of service attacks. In Proceedings of the 2000 IEEE International Conference on Systems, Man and Cybernetics, Nashville, TN, USA, 8–11 October 2000; Volume 3, pp. 2275–2280, [CrossRef]
32. Parvez, I.; Islam, A.; Kaleem, F. A key management-based two-level encryption method for AMI. In Proceedings of the 2014 IEEE PES General Meeting | Conference & Exposition, National Harbor, MD, USA, 27–31 July 2014, pp. 1–5.
33. Thomas, M.S.; Ali, I.; Gupta, N. A secure way of exchanging the secret keys in advanced metering infrastructure. In Proceedings of the 2012 IEEE International Conference on Power System Technology (POWERCON), Auckland, New Zealand, 30 October–2 November 2012; pp. 1–7.
34. Zhang, F.; Mahler, M.; Li, Q. Flooding attacks against secure time-critical communications in the power grid. In Proceedings of the 2017 IEEE International Conference on Smart Grid Communications (SmartGridComm), Dresden, Germany, 23–27 October 2017, pp. 449–454.
35. Lu, Z.; Wang, W.; Wang, C. Modeling, evaluation and detection of jamming attacks in time-critical wireless applications. *IEEE Trans. Mob. Comput.* **2013**, *13*, 1746–1759. [CrossRef]
36. DiChristopher, Tom, K.F. An Alarmingly Simple Cyberattack Hit Electrical Systems Serving LA and Salt Lake, but Power Never Went Down. 2019. Section: Cybersecurity. Available online: https://finance.yahoo.com/news/alarmingly-simple-cyberattack-hit-electrical-193034191.html (accessed on 4 December 2022).
37. Mallik, A.; Ahsan, A.; Shahadat, M.M.Z.; Tsou, J.C. Understanding Man-in-the-middle-attack through Survey of Literature. *Indones. J. Comput. Eng. Des.* **2019**, *1*, 44–56. [CrossRef]
38. Psiaki, M.L.; Humphreys, T.E. GNSS Spoofing and Detection. *Proc. IEEE* **2016**, *104*, 1258–1270. [CrossRef]
39. Schuckers, S.A. Spoofing and anti-spoofing measures. *Inf. Secur. Tech. Rep.* **2002**, *7*, 56–62. [CrossRef]
40. ARP Poisoning. Available online: https://www.radware.com/security/ddos-knowledge-center/ddospedia/arp-poisoning/ (accessed on 17 November 2021).
41. Conti, M.; Dragoni, N.; Lesyk, V. A survey of man in the middle attacks. *IEEE Commun. Surv. Tutor.* **2016**, *18*, 2027–2051. [CrossRef]
42. Callegati, F.; Cerroni, W.; Ramilli, M. Man-in-the-Middle Attack to the HTTPS Protocol. *IEEE Secur. Priv.* **2009**, *7*, 78–81. [CrossRef]
43. Cheng, K.; Gao, M.; Guo, R. Analysis and research on HTTPS hijacking attacks. In Proceedings of the 2010 Second International Conference on Networks Security, Wireless Communications and Trusted Computing, Austin, TX, USA, 24–25 April 2010; Volume 2, pp. 223–226.
44. What Is HTTPS Spoofing MitM? Secret Double Octopus. Available online: https://doubleoctopus.com/security-wiki/threats-and-tools/https-spoofing/ (accessed on 17 November 2021).
45. Verizon Data Breach Investigations Report. Verizon, 2019. Available online: https://www.verizon.com/business/resources/reports/dbir/ (accessed on 17 November 2021).
46. Josh Fruhlinger. CSO Online. Equifax Data Breach: What Happened, Who Was Affected, What Was the Impact? Available online: https://www.csoonline.com/article/3444488/equifax-data-breach-faq-what-happened-who-was-affected-what-was-the-impact.html (accessed on 7 May 2022).

47. Four Members of China's Military Indicted over Massive Equifax Breach. Wall Street Journal, 11 February 2020. Available online: https://www.wsj.com/articles/four-members-of-china-s-military-indicted-for-massive-equifax-breach-11581346824 (accessed on 7 May 2022).
48. Stavroulakis, P.; Stamp, M. *Handbook of Information and Communication Security*; Springer Science & Business Media: Berlin/Heidelberg, Germany, 2010.
49. IC3. *Cyber Crime Report*; Federal Bureau of Investigation, Internet Crime Complaint Center: Washington, DC, USA, 2020.
50. Lanna Deamer. The DDoS Threat for Energy and Utility Companies. ElectronicSpecifier. 19 January 2018. Available online: https://www.electronicspecifier.com/products/cyber-security/the-ddos-threat-for-energy-and-utility-companies (accessed on 17 November 2021).
51. Spearphishing via Service, Technique T1194—Enterprise | MITRE ATT&CK®. Available online: https://attack.mitre.org/techniques/T1566/003/ (accessed on 14 July 2022).
52. Alkhalil, Z.; Hewage, C.; Nawaf, L.; Khan, I. Phishing Attacks: A Recent Comprehensive Study and a New Anatomy. *Front. Comput. Sci.* **2021**, *3*, 563060. [CrossRef]
53. Vishing. In Proceedings of the 5th Annual Conference on Information Security Curriculum Development, Kennesaw, GA, USA, 26–27 September 2008.
54. Static detection of cross-site scripting vulnerabilities. In Proceedings of the 2008 ACM/IEEE 30th International Conference on Software Engineering, Leipzig, Germany, 10–18 May 2008.
55. Sundararajan, A.; Chavan, A.; Saleem, D.; Sarwat, A.I. A Survey of Protocol-Level Challenges and Solutions for Distributed Energy Resource Cyber-Physical Security. *Energies* **2018**, *11*, 2360. [CrossRef]
56. Ameli, A.; Hooshyar, A.; El-Saadany, E.F.; Youssef, A.M. Attack detection and identification for automatic generation control systems. *IEEE Trans. Power Syst.* **2018**, *33*, 4760–4774. [CrossRef]
57. Pratama, A.; Rafrastara, F.A. Computer worm classification. *Int. J. Comput. Sci. Inf. Secur.* **2012**, *10*, 21.
58. Nissim, N.; Moskovitch, R.; Rokach, L.; Elovici, Y. Detecting unknown computer worm activity via support vector machines and active learning. *Pattern Anal. Appl.* **2012**, *15*, 459–475. [CrossRef]
59. Kerr, P.K.; Rollins, J.; Theohary, C.A. *The Stuxnet Computer Worm: Harbinger of an Emerging Warfare Capability*; Congressional Research Service: Washington, DC, USA, 2010.
60. Chandel, S.; Yu, S.; Yitian, T.; Zhili, Z.; Yusheng, H. Endpoint protection: Measuring the effectiveness of remediation technologies and methodologies for insider threat. In Proceedings of the 2019 International Conference on Cyber-Enabled Distributed Computing and Knowledge Discovery (Cyberc), Guilin, China, 17–19 October 2019; pp. 81–89.
61. Clark, J.; Leblanc, S.; Knight, S. Risks associated with USB hardware trojan devices used by insiders. In Proceedings of the 2011 IEEE International Systems Conference, Montreal, QC, Canada, 4–7 April 2011; pp. 201–208.
62. Kaspersky, J. *BlackEnergy APT Attacks in Ukraine*; Kaspersky Co.: Moscow, Russia, 2015.
63. Bhunia, S.; Hsiao, M.S.; Banga, M.; Narasimhan, S. Hardware Trojan attacks: Threat analysis and countermeasures. *Proc. IEEE* **2014**, *102*, 1229–1247. [CrossRef]
64. Konstantinou, C.; Keliris, A.; Maniatakos, M. Taxonomy of firmware trojans in smart grid devices. In Proceedings of the 2016 IEEE Power and Energy Society General Meeting (PESGM), Boston, MA, USA, 17–21 July 2016; pp. 1–5.
65. Miller, B.; Rowe, D. A Survey SCADA of and Critical Infrastructure Incidents. In Proceedings of the Proceedings of the 1st Annual Conference on Research in Information Technology, New York, NY, USA, 4–7 October 2012; pp. 51–56. [CrossRef]
66. Maglaras, L.A.; Kim, K.H.; Janicke, H.; Ferrag, M.A.; Rallis, S.; Fragkou, P.; Maglaras, A.; Cruz, T.J. Cyber security of critical infrastructures. *ICT Express* **2018**, *4*, 42–45. [CrossRef]
67. Alcaide, J.I.; Llave, R.G. Critical infrastructures cybersecurity and the maritime sector. *Transp. Res. Procedia* **2020**, *45*, 547–554. [CrossRef]
68. Liu, X.; Qian, C.; Hatcher, W.G.; Xu, H.; Liao, W.; Yu, W. Secure Internet of Things (IoT)-Based Smart-World Critical Infrastructures: Survey, Case Study and Research Opportunities. *IEEE Access* **2019**, *7*, 79523–79544. [CrossRef]
69. Sundararajan, A.; Wei, L.; Khan, T.; Sarwat, A.I.; Rodrigo, D. A Tri-Modular Framework to Minimize Smart Grid Cyber-Attack Cognitive Gap in Utility Control Centers. In Proceedings of the 2018 Resilience Week (RWS), Denver, CO, USA, 20–23 August 2018, pp. 117–123. [CrossRef]
70. MITRE ATTCK. ATTCK Matrix for Enterprise. 2022. Available online: https://attack.mitre.org (accessed on 5 November 2022).
71. Kifayat, K.; Merabti, M.; Younis, Y.A. Secure Cloud Computing for Critical Infrastructure: A Survey. In Proceedings of the 14th Annual Post Graduate Symposium on The Convergence of Telecommunications, Networking and Broadcasting, Liverpool, UK, 24–25 June 2013; pp. 599–610.
72. Ozgur, U.; Nair, H.T.; Sundararajan, A.; Akkaya, K.; Sarwat, A.I. An efficient MQTT framework for control and protection of networked cyber-physical systems. In Proceedings of the 2017 IEEE Conference on Communications and Network Security (CNS), Las Vegas, NV, USA, 9–11 October 2017; pp. 421–426. [CrossRef]
73. Stopel, D.; Boger, Z.; Moskovitch, R.; Shahar, Y.; Elovici, Y. Application of artificial neural networks techniques to computer worm detection. In Proceedings of the The 2006 IEEE International Joint Conference on Neural Network Proceedings, Vancouver, BC, Canada, 16–21 July 2006; pp. 2362–2369.

74. Sundararajan, A.; Khan, T.; Aburub, H.; Sarwat, A.I.; Rahman, S. A Tri-Modular Human-on-the-Loop Framework for Intelligent Smart Grid Cyber-Attack Visualization. In Proceedings of the SoutheastCon 2018, St. Petersburg, FL, USA, 19–22 April 2018; pp. 1–8. [CrossRef]
75. Sundararajan, A.; Riggs, H.; Jeewani, A.; Sarwat, A.I. Cluster-based Module to Manage Smart Grid Data for an Enhanced Situation Awareness: A Case Study. In Proceedings of the 2019 Resilience Week (RWS), San Antonio, TX, USA, 4–7 November 2019; Volume 1, pp. 81–87. [CrossRef]
76. Kunta, H.; Induri, B.; Bourgeois, A.G.; Maimon, D.; Ashok, A. Towards an Experimental Testbed to Study CyberWorm Behaviors in Large Scale Networks. In Proceedings of the 14th International Workshop on Wireless Network Testbeds, Experimental evaluation & Characterization, London, UK, 25 September 2020; pp. 120–121.
77. Fuentes-García, M.; Camacho, J.; Maciá-Fernández, G. Present and future of network security monitoring. *IEEE Access* **2021**, *9*, 112744–112760. [CrossRef]
78. Nisioti, A.; Mylonas, A.; Yoo, P.D.; Katos, V. From intrusion detection to attacker attribution: A comprehensive survey of unsupervised methods. *IEEE Commun. Surv. Tutor.* **2018**, *20*, 3369–3388. [CrossRef]
79. David A. Wheeler, G.N.L. *Techniques for Cyber Attack Attribution*; IDA Paper P-3792; Institute for Defense Analyses: Alexandria, VA, USA, 2003.
80. Delbert Tran. The Law of Attribution: Rules for Attribution the Source of a Cyber-Attack. Available online: https://yjolt.org/law-attribution-rules-attributing-source-cyber-attack (accessed on 22 May 2022).
81. Skopik, F.; Pahi, T. Under false flag: Using technical artifacts for cyber attack attribution. *Cybersecurity* **2020**, *3*, 8. [CrossRef]
82. Geers, K. The challenge of cyber attack deterrence. *Comput. Law Secur. Rev.* **2010**, *26*, 298–303. [CrossRef]
83. Payne, C.N.; Finlay, L. Addressing Obstacles to Cyber-Attribution: A Model Based on State Response to Cyber-Attack. *Georg. Wash. Int. Law Rev.* **2017**, *49*, 535.
84. Cichonski, P.; Millar, T.; Grance, T.; Scarfone, K. *Computer Security Incident Handling Guide*; NIST Special Publication; NIST: Gaithersburg, MD, USA, 2012; Volume 800, pp. 1–147.
85. Sundararajan, A.; Khan, T.; Moghadasi, A.; Sarwat, A.I. Survey on synchrophasor data quality and cybersecurity challenges, and evaluation of their interdependencies. *J. Mod. Power Syst. Clean Energy* **2019**, *7*, 449–467. [CrossRef]
86. Kemp, R. ISO 27018 and personal information in the cloud: First year scorecard. *Comput. Law Secur. Rev.* **2015**, *31*, 553–555. [CrossRef]
87. Pillitteri, V.Y.; Brewer, T.L. *Guidelines for Smart Grid Cybersecurity*; NIST Special Publication; NIST: Gaithersburg, MD, USA, 2014.

sensors

Article

An Efficient Framework for Securing the Smart City Communication Networks

Faisal Abdulaziz Alfouzan [1,*], Kyounggon Kim [2] and Nouf M. Alzahrani [3]

[1] Department of Forensic Sciences, College of Criminal Justice, Naif Arab University for Security Sciences, Riyadh 14812, Saudi Arabia
[2] Center of Excellence in Cybercrime and Digital Forensics, College of Criminal Justice, Naif Arab University for Security Sciences, Riyadh 14812, Saudi Arabia; kkim@nauss.edu.sa
[3] Information Technology Department, Collage of Computer Science and Information Technology, Al Baha University, Al Bahah 65731, Saudi Arabia; noufalzahrani@bu.edu.sa
* Correspondence: falfouzan@nauss.edu.sa; Tel.: +966-(0)-11-246-3444 (ext. 1515)

Abstract: Recently, smart cities have increasingly been experiencing an evolution to improve the lifestyle of citizens and society. These emerge from the innovation of information and communication technologies (ICT) which are able to create a new economic and social opportunities. However, there are several challenges regarding our security and expectation of privacy. People are already involved and interconnected by using smart phones and other appliances. In many cities, smart energy meters, smart devices, and security appliances have recently been standardized. Full connectivity between public venues, homes, cares, and some other social systems are on their way to be applied, which are known as Internet of Things. In this paper, we aim to enhance the performance of security in smart city communication networks by using a new framework and scheme that provide an authentication and high confidentiality of data. The smart city system can achieve mutual authentication and establish the shared session key schemes between smart meters and the control center in order to secure a two-way communication channel. In our extensive simulation, we investigated and evaluated the security performance of the smart city communication network with and without our proposed scheme in terms of throughput, latency, load, and traffic received packet per seconds. Furthermore, we implemented and applied a man-in-the-middle (MITM) attack and network intrusion detection system (NIDS) in our proposed technique to validate and measure the security requirements maintaining the constrained resources.

Keywords: smart city; cyber security for smart cities; communication wireless network; man-in-the-middle (MITM) attack; network intrusion detection system (NIDS)

Citation: Alfouzan, F.A.; Kim, K.; Alzahrani, N.M. An Efficient Framework for Securing the Smart City Communication Networks. *Sensors* **2022**, *22*, 3053. https://doi.org/10.3390/s22083053

Academic Editor: Juan M. Corchado

Received: 21 March 2022
Accepted: 14 April 2022
Published: 15 April 2022

1. Introduction

The rapid rise of information and communication technology (ICT) has changed not only modern society, but also the automation system in industries, including the electric power system, to a more convenient and dependable system. Electric power system reliability is one of our society's most basic necessities, and in today's technologically oriented world, an efficient and dependable electric power system is heavily reliant on ICT [1,2]. In recent years, there has been a lot of interest in upgrading traditional power systems to smart grid systems, which has boosted the integration of power systems with ICTs to ensure a dependable system that can overcome the issues that traditional power systems face. Increased consumption is one of the issues, as is the incorporation of cutting-edge technologies into power networks, such as renewable energy generation, electric vehicle charging, and smart meters. The ever-increasing reliance on electricity, as well as the demand for high-quality power, have necessitated smarter power delivery, more accessible pricing, and faster power restoration [3].

Although a new city that connects everything through the internet smartly and electrically does not exist yet, many cities are on their way to develop the connectivity to be fully smart. However, the increasing number of security concerns have demonstrated that cyber attacks in smart cities are ubiquitous and can occur at any time. For critical infrastructure, complex cyber attacks can occur that may paralyze industrial control systems, which results in serious damage and terrible consequences. For humans, smart wearables, such as mobile ransomware and communications hijacking, are able to steal users' data and personal identity information (PII). For society, changing or manipulation of these data can consequently lead to widespread panic and public opinion that can also cause threaten social management. Furthermore, the cyber security risks have significantly become the most serious threats in terms of developing the interconnect facilities in smart cities and the self-driven car industry, where accident liability will also become a particularly important and hot topic [4].

Therefore, communication trust and security challenges in the deployment of smart city systems have recently become a source of concern. In particular, a smart electrical meter distributed in multiple hierarchical networks can significantly achieve mutual authentication and establish the exchange keys. In order to enhance the performance of the smart city communication network, the OPNET modeler will be used to implement both (1) the secure scenario with authentication and key session techniques and (2) another scenario without these techniques. Various objectives should be processed and carried out in order to build our contribution. These include studying and investigating the security methods of authentication and confidentiality data relating to smart city technology and the impact of its integration with various emerging technologies, as well as selecting and designing authentication and a key exchange schemes approach in order to secure smart city data. In turn, this will include collecting data and conducting analysis of a secure control network in smart city communication, comparing and determining secure and insecure traffic parameters using authentication and key session schemes, and eventually evaluating the performance of a secure control network in smart city communication using authentication and key session schemes.

In this paper, we provide a new framework/scheme for enhancing secure control in smart city communication networks by boosting security performance. This works by enforcing data secrecy and authentication using efficient key exchange mechanisms. In order to secure a two-way communication channel, a smart city system can achieve mutual authentication and build shared session key schemes between smart meters and the control center.

The remainder of this paper is organised as follows. In Section 2, we review the related work. Section 3 describes our proposed cyber security framework for smart cities in detail. Section 4 presents and discusses the results of our simulation study. Finally, Section 5 concludes the paper.

2. Related Work

This section depicts the relevant background information and materials for this investigation. It will be described how to secure a smart city communication network as well as information and communication technology (ICT). This is followed by a literature review that focuses on reviews related to the security networks in smart city communication by implementing authentication and key exchange schemes. In addition, certain security strategies that can be applied as security requirements in smart city communication networks will be discussed in detail.

2.1. Securing Smart City Communication Network

Smart cities have piqued the interest of numerous disciplines, including scholars, enterprises, and governments, as a result of the rapid growth of ICT. Because a lack of proper cyber security can result in the theft of a user's sensitive data, utility fraud, and grid instability, cyber security is a major concern in the implementation and adoption of smart cities [5]. The United States' National Institute of Standards and Technology (NIST) established three critical needs for smart city security. These requirements are based on information availability, integrity, and confidentiality [6].

2.2. Information and Communication Technology (ICT) in a Smart City

As presented in Figure 1, the smart city system has been divided into four layers, each with its own set of needs in order to be more efficient, reliable, and intelligent. Furthermore, efficiency, reliability, scalability, and intelligence have become increasingly vital in order to be fast, better, secure, and resilient controls and communication. The deployment of ICT in smart grid has involved different applications such as advanced metering infrastructure, wide area measurement system, substation automation system, and common information models [7].

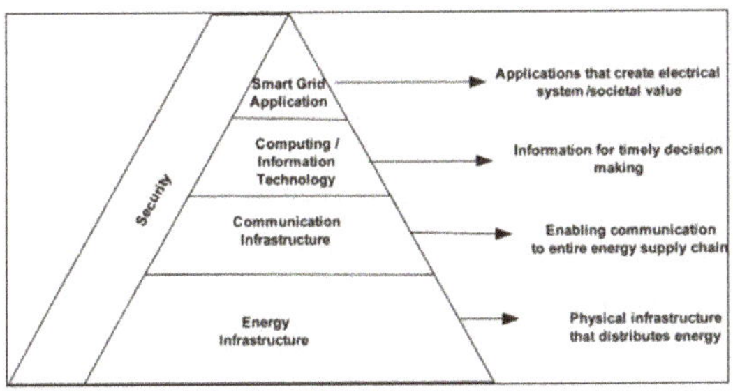

Figure 1. Presents the smart city definition.

In a nutshell, overall, the primary purpose of a smart city is to focus on urban residents, not just to meet current requirements but also to ensure that future generations are kept in mind with regards to cultural, social, economic, and environmental aspects.

2.3. Offensive Cyber Security Framework

To examine the goal and flow of cyber attacks, we systematized aspects of the offensive cyber security framework. Individual hackers, cyber crime organizations, and nation-state hackers, as indicated in Figure 2, perform cyber attacks to achieve their goals, such as financial gain or system destruction. The framework is designed depending on the internet/network, treat actors, and targets. Moreover, individuals, nation states, and cyber crime organizations are all dangerous players in cyber attacks. In order to gain access to the attack target, the Internet and network are employed, with Public Networks, Proxies, Virtual Private Networks (VPNs), and the Darknet/Deepweb being used. Organizations and CPS are the most common targets of an assault.

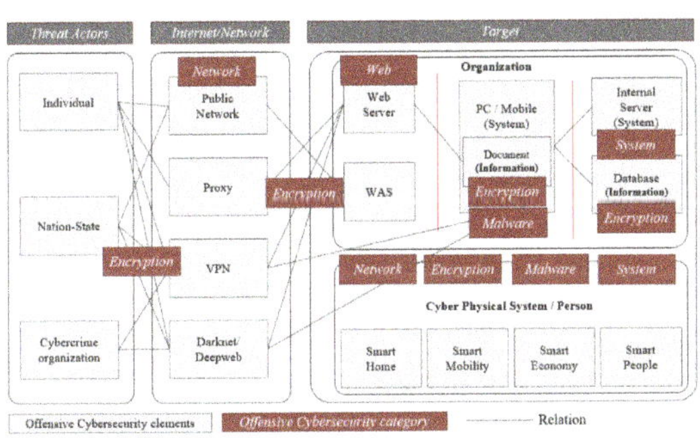

Figure 2. Offensive cyber security framework [8].

There are open to the outside Web Servers and Web Application Servers (WAS) within the firm, as well as personal computers and mobile phones with documents and information stored on them. Internal systems and databases carrying critical information are also used by businesses. CPS and Persons have both been targeted in recent attacks. Smart Homes, Smart Mobility, Smart Economy, and Smart People are among the CPS's detailed attack targets [9,10]. The assault categories that cyber attackers must breach, such as encryption, networks, web, malware, and systems, are referred to as offensive cyber security [8].

2.4. Literature Review

The integration of information and communication technologies in numerous application domains, such as transportation, academics, industries, medicine, and energy, has a significantly positive impact on society; however, systems' inability to cope with vulnerable attacks and loss of confidentiality, integrity, and authenticity has stymied this integration [11]. All application domains with communication infrastructure are vulnerable to cyber attacks, with the smart city being the most serious and vulnerable. Sensing, communication, control, and actuation systems work together to enable bi-directional pervasive communication for a variety of applications, including monitoring real-time energy consumption, providing real-time information to consumers, smart monitoring and tracking, control system commands, and more. One of the most essential roles of SCADA systems for remotely monitoring the grid's physical processes is power state system estimation [12]. The power system model and telemetered data help provide the best potential state of the system and can aid in ensuring appropriate power station performance through various applications. Because the system's state is based on a power state system estimation, a value that is invalid, corrupt, or deceptive as a result of a vulnerable attack might cause the system's state, management, and analysis to be diverted. Due to the risk to business-critical information, the bi-directional flow of information has generated a number of security and privacy issues; as a result, an efficient strategy is necessary to reduce complexity, computational cost, and processing time for immediate communication [13]. Although a large amount of research has been conducted in the realm of traditional cyber security, many approaches do not effectively meet the security requirements of grid systems. The following are some of the restrictions:

- Lifespan of grid systems;
- Proprietary system's dependency;
- Remotely located resources;
- Lack of physical protection;
- Limited computational resources.

The need of key management has been acknowledged in relation to advanced metering infrastructure, but no plan to combat threats has been proposed [14]. In the context of AMI, a key management system (KMS) for unicast, multicast, and broadcast transmission has been developed to improve efficiency, key storage, computation, and administration of keys with forward and backward security based on key graphs [15]. The authors have taken into account key creation and key freshness, authentication and integrity, and forward and backward security, but they have left out the process of key distribution, key destruction, key renewal/revocation, and the node replacement phase. Furthermore, because storage costs have become a concern, SELINDA has been proposed as a key establishment and data collection protocol to allow power operators to initiate shared keys with various measurement devices through an untrusted data relaying data collector unit [16]. In order to assure security during data collection, the data collector unit was not deemed to hold any key established between the power operator and measuring devices. Because the power operator controls all public keys for all measuring devices, unauthorized access or a little breach can put the entire system at risk. Furthermore, untrusted data replay nodes can expose data to a man-in-the-middle attack. The authors of [17] investigated a trust anchor so that a data collector and a device may do mutual authentication and key establishment. The suggested technique considers man-in-the-middle and replay attacks and is based on the public key and Needham–Schroeder authentication protocol. The key benefit of this method is that it reflects high levels of security, fault tolerance, and accessibility [18].

However, the complexity of resource constraint nodes in a smart grid is great due to the integration of both PKI and trustworthy anchors. The approach is not scalable because as the number of devices grows, the trust anchor's ability to perform mutual authentication and key generation may be limited. The authors of [13] demonstrated that a key management scheme for unicast, multicast, and broadcast has been discussed using the Iolus framework, which is based on a secure distribution tree using group security controllers (GSC) to manage top level subgroups and group security intermediaries to manage the remaining subgroups. The GSC is in charge of all subgroup keys, whereas subgroups are in charge of all node keys. In terms of scalability, the Iolus framework can be advantageous because changes in membership within a group will not affect other subgroups. The Iolus-based technique offers limited multicasting functionality as well as the production of a larger number of keys stored in remote terminal units, but at the cost of higher computational overhead and complexity [19].

From the above literature review, it should be emphasized that one of the most basic requirements for grid systems to enable appropriate resilience to attacks is an improved security system. Smart grid systems are expected to use a variety of authentication and secure key management mechanisms. Most of these secure key management systems and practices, on the other hand, are either incomplete or do not correspond to real-world scenarios and emerging technologies. As a result, it is crucial to understand that security requirements differ from one system to the next. As a result, in terms of secrecy, a planned design is required. However, [20] estimated that the D-H key exchange algorithm provides security for unprotected channels by exchanging secret keys over unprotected channels, whereas [21] argued that because the traffic between smart meters and utilities in the smart grid system is predicted to be massive, the most secure authentication and encryption solutions may not be the fastest. As a result, it is apparent that secure authentication and session key exchange systems have an impact on the smart grid communication network's performance. Therefore, the goal of this paper is to enhance the performance of security in smart city communication networks by using a new framework and scheme that provide an authentication and high confidentiality of data by implementing them along with session key exchange schemes through the OPNET simulator modeler.

3. Description of Our Proposed Cyber Security Framework for Smart Cities

This section briefly gives an overview of our proposed framework followed by a description of each operational stage in detail. We finally analyze and evaluate our proposed framework by using a cyber attack and intrusion detection system.

3.1. Overview

The purpose of our proposed cyber security framework is to enhance the communication efficiency as well as to reduce any negative impacts of security. In other words, we aim to improve the security system, which is one of the fundamental requirements for smart city systems, and also to ensure the sufficient resilience between two parties without attacks. Notably, the security requirements can vary from system to system. Hence, the planned design is required in order to overcome these challenges at an early stage.

To deploy our proposed framework along with the key exchange Algorithm 1, two different tasks are used; namely, authentication and encryption between every smart meter and the control center in order to create a two-way secure communication.

Algorithm 1: Modifying the process of D-H keys exchange algorithm.

Procedure Exchanging a random number (x, y)
Select (p, g, x, y)
for *user A* **do**
  $A = g^x \; \mathbf{mod} \; p$
for *user B* **do**
  $B = g^y \; \mathbf{mod} \; p$
*// **Both users exchange keys in order to calculate** K1* and *K2*
$K1 = B^x \; \mathbf{mod} \; p$
$K2 = A^y \; \mathbf{mod} \; p$
Select (t, s, m)
*// **Both users choose** m **as secret number "log"***
for *user A* **do**
  $C = log_m \; (t, K1)$
for *user B* **do**
  $D = log_m \; (s, K2)$
*// **Both users exchange their final keys***
$FK_A = C * D$
$FK_B = D * C$
// $FK_A = FK_B = FK$ ***are shared symmetric keys***
End procedure

3.2. Secure Two-Way Communication by Using an Authentication and Key Exchange Scheme

The authentication and session key discovery portion of secure two-way communication work is presented in Figure 3. Every message, m, will be authenticated and encrypted with key K. Now, every smart meter transmits m to the control center. This m is a trusted communication, which requests a session key for its communication with the control center. A control center generates a K_{SMP}, which includes a smart meter's public key K_{SMCC} and a ticket generated by the control center to secure the communication between them T_{CC}. Upon receiving the K_{SMP}, a smart meter immediately decrypts it. Another message, K_{SMCC}, is sent by the smart meter to the control center, which is a key session between them. This message contains T_{CC} and random secret number S_{SM}. When the message is received, the control center decrypts both T_{CC} and S_{SM}; then, it sends the message back to the smart meter as $S_{SM} + 1$, which is a value of that message encrypted between them in order to protect the two-way communication.

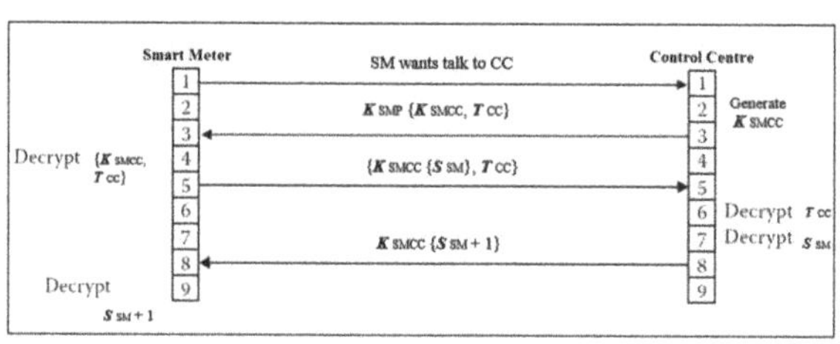

Figure 3. Secure two-way communication by providing an authentication and key session algorithm.

For the authentication and key exchange scheme, which is used between a smart meter and the control center to securely provide a two-way communication as outlined above, two tasks configurations, authentication and encryption tasks, are illustrated below in detail.

3.3. Authentication and Encryption Tasks

The authentication task includes two phases, which are described as follows. In the first phase, a smart meter sends a single message with a specific size to the control center. None of the smart meters spend any initialization time to generate this message. In order to send a response message for every individual smart meter's request, the control center spends only 0.5 s of processing time. The second phase involves confirming the authentication process in each smart meter, and it takes 0.2 s to process the control center's return message.

During the second task, encryption, of securing two-way communication between each smart meter and the control center, two phases are applied to achieve the processing deployment. The first phase consists of encrypting messages before delivering them individually from each smart meter to the control center in order to ensure the connection between these two parties. These encrypted messages are uniformly distributed within a size of 1 to 10 kilobytes, and the process of the encrypting takes only 0.2 s. The second phase encrypts messages before sending them from the control center to each smart meter, as well as ensuring secure connection between them. Table 1 shows the authentication and encryption tasks, as well as the configurations for each phase, each of which has two phases that drive traffic from each smart meter to the control center, including two-way communication, authentication, and encryption.

Table 1. The phases' configuration.

Task Name	Phase Name	Start After	Source	Destination
Authentication task	Smart Meter to Control Center	Application Starts	Smart Meter	Control Center
	Verify Control Centre	Previous Phase Ends	Smart Meter	Not Applicable
Encryption task	Smart Meter to Control Center	Application Starts	Smart Meter	Control Center
	Control Center to Smart Meter	Application Starts	Control Center	Smart Meter

3.4. Key Exchange Scheme

A cryptographic key exchange system is a method of exchanging cryptographic keys. We modified the (D-H) key exchange technique, which enables two-way communication between transmitters with no prior knowledge of one another to join and construct a shared secret key across unsecured network communications. As a result, using a symmetric key, this key may be used to encrypt subsequent transmission, which the destination can then decrypt to safely receive communications. The collaboration of Hitfield Diffie and Martin Hellman produced the first practical technique for establishing shared secret keys between two-side communications (such as A and B) via an unsecured communication network, as shown in our Algorithm 1. The modification of the D-H keys exchange technique is illustrated in the Algorithm 1. In this pseudocode, 'p' denotes a prime number and 'g' indicates a primitive root such that $g < p$. As 'g' is a primitive root of 'p', the numbers $g \ mod \ p, g^2 \ mod \ p, \ldots, g^x \ mod \ p$ will produce all numbers from 1 to $p - 1$. A and B are denoted as Alice and Bob, respectively, and $K1$ and $K2$ indicated the secret key of each. They select the arbitrary numbers 't' and 's' such that 0 < t, s < p. Now, the first user selects a random number (x) as its private key. Its public key is calculated as $g^x \ mod \ p$. Similarly, the second user user selects a random number (y) as its private key. Its public key is calculated as $g^y \ mod \ p$. The public keys are exchanged over a public channel in order to calculate $K1$ and $K2$. Both users are now required to choose m as a secret number. Thereafter, both users exchange their final keys, and then shared the symmetric keys.

3.5. Cyber Attack Evaluation

In this subsection, the analysis and evaluation of our proposed framework requires first to know the nature of the cyber attack. Because of this nature, malicious malware incorporates increasingly complex strategies. As a result, we undertook a more in-depth study of the harmful activities while computing the cyber attack score. The 12 phases of the MITRE ATT&CK were subjected to an analysis, and the findings are displayed.

Each layer, in Figure 4, depicts the steps of a cyber attack in MITRE ATT&CK. The flow of an attack is determined by the layer order. The red line demonstrates a link between a malicious code technique and the previous approaches used. The color of the blue plane indicates the cyber attack strategies at each level. Furthermore, the circle's hue shows the number of cyber attacks that have used it. For example, if there are five or more cyber attack kinds, the circle is dark brown; if there are three or four, it is orange; and if there are two or fewer, it is apricot [8,22,23].

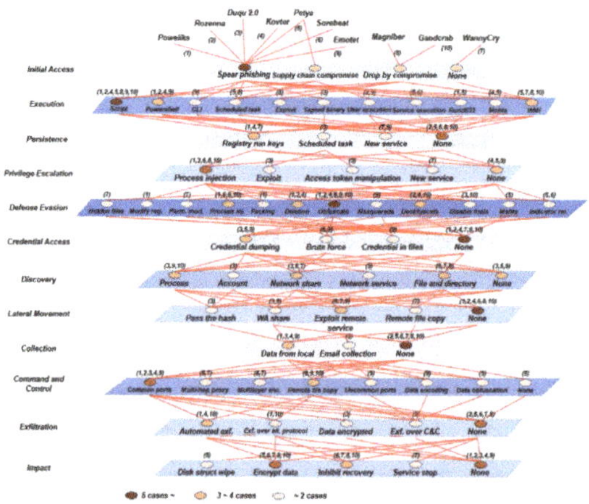

Figure 4. Attack techniques of a fileless cyber attack [8].

Due to these potential cyber attacks that may occur in smart city communication networks, we implemented and applied a man-in-the-middle (MITM) attack and network intrusion detection system (NIDS) in our proposed technique to validate and measure the security requirements maintaining the constrained resources.

Signature-based systems (SBS) and anomaly-based systems (ABS) are examples of NIDS systems that can dynamically monitor and analyze system events to identify whether they are assaults or authorized accesses [24]. The goal of NIDS is to detect malicious behavior related to messages sent across the smart city communication network between a smart meter and the control center. Different sorts of assaults, such as replay and man-in-the-middle attacks, can be detected using the simultaneous network intrusion detection monitoring method. The given topology network is used to investigate the performance of the smart city wireless communication network by implementing authentication and session key exchange schemes, and also by preventing the man-in-the-middle attacks in order to measure our proposed security concerns over the smart city system.

As seen in Figure 5, an MITM attack is characterized as follows: suppose entity "A" wants to monitor the messages sent by entity "B" as a smart meter to entity "C" as the control center server. A man-in-the-middle attack using keys can be conducted in this case to intercept messages between entities "B" and "C". Thereafter, the interceptor can replace the message and deceive entity "A". Entity "A" can be a mobile node observing data between a smart meter and a capturing device in the context of grid computing. Eavesdropping, replay attacks, and physical jamming attacks can all be treated in the same way as man-in-the-middle attacks. Consequently, Figure 6 shows how a man-in-the-middle attack can be detected and messages intercepted across the smart city communication network.

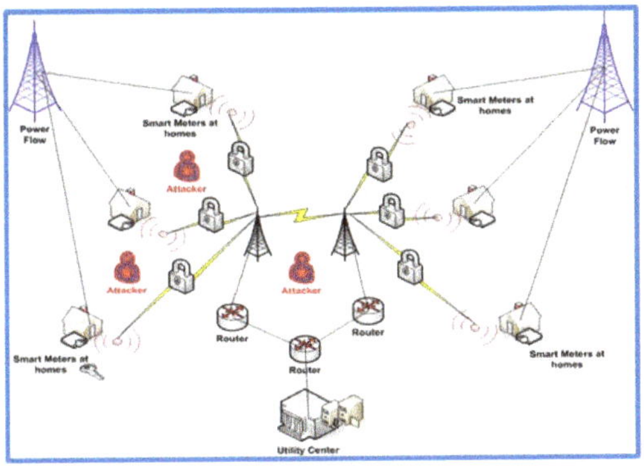

Figure 5. Implementing and applying a man-in-the-middle (MITM) attack and network intrusion detection system (NIDS) in our topology.

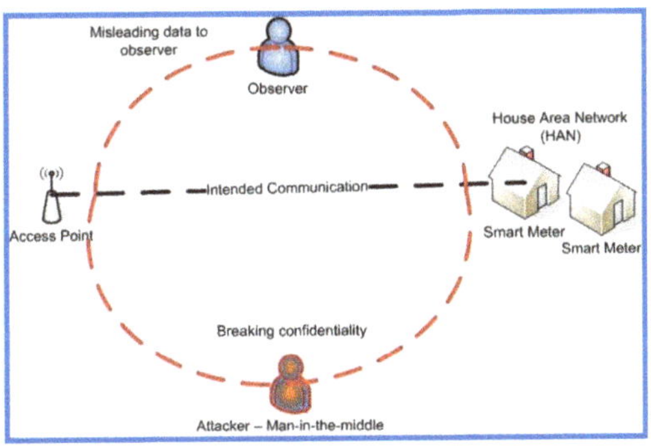

Figure 6. How a man-in-the-middle attack intercepted messages passing via the smart city communication network.

4. Simulation Network and Setup

In this section, we first describe our used simulation tool called OPNET. We then discuss the simulation setup of our proposed framework for enhancing the security control in smart city communication networks. We also use the OPNET simulator to study the performance of the smart city communication network with and without authentication and key exchange mechanisms. Various performance measures are defined in this study such as throughput (packet per seconds), delay (packet per seconds), load, and traffic received (packet per seconds). We finally present and analyze the simulation evaluation.

4.1. Network Simulator

A network simulator is a program that simulates networks with a variety of nodes. To create real-time network experiments, simulation tools are practically required. Several simulation tools, such as the OPNET modeler, and other open source simulators, such as NS-2, GNS2, GNS3, and OMNET, are available.

One of the most widely used data network modeling and network simulation technologies is the OPNET modeler. It is a comprehensive network modeling tool with a slew of useful features. The OPNET modeler, in particular, allows for the simulation of heterogeneous networks using various scenarios, as well as the simulation of specifically planned network architecture and analysis in order to compare different types of traffic and situations [25]. It is also used to support a variety of operating systems, including Linux and Microsoft. Because of its real-time configuration and operation capabilities, the fundamental advantage of using OPNET is the ability to evaluate and analyze networks in realistic experiments. As a result, OPNET allows for the comparison of network improvements based on multiple protocols, as well as the development of new protocols and technologies. OPNET also supports the following simulation technologies: Discreet Event Simulator (DES), Flow Analysis Simulation, ACE Quick Predict Simulation, and Hybrid Simulation [26]. Therefore, DES will be the simulation technology used in this study.

As shown in Table 2, we use the following parameters for both authentication and encryption tasks. Each has two phases, as mentioned above. In the first phase of the authentication task (Smart Meter to Control Center), we set up the requests to be in initialization time = 0 second, request count = 1 (constant), request packet size = 3 Kbytes (constant). For the response of this phase, the processing time = 0.5 s, number of replies = 1 (constant), request packet size = 5 kbytes (constant). In the second phase of the authentication task (Verify Control Center), we also set the requests in the initialization time = 0.2 s, request count = 0 (constant), whereas there is no response in this phase.

Table 2. Simulation parameters.

Parameter	Value
Initialization time of the first phase	0 s
Request count	1 (constant)
Request packet size	3 Kbytes (constant)
Processing time	0.5 s
Number of replies	1 (constant)
Request packet size of the response	5 kbytes (constant)
Initialization time of the second phase	0.2 s
Request count of the second phase	0 (constant)
Initialization time of encryption task	0.2 s
Request count of encryption task	1000 (constant)
Inter-request time	1.2 s (exponential)
Request packet size of encryption task	1–10 kbytes (uniform_int)
Simulation time for each run	3600 s

For the second task (encryption), the first phase (Smart Meter to Control Center) includes the following in the request field: initialization time = 0.2 s, request count = 1000 (constant), inter-request time = 1.2 s (exponential), request packet size = 1–10 kbytes (uniform_int). The request of the second phase (Control Centre to Smart Meter) is exactly the same as that in the first phase of the encryption task. Now, both requests in both phases have no responses.

All the results are averaged over 50 runs for randomly generated topologies, while the simulation time for each run is set to 3600 s.

4.2. Simulation Evaluation

In this set of simulations, we evaluate the throughput packets per seconds, delay, load, and traffic received. As outlined previously, a smart meter is a two-way data transmission device. As a result, a smart meter generates packets and assigns a destination address to each one, which is subsequently sent over an Ethernet switch and then a router to the control center server. Figures 7–12 show the point-to-point throughput packets per second, delay, load, and traffic received between a smart meter and the control center; from the smart meter to the Ethernet switch, the Ethernet switch to the router, and the router to the control center server, respectively.

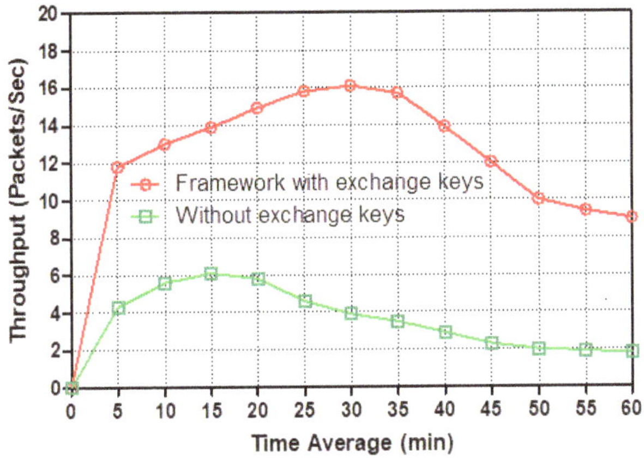

Figure 7. Throughput (packets/s) from a smart meter to the switch for different tasks.

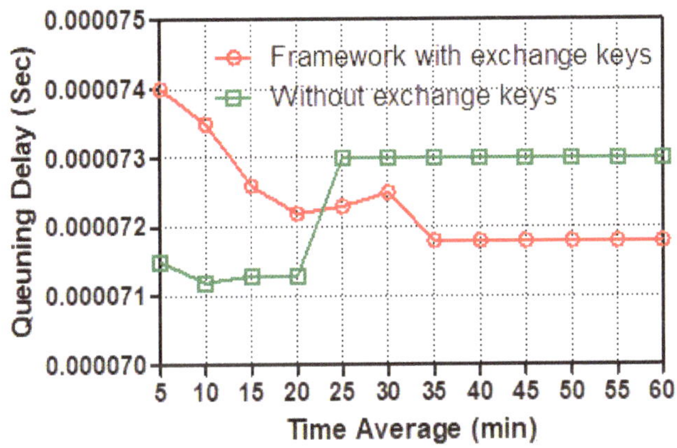

Figure 8. Queuing delay (s) from a smart meter to the switch for different tasks.

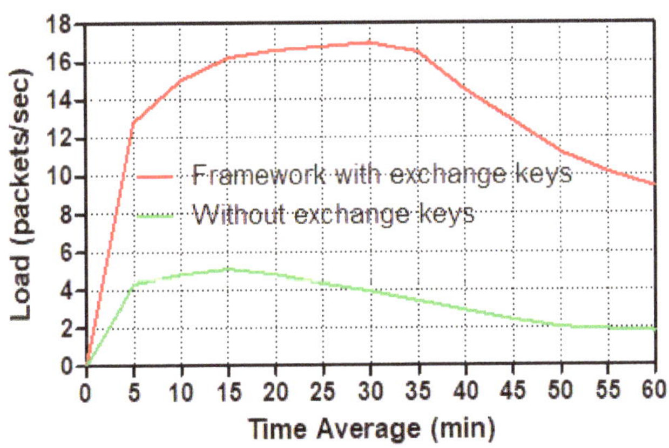

Figure 9. Load (packets/s) in a smart meter for different tasks.

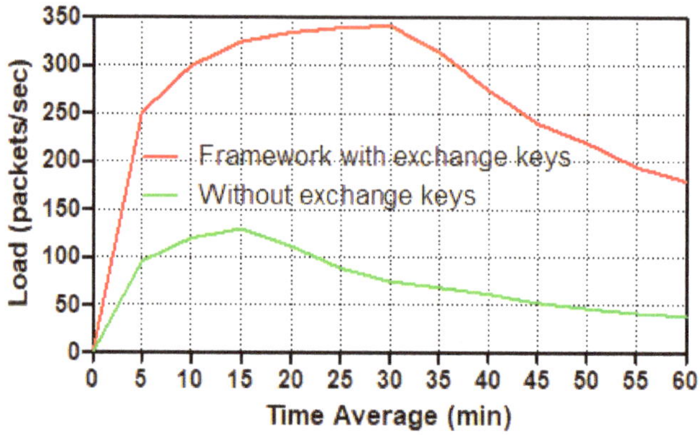

Figure 10. Load (packets/s) in the control center for different tasks.

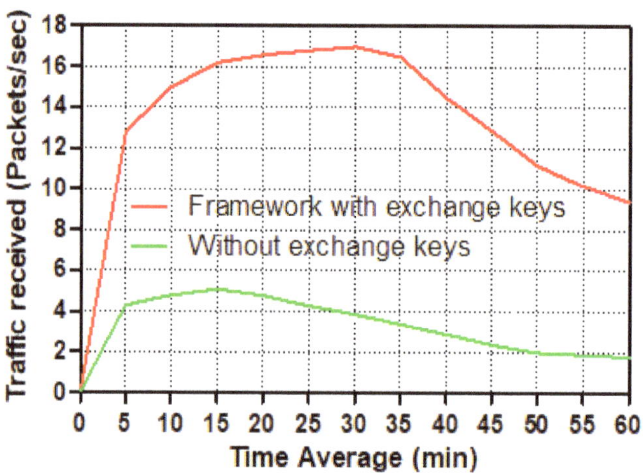

Figure 11. Traffic received (packets/s) in a smart meter for different tasks.

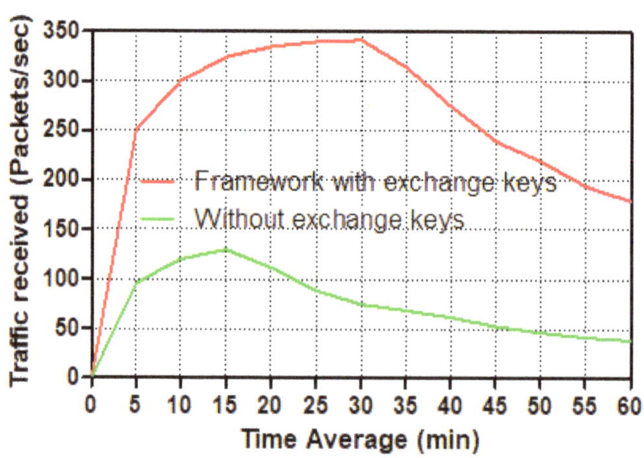

Figure 12. Traffic received (packets/s) in the control center for different tasks.

Figure 7 plots the point-to-point throughput packets per second from a smart meter to an Ethernet switch. It is important to understand the difference between the two line graphs: the red line depicts secure two-way communication from a smart meter to a switch via authentication and key exchange schemes, whereas the green line depicts normal traffic without authentication and key exchange sessions as insecure two-way communication traffic. The temporal average has been used to evaluate point-to-point throughput packets per second between a smart meter and a switch in this diagram. Both line graphs began abruptly from zero to twenty mins with different amounts of packets per second; the green line for insecure traffic nearly reached seven packets per second, whereas the red line for securing traffic nearly doubled the green line, with fifteen packets per second as the point to point throughput. During the 30-min experiment, the red line had the maximum point-to-point throughput packets per second, around seventeen, while the green line had declined somewhat, to four packets per second. Following this, the red and green lines dropped drastically until they reached ten and two packets/second, respectively, at the end of our simulation time (60 mins). Overall, secure transmission employing authentication and session keys outperformed insecure traffic from a smart meter to an Ethernet switch in terms of packets per second.

Figure 8 shows that the best effort between secure and insecure two-way communication was started with a significant point of queuing delay by 0.77 s and 0.69 s, respectively. Both graphs illustrate that the number of seconds has fluctuated significantly after five mins of simulation time. After 10 mins, the seconds of queuing delay in trusted communication are expected to decrease continuously, whereas the seconds of queuing delay in un-trusted communication are expected to grow significantly. Both graphs converge significantly between 15 and 19 mins of simulation time. After that, the un-trusted traffic line (green) stayed in a steady state for about 0.73 seconds, which was almost 0.2 seconds longer than the trusted traffic line of queuing delay until the simulation period was up. To summarize, secure communication traffic between a smart meter and a switch has more service requests and key session establishment than un-secure communication traffic. When services are requested and key sessions are established, the queuing latency in trusted communication appears to be less than in un-trusted communication from a smart meter to the switch.

The load refers to how much load it takes to move packets from a smart meter to a control center or the other way around in the network. For various circumstances, Figures 9 and 10 depict how much load the smart meter and control center require to transport packets across the network.

These two figures show that the number of packets per second for the trusted communication scenario in the smart meter and control center has increased dramatically. During the simulation duration, the number of packets per second peaked at around 34 mins for the smart meter and 36 mins for the control center. Because the control center sent and received packets from twenty smart meters, and the smart meter only sent and received packets from the control center, the control center load had a higher amount of packets per second than the smart meter. At the secure communication scenario in the control center, the load spiked to almost 340 packets per second after 36 mins, then dropped substantially (by roughly 50) to nearly 170 packets per second at the end. Meanwhile, the load in the unsafe scenario increased to just over 120 packets per second at 18 mins, then decreased significantly to roughly 40 packets per second at the conclusion of the simulation duration. The load in the smart meter, on the other hand, experienced a sharp increase in the number of packets per second, reaching over 17 packets at 34 mins; thereafter, the load has reduced dramatically to around 9 packets per second. The load increased dramatically from 1 to 18 mins in the insecure smart meter scenario, reaching 6 packets per second. Following that, the amount of load gradually decreased until it reached around 2 packets per second at the end of the period time. To summarize, because of the requirements of services and the establishment of keys exchange in order to connect securely, both the smart meter and the control center loaded many more packets per second in the secure scenario than in the un-secure situation. As a result, the load for trusted communication exchange was nearly double that of un-trusted communication exchange.

To evaluate the performance of receiving packets per second across the smart grid communication network, the volume of traffic received in packets per second is shown for both smart meters and control centers.

Two distinct possibilities among a smart meter and the control center server are depicted in Figures 11 and 12. From 2 to 38 mins of simulation time, the traffic received packets per second among trusted communication traffic in the smart meter and the control center has increased significantly. Individually, the control center's trusted scenario achieves a peak of around 330 packets per second at 38 mins, following which the traffic received for the trustworthy scenario reduces substantially to just under 170 packets per second at the end. Although the control center's level rose to 120 packets per second between 2 and 19 mins in the un-trusted scenario, the receiving traffic level declined slightly to around 35 packets per second at the end. According to the smart meter, the traffic received in the trusted communication scenario spikes at roughly 38 mins, hitting 17 packets per second, before plummeting to barely 9 packets per second at the end.

Meanwhile, just before 20 mins, the number of packets received in the un-trusted communication situation increased to 6 packets per second. After 20 mins, the amount of traffic received in packets per second rapidly decreased until it reached only 2 packets per second. The control center, on the other hand, has received traffic from 20 smart meters, but according to the smart grid communication system's design, the smart meter can only accept traffic from the control center server. Because it deals with twenty smart meters, the control center has received significantly more traffic packets per second in both scenarios, especially in the scenario with authentication and session key exchange schemes, which required a secure two-way communication channel from each smart meter to the control center. In this situation, the control center receives many more packets per second than in the other scenario, which does not use authentication or key exchange systems.

5. Conclusions

This paper has been enhancing and investigated the performance of securing two-way communication in smart city communication networks by using a new framework and scheme that provide an authentication and high confidentiality of over two different tasks over the OPNET simulator. In order to secure a two-way communication channel, the smart city system can achieve mutual authentication and build shared session key schemes between smart meters and the control center. We researched and assessed the security performance of the smart city communication network with and without our proposed algorithm in terms of throughput and end-to-end delay in our extensive simulation.

The results of this study reveal that raising the time average in all measures within the authentication and session key exchange schemes scenario degrades performance when compared to the situation without authentication and encryption. There is a comparison of different smart city communication network tasks to investigate authentication and session key exchange schemes in smart city communication networks, as well as other reasons for lower performance when using authentication and session key exchange schemes in smart city communication networks.

In the future, instead of considering a handshaking algorithm between neighborhoods, it would be interesting to increase and enhance the cyber security systems by using the proposed framework for underwater sensor networks in [27–29]. Another area of future interest is verifying the protocol's reliability when employing the mobile Autonomous Underwater Vehicle (AUV) in a distributed way in order to develop an efficient cyber security framework [30–33].

Author Contributions: Conceptualization, F.A.A.; Data curation, F.A.A.; Formal analysis, F.A.A.; Funding acquisition, K.K.; Investigation, F.A.A., K.K. and N.M.A.; Methodology, F.A.A.; Project administration, F.A.A.; Resources, F.A.A., K.K. and N.M.A.; Software, F.A.A.; Validation, F.A.A.,

K.K. and N.M.A.; Visualization, F.A.A.; Writing—original draft, F.A.A.; Writing—review and editing, F.A.A., K.K. and N.M.A. All authors have read and agreed to the published version of the manuscript.

Funding: This research work has received funding from the Security Research Center of Naif Arab University for Security Sciences, under grant agreement no PR2-05.

Institutional Review Board Statement: Not applicable.

Informed Consent Statement: Not applicable.

Data Availability Statement: Not applicable.

Conflicts of Interest: The authors declare no conflict of interest.

References

1. Elmaghraby, A.S.; Losavio, M.M. Cyber security challenges in Smart Cities: Safety, security and privacy. *J. Adv. Res.* **2014**, *5*, 491–497. [CrossRef] [PubMed]
2. Yan, J. *Modelling and Analysis on Smart Grid Against Smart Attacks*; University of Rhode Island: Kingston, RI, USA, 2013.
3. Baig, Z.A.; Szewczyk, P.; Valli, C.; Rabadia, P.; Hannay, P.; Chernyshev, M.; Johnstone, M.; Kerai, P.; Ibrahim, A.; Sansurooah, K.; et al. Future challenges for smart cities: Cyber-security and digital forensics. *Digit. Investig.* **2017**, *22*, 3–13. [CrossRef]
4. Kimani, K.; Oduol, V.; Langat, K. Cyber security challenges for IoT-based smart grid networks. *Int. J. Crit. Infrastruct. Prot.* **2019**, *25*, 36–49. [CrossRef]
5. Jain, R.; Nagrath, P.; Thakur, N.; Saini, D.; Sharma, N.; Hemanth, D.J. Towards a Smarter Surveillance Solution: The Convergence of Smart City and Energy Efficient Unmanned Aerial Vehicle Technologies. In *Development and Future of Internet of Drones (IoD): Insights, Trends and Road Ahead*; Springer: Berlin/Heidelberg, Germany, 2021; pp. 109–140.
6. Ghosal, A.; Conti, M. Key management systems for smart grid advanced metering infrastructure: A survey. *IEEE Commun. Surv. Tutorials* **2019**, *21*, 2831–2848. [CrossRef]
7. Gao, J.; Xiao, Y.; Liu, J.; Liang, W.; Chen, C.P. A survey of communication/networking in smart grids. *Future Gener. Comput. Syst.* **2012**, *28*, 391–404. [CrossRef]
8. Kim, K.; Alfouzan, F.A.; Kim, H. Cyber-Attack Scoring Model Based on the Offensive Cybersecurity Framework. *Appl. Sci.* **2021**, *11*, 7738. [CrossRef]
9. Kim, K.; Cho, K.; Lim, J.; Jung, Y.H.; Sung, M.S.; Kim, S.B.; Kim, H.K. What is your protocol: Vulnerabilities and security threats related to Z-Wave protocol. *Pervasive Mob. Comput.* **2020**, *66*, 101211. [CrossRef]
10. Kim, K.; Kim, J.S.; Jeong, S.; Park, J.H.; Kim, H.K. Cybersecurity for autonomous vehicles: Review of attacks and defense. *Comput. Secur.* **2021**, *103*, 102150. [CrossRef]
11. Axelrod, C.W. Applying lessons from safety-critical systems to security-critical software. In Proceedings of the 2011 IEEE Long Island Systems, Applications and Technology Conference, Farmingdale, NY, USA, 6 May 2011; pp. 1–6.
12. Dán, G.; Sandberg, H.; Ekstedt, M.; Björkman, G. Challenges in power system information security. *IEEE Secur. Priv. Mag.* **2012**, *10*, 62–70. [CrossRef]
13. Long, X.; Tipper, D.; Qian, Y. An advanced key management scheme for secure smart grid communications. In Proceedings of the 2013 IEEE International Conference on Smart Grid Communications (SmartGridComm), Vancouver, BC, Canada, 21–24 October 2013; pp. 504–509.
14. Shein, R. Security measures for advanced metering infrastructure components. In Proceedings of the 2010 Asia-Pacific Power and Energy Engineering Conference, Chengdu, China, 28–31 March 2010; pp. 1–3.
15. Liu, N.; Chen, J.; Zhu, L.; Zhang, J.; He, Y. A key management scheme for secure communications of advanced metering infrastructure in smart grid. *IEEE Trans. Ind. Electron.* **2012**, *60*, 4746–4756. [CrossRef]
16. Dán, G.; Lui, K.S.; Tabassum, R.; Zhu, Q.; Nahrstedt, K. SELINDA: A secure, scalable and light-weight data collection protocol for smart grids. In Proceedings of the 2013 IEEE International Conference on Smart Grid Communications (SmartGridComm), Vancouver, BC, Canada, 21–24 October 2013; pp. 480–485.
17. Wu, D.; Zhou, C. Fault-tolerant and scalable key management for smart grid. *IEEE Trans. Smart Grid* **2011**, *2*, 375–381. [CrossRef]
18. Alzahrani, N.M.; Alfouzan, F. Augmented Reality (AR) and Cyber-security for Smart Cities. *Sensors* **2022**, *22*, 2792. [CrossRef] [PubMed]
19. Chen, D.; Wawrzynski, P.; Lv, Z. Cyber security in smart cities: A review of deep learning-based applications and case studies. *Sustain. Cities Soc.* **2021**, *66*, 102655. [CrossRef]
20. Ibrahem, M.K. Modification of Diffie-Hellman key exchange algorithm for Zero knowledge proof. In Proceedings of the 2012 International Conference on Future Communication Networks, Baghdad, Iraq, 2–5 April 2012; pp. 147–152.
21. Mahmood, K.; Chaudhry, S.A.; Naqvi, H.; Shon, T.; Ahmad, H.F. A lightweight message authentication scheme for smart grid communications in power sector. *Comput. Electr. Eng.* **2016**, *52*, 114–124. [CrossRef]
22. Qureshi, K.N.; Ahmad, A.; Piccialli, F.; Casolla, G.; Jeon, G. Nature-inspired algorithm-based secure data dissemination framework for smart city networks. *Neural Comput. Appl.* **2021**, *33*, 10637–10656. [CrossRef]
23. Lee, G.; Shim, S.; Cho, B.; Kim, T.; Kim, K. Fileless cyberattacks: Analysis and classification. *ETRI J.* **2021**, *43*, 332–343. [CrossRef]

24. Obimbo, C.; Zhou, H.; Wilson, R. Multiple SOFMs working cooperatively in a vote-based ranking system for network intrusion detection. *Procedia Comput. Sci.* **2011**, *6*, 219–224. [CrossRef]
25. Saputro, N.; Akkaya, K.; Uludag, S. A survey of routing protocols for smart grid communications. *Comput. Netw.* **2012**, *56*, 2742–2771. [CrossRef]
26. Christhu, M.; Marium, N.; Major, J.; Shibin, D. A comprehensive overview on different network simulators. *Int. J. Eng. Technol. (IJET)* **2013**, *5*, 325–332.
27. Alfouzan, F.; Shahrabi, A.; Ghoreyshi, S.; Boutaleb, T. An Efficient Scalable Scheduling MAC Protocol for Underwater Sensor Networks. *Sensors* **2018**, *18*, 2806. [CrossRef]
28. Alfouzan, F.A.; Shahrabi, A.; Ghoreyshi, S.M.; Boutaleb, T. A Collision-Free Graph Coloring MAC Protocol for Underwater Sensor Networks. *IEEE Access* **2019**, *7*, 39862–39878. [CrossRef]
29. Alfouzan, F.A.; Shahrabi, A.; Ghoreyshi, S.M.; Boutaleb, T. An Energy-Conserving Collision-Free MAC Protocol for Underwater Sensor Networks. *IEEE Access* **2019**, *7*, 27155–27171. [CrossRef]
30. Alfouzan, F.A.; Ghoreyshi, S.M.; Shahrabi, A.; Ghahroudi, M.S. An AUV-Aided Cross-Layer Mobile Data Gathering Protocol for Underwater Sensor Networks. *Sensors* **2020**, *20*, 4813. [CrossRef] [PubMed]
31. Alfouzan, F.A. Energy-efficient collision avoidance MAC protocols for underwater sensor networks: Survey and challenges. *J. Mar. Sci. Eng.* **2021**, *9*, 741. [CrossRef]
32. Alfouzan, F.; Shahrabi, A.; Ghoreyshi, S.M.; Boutaleb, T. An energy-conserving depth-based layering mac protocol for underwater sensor networks. In Proceedings of the 2018 IEEE 88th Vehicular Technology Conference (VTC-Fall), Chicago, IL, USA, 27–30 August 2018; pp. 1–6.
33. Alfouzan, F.; Shahrabi, A.; Ghoreyshi, S.; Boutaleb, T. A comparative performance evaluation of distributed collision-free MAC protocols for underwater sensor networks. In Proceedings of the 8th International Conference on Sensor Networks (SENSORNETS 2019), Setubal, Portugal, 26–27 February 2019; pp. 1–9.

sensors

MDPI

Article

Cyber Threat Intelligence-Based Malicious URL Detection Model Using Ensemble Learning

Mohammed Alsaedi [1], Fuad A. Ghaleb [2,*], Faisal Saeed [1,3], Jawad Ahmad [4] and Mohammed Alasli [1]

[1] College of Computer Science and Engineering, Taibah University, P.O. Box 344, Medina 41411, Saudi Arabia; masadi@taibahu.edu.sa (M.A.); fsaeed@taibahu.edu.sa (F.S.); masali@taibahu.edu.sa (M.A.)
[2] School of Computing, Faculty of Engineering, Universiti Teknologi Malaysia, Johor Bahru 81310, Malaysia
[3] DAAI Research Group, Department of Computing and Data Science, School of Computing and Digital Technology, Birmingham City University, Birmingham B4 7XG, UK
[4] School of Computing, Edinburgh Napier University, Edinburgh EH10 5DT, UK; j.ahmad@napier.ac.uk
* Correspondence: abdulgaleel@utm.my

Abstract: Web applications have become ubiquitous for many business sectors due to their platform independence and low operation cost. Billions of users are visiting these applications to accomplish their daily tasks. However, many of these applications are either vulnerable to web defacement attacks or created and managed by hackers such as fraudulent and phishing websites. Detecting malicious websites is essential to prevent the spreading of malware and protect end-users from being victims. However, most existing solutions rely on extracting features from the website's content which can be harmful to the detection machines themselves and subject to obfuscations. Detecting malicious Uniform Resource Locators (URLs) is safer and more efficient than content analysis. However, the detection of malicious URLs is still not well addressed due to insufficient features and inaccurate classification. This study aims at improving the detection accuracy of malicious URL detection by designing and developing a cyber threat intelligence-based malicious URL detection model using two-stage ensemble learning. The cyber threat intelligence-based features are extracted from web searches to improve detection accuracy. Cybersecurity analysts and users reports around the globe can provide important information regarding malicious websites. Therefore, cyber threat intelligence-based (CTI) features extracted from Google searches and Whois websites are used to improve detection performance. The study also proposed a two-stage ensemble learning model that combines the random forest (RF) algorithm for preclassification with multilayer perceptron (MLP) for final decision making. The trained MLP classifier has replaced the majority voting scheme of the three trained random forest classifiers for decision making. The probabilistic output of the weak classifiers of the random forest was aggregated and used as input for the MLP classifier for adequate classification. Results show that the extracted CTI-based features with the two-stage classification outperform other studies' detection models. The proposed CTI-based detection model achieved a 7.8% accuracy improvement and 6.7% reduction in false-positive rates compared with the traditional URL-based model.

Keywords: malicious URLs; cyber threat intelligence; ensemble learning; internet security; cybersecurity

Citation: Ghaleb, F.A.; Alsaedi, M.; Saeed, F.; Ahmad, J.; Alasli, M. Cyber Threat Intelligence-Based Malicious URL Detection Model Using Ensemble Learning. *Sensors* **2022**, *22*, 3373. https://doi.org/10.3390/s22093373

Academic Editor: Jiankun Hu

Received: 7 March 2022
Accepted: 20 April 2022
Published: 28 April 2022

1. Introduction

Recently, the number of users surfing the Internet has increased exponentially. Due to the proliferation of mobile devices, ad hoc networks, smart sensors, and the Internet of Things technologies fueled by the imposed lockdown to mitigate the COVID-19 pandemic, the Internet has become an essential part of people's daily lives and activities worldwide [1–4]. Most businesses shifted online due to the availability of reliable infrastructures such as cloud storage, cost-effective platforms, and a large target market. However, the Internet brings many cyber threats such as malware, spamming, phishing, financial fraud, information theft, and data sabotage [3–6]. Malicious websites are the primary attack vector that is used by cybercriminals to spread malware and archive attackers' objectives [1]. A malicious website contains content that can be harmful such as malware or phishing attacks infecting the visitors' smart devices with malware without user interaction, such as clicking or downloading, with the website.

According to [5], 18.5 million websites are infected by malware. Moreover, according to Google's safe browsing report [6], there were two million phishing websites in September 2020, an increase of nearly 2800% compared with the number in September 2010. Attackers spread fake information and advertisements to attract users to visit malicious websites. Once a victim visits a malicious website, attackers use different strategies to infect users browsing devices with malicious payloads or deceive victims into interacting with the attackers for financial fraud or other types of attacks. Many harmful websites are not intended to be malicious by the developers. Attackers can exploit vulnerable websites to perform malicious intent. For example, an attacker can inject cross-site scripting into a vulnerable website to steal a visitor victim's sensitive information or perform a phishing attack [7].

The problem with detecting malicious websites has been around since early 2004 [8–18]. Many solutions have been proposed to accurately detect these websites. These solutions can be divided into three categories by their source of investigation: URL-based [8–16], web content-based [19–21], and script-based [17,18]. URL-based detection is the most investigated approach followed by content-based detection, while little research has been investigated on script-based detection. URL-based detection is preferable because it is a proactive and safe approach for the detection machines as it can detect the malicious URLs before it is visited by the user. Moreover, detecting malicious URLs is more efficient for real-time detection and resource-constrained applications such as mobile and Internet of Things (IoT) devices.

Various techniques have been suggested to detect malicious websites and harmful content by extracting features from their URLs [15,16,22–28]. Most of these techniques rely on humans to derive the features [16,22–26] while few solutions used deep learning techniques to automate the features [15,27,28]. Many sets of features were extracted and used for the detection including host information features such as country name and host sponsor, domain features such as .com and .tk, and lexical features such as the number of dots in URL and URL length. However, the URL-based features are subject to manipulation by attackers and can be dynamically changed, and may be insufficient for effective representation. Attackers can use evasive techniques to bypass the security countermeasures. Accordingly, any features extracted from these URLs can be misleading as attackers can manipulate them to hide the malicious intent and malicious patterns of the website. Therefore, features that are out of attackers' control will be beneficial for improving detection accuracy and reducing the false alarm rate.

The CTI feature can be used to enrich URL-based features to improve detection performance. Cybersecurity analysts, users' experiences, and website reputations can be important sources of information. People usually share knowledge regarding malicious websites in discussion forums, social media, and news websites. Cyber threat intelligence can be safer, more efficient, and provide more accurate results than investigating the website content. This study designed and developed a malicious URL detection model that utilizes cyber threat intelligence-based features to improve classification performance. The proposed model, called the Cyber Threat Intelligence-based Malicious URL Detection model (CTI-MURLD), consists of three main components. The first component is for feature collection. Three types of features are extracted: URL-based features, Whois information-based features, and cyber threat intelligence features. The threat factors are researched using Google searches and Whois information. The second component contains data processing, including feature extraction, representation, and selection. N-gram is used for feature extraction, the Term Frequency-Inverse Document Frequency (TF-IDF) technique is used for feature representation, and mutual information (information gain) is used for feature selection. The third component is classification and decision making. The RF algorithm was used to train three ensemble classifiers. Each classifier was trained using different features. The probabilistic outputs of the decision tree classifiers in each forest were aggregated and used to train a multilayer perceptron-based classifier for decision making. The multilayer perceptron-based classifier could learn the hidden patterns that map the classifier's output with the correct class of URLs. We hypothesize that the multilayer perceptron-based classifier can be more effective than the random forest classifiers' three independent majority voting schemes. The results of the experiments were validated employing commonly used performance measures and benchmarked using a widely accepted dataset that contains benign and malicious URLs. Additionally, a comparison of related studies was carried out that shows the superiority of the proposed work. The results show a significant improvement in the proposed model's performance compared with the state-of-the-art models. This study makes the following contributions:

1. A malicious URL-detection model based on CTI was designed and developed. Both the URL and web content are subject to obfuscation; an independent source of features that are outside of the attacker's control was needed to strengthen the model's performance. Thus, cyber threat intelligence-based features were extracted from a Google search and Whois information and used as new knowledge to train the proposed detection model.
2. The study designed and developed an ensemble learning-based model that combines three random forest-based predictors such as predetection and feature extractions with multilayer perceptron-based classifiers for the final decision. Three RF classifiers were trained using different feature dimensions extracted from URLs, Whois, and Google-based CTI. The three majority voting schemes that were used by the trained RF classifiers were replaced by the trained multilayer perceptron-based classifiers for accurate detection.
3. Several machine learning algorithms have been investigated, including deep learning techniques such as the convolutional neural network (CNN) model and sequential deep learning model, which were trained to distinguish between malicious and benign patterns. Results demonstrated that the cyber threat intelligence collected from Google improves the detection performance of malicious websites.

The remainder of the manuscript is organized as follows: Section 2 reviews the related work; Section 3 describes the proposed model; Section 4 explains the experimental design; Section 5 presents the results with a detailed discussion; Section 6 presents the conclusion and future work.

2. Related Work

For many decades, malicious URL detection has been a major concern for cybersecurity specialists [8–14]. Several solutions have been proposed to detect malicious URLs and protect users from being victims of an attack. These solutions can be categorized based on the type of detection into feature-based detection or blacklist-based detection [23]. In feature-based detection, the features that represent the URLs are extracted and automatically analyzed while blacklist-based detection relies on user reports and expert analysis. The centralized blacklist is the most widely used detection method in practice. The Internet Protocol (IP) address of the malicious website is stored in a database through matching detection. The feature-based detection can be further categorized into URL based features or web content-based features. In the former, the features are extracted from the URL's characters using N-gram techniques or derived directly from the URL (i.e., the length of the URL, whether it contains a file, the status, request protocol, IP, domain name, and registrar information). Meanwhile, in the latter, the features are crawled from the web content in terms of text, HTML code, and programs scripts. Detecting malicious URLs is crucial as many attackers spread malicious links to legitimate websites such as social media platforms and e-mails. Moreover, some malicious URLs are spread by downloading malware which can infect the detection machine during the crawling. Furthermore, detecting malicious URLs is more efficient and accurate than detecting web content due to the high similarity of some malicious web content with legitimate content, for example, phishing and fraudulent websites. Accordingly, this study focuses on reviewing URL-based detection solutions.

In [26], the authors proposed an improved malicious URL-detection model based on a two-stage distance-metric learning approach, namely singular value decomposition and linear programming for feature extraction. A set of 62 features were extracted from the URLs including information from Whois such as top-level domain names (TLDs), registrar information, lexical features such as the number of dots, keywords, and reputation-based features. A dataset consisting of 33,1622 URLs was collected from "PhishTank" and used to train three machine learning classifiers for the evaluation, namely K-nearest neighbor, support vector machine, and neural networks. Results showed that the improvement of the proposed feature extraction method was significant. However, the results showed that the false alarm (false positive) and misrate (false negative) were still high.

Rakesh and Muthurajkumar [22] modified the C4.5 algorithm to detect cross-site request forgery. Authors in [23] analyzed malicious URLs to extract common features regarding attacker behavior. A similarity matching technique was used to detect attackers' habitual behavior. A small set of features were extracted from the URLs. Chiramdasu and Srivastava [16] proposed a malicious-URL-detection model using logistic regression. Three sets of features were extracted, host information features such as country name and host sponsor, domain features such as .com and .tk, and lexical features such as the number of dots in the URL and URL length. He and Li [24] focused on the class imbalance issue and then trained a model using XGBoost with cost-sensitive learning for detecting malicious URLs. A total of 28 features were extracted from the domain name, Whois information, geographic information, and suspicious words. Despite the results demonstrating that the proposed model outperformed related studies, the poor sensitivity achieved is the main limitation of this model. Authors in [29] proposed ensemble learning using a support vector machine (SVM) and a neural network to identify the command and control (C&C) server. The classifiers were trained based on features extracted from Whois and the DNS of domains of C&C servers. Another study [25] extracted 117 features from URL features, lexical features, domain name features, webpage source features, and short URL features. Then, various decision-tree-based learning algorithms were studied including J48 decision tree, simple CART, random forest (RF), random tree, ADTree, and REPTree for detecting malicious URLs. Results showed that the random forest-based classifier outperformed other constructed classifiers. In [30], two classifiers were trained using naïve Bayes and logistic regression. Different sets of features were extracted including lexical features and textual features represented by terms frequency/inverse documents frequency (TF-IDF). In their experiments, logistic regression outperformed the naïve Bayes algorithm.

The performance of various deep learning techniques in detecting malicious URLs was evaluated in [28]. The evaluated techniques included the recurrent neural network (RNN), identity-recurrent neural network (I-RNN), long short-term memory (LSTM), convolution neural network (CNN), and convolutional neural network-long short-term memory (CNN-LSTM). The model constructed using LSTM and the hybrid network of CNN and LSTM outperformed other studied models.

To summarize, many solutions have been proposed for detecting malicious URLs [16,22,26]. Most of these solutions utilize supervised-based machine learning techniques for classification [16,26,29]. The deep learning approach has also been investigated [28]. However, most of these solutions extract the features solely from URLs such as lexical features, textual features, and host features. It is commonly agreed among researchers that obfuscated URLs and web content hinder effective detection. CTI has not yet been investigated for improving detection performance. Therefore, this study proposed a malicious URL detection model that utilized CTI to extract features safely without crawling the actual malicious websites. User expertise regarding malicious URLs can be used for the early detection of URLs without the need for intensive analysis of the websites. Due to their classification performance and ability to extract effectiveness patterns from textual-based features, the RF algorithm and the multilayer perceptron were combined to improve the classification performance. A detailed description of the proposed model is provided in the following section.

3. The Proposed CTI-MURLD Model

Figure 1 shows the proposed Cyber Threat Intelligence-based Malicious URL Detection (CTI-MURLD) model. The proposed CTI-MURLD model consists of seven phases: data collection, feature preprocessing, feature extraction, feature representation, feature selection, ensemble learning-based prediction, and decision making. In the first six phases, three ensemble-learning-based predictors were constructed using the random forest (RF) algorithm. Each RF-based predictor was trained using different feature sets, URL-based, Google-based CTI, and Whois-based features. Each RF classifier had two probabilistic outputs. The first output represents the belief that a sample is a malicious URL (p_0 in Figure 1), and the second output is the amount of belief that the URL is benign (p_1 in Figure 1). In the last phase, an artificial neural network (ANN) classifier was built for decision making. The probabilistic outputs of the three RF classifiers were used to train the ANN classifier for the final decision. As shown in Figure 1, the URLs requested by users were intercepted, and three types of features were extracted. Each type of feature set was preprocessed to remove the noise. Then, more features were extracted using the N-gram technique and then represented by the TF-IDF technique. Then, the most representative features in each set were selected from each feature set (denoted by f_1 to f_n) using information gain. Each feature set is passed to its specifically trained RF predictor. The probabilistic outputs (two probabilistic outputs for each predictor) of these predictors were fed into the ANN classifier for the final decision about the URL class, whether it was malicious or benign. A detailed description of each phase is presented in the following subsections.

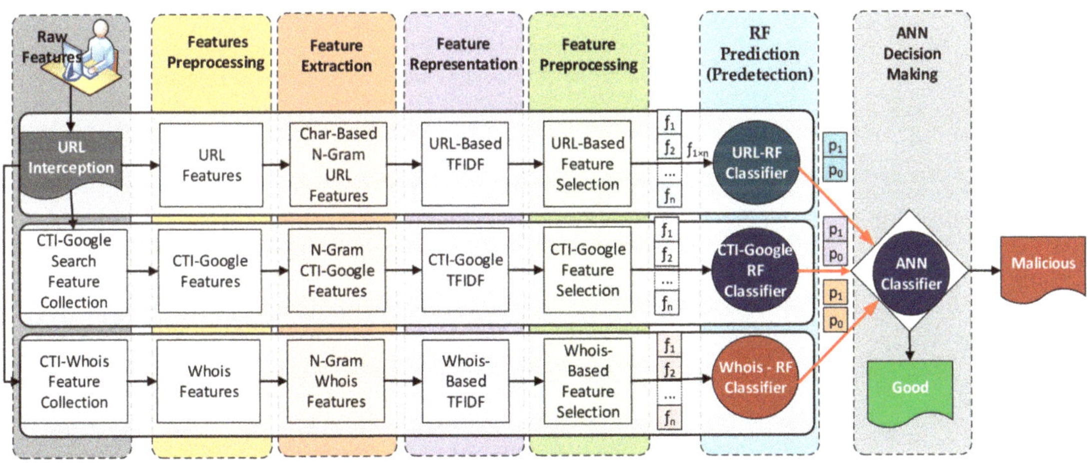

Figure 1. The proposed CTI-MURLD model.

3.1. Phase 1: Data Collection Phase

In this phase, three types of features were collected, namely, URL content features, cyber threat intelligence data crawled using a web search (Google-based CTI), and data related to the domain owners crawled from Whois lookup (Whois-based CTI). The URL data were collected by intercepting user HTTP requests in the application layer. To collect the Google-based CTI, first, the domain name was extracted from the URL and the IP address of the domain was searched, and then data related to the domain and its IP were crawled from a Google search. The Whois-based CTI was crawled from a Whois search which included the website owner, creation date, contacts, domain status, registrant email, and registrant country.

3.2. Phase 2: Data Preprocessing

In this phase, the textual data that were collected in the previous phase were sanitized and normalized. The URL was preprocessed by removing symbols while the Google-based CTI and Whois-based CTI data were preprocessed using natural language processing (NLP) text-preprocessing techniques. Since Google-based CTI and Whois-based CTI data were crawled from websites, unwanted text such as HTML codes, symbols, and punctuation were removed to reduce feature complexity and enhance the classification performance. The collected text data were converted to lower case and then normalized. The normalization process aims were two-fold. Firstly, to convert the text from unstructured data to a structured word vector. Secondly, to reduce the scarcity of the feature vectors by removing unnecessary words and reducing the number of words by rooting the words to their originals. The normalization started with tokenization, then the removal of stop words, lemmatization, stemming, and finally converting the words to their equivalent numerical format. Tokenization is the process of representing the text sample by a list of words that construct the URL data sample. Stemming is converting the words into their roots e.g., removing "s" from the plural words and removing "ing" from the word. Lemmatization is the process of converting the words into base form by rooting the verbs to their root using lexical knowledge base e.g., 'took' to 'take'.

3.3. Phase 3: N-Gram Feature Extraction

The N-gram technique [31] was used to enrich the feature sets and create more representative features. N-gram has been a commonly used method for malicious URL detection and text analysis due to its effectiveness in improving the classification accuracy as reported by previous researchers [32–36]. Both word N-gram and character N-gram were used in this study. The character N-gram was used to extract features from the URL while the word N-gram was used for Google-based CTI and Whois-based CTI data. The URL data were converted to vectors of words each consisting of three, four, or five characters. To reduce feature complexity the word bi-gram technique was used for Google-based CTI and Whois-based CTI data. Each subsequent word was considered one additional feature. The output of this phase was three feature vectors each consisting of sets of words called tokens.

3.4. Phase 4: TF-IDF Feature Representations

To convert the words (the tokens) to their equivalent numerical values, a corpus that contained the list of unique tokens was constructed based on their frequency of occurrence in each class. Then, the statistical-based text representation, namely TF-IDF was calculated using the following equation:

$$tf\_idf = tf.log\frac{N}{df} \tag{1}$$

where tf is the term frequency of the word in a specific instance, df is the document frequency for the word, N is the number of samples in the dataset. The term frequency tf is the number of times a word has occurred in the sample while the inverse document frequency idf refers to the inverse number of documents where the word has occurred. The higher the tf_idf of a word in a document, the more relevant the document. The output of this phase was three numerical vectors for each sample.

3.5. Phase 5: Feature Selections

In this phase, the features that represented the URL well were selected using information gain (mutual information). As the CTI features were collected from Google, a huge number of irrelevant features were included. These irrelevant features hindered the ability to differentiate between benign and malicious URLs due to the high dimensionality of the features. Thus, the learning task became complex, leading to poor training accuracy [28,37]. Similarly, the Whois information and URL features also contained irrelevant features, especially when the N-gram was used. The features were doubled based on the n-value of the N-gram. Moreover, feature selection is common research procedure for text-based features [12,28,37,38]. Therefore, feature selection is important in this study. However, this study selected the top five thousand features to minimize the probability of losing some information while maximizing the generalizability of the trained models.

The features with low probabilities (the uncommon features) have more information compared to features with high probabilities (the common features). The mutual information-based feature selection uses entropy to measure the impurity of the features when it is used to split the target variable. The entropy can be calculated using Equation (2). The higher the entropy the more information. Mathematically, the entropy is written as:

$$E(p) = -\sum_{i=1}^{n} p_i \log(p_i) \tag{2}$$

where n is the target class, p_i the probability of a feature split the class i. The information gain which represents the quality of the split can be calculated using the following equation.

$$Gain = 1 - E(p) \tag{3}$$

where n is the target class of the entropy and the $Gain$ is the quality of the split. A feature is important for classification if it has a high gain. The higher the gain, the lower the entropy. If the entropy is zero, the less impure the split. The output of this phase is a feature vector with only high-gain features selected.

3.6. Phase 6: RF Ensemble-Based Prediction

Three predictors were constructed and grouped using the RF algorithm in this phase. A predictor was trained for each type of feature, namely URL, Google-CTI, and Whois-CTI. A random forest algorithm was selected to construct these predictors. RF was selected for two reasons: firstly, for its diversity, which fits the diverse nature of the features in our collected datasets, and secondly for its effectiveness with high-dimensional data. Even after selecting a subset of important features, a high-dimensional vector consisting of 5000 elements was selected so that we did not lose the valuable features and generalizability. RF is a supervised machine learning algorithm that trains ensembles of weak classifiers using decision trees and bagging methods. The RF algorithm searches for the best split in a random subset of features before the tree is constructed. Thus, diverse trees were constructed that would improve the model's performance. The RF classifier was constructed using 100 decision tree classifiers. Each decision tree classifier was trained based on a random subset of the original features with a random subset of the training dataset. The results were three forests of weak but diverse classifiers. The probabilistic outputs of these weak classifiers were averaged to be used as the RF decision about the class of the sample. The output of a tree was a real number between zero and one for each class. When the output value approaches zero, it means a low probability that the sample belongs to that class. Because the trees in the three RF classifiers were trained based on three different datasets, the results were more diverse and thus the probabilistic output. Instead of using the majority voting as the RF, in this study, the probabilistic outputs of the ensemble classifiers are used as input to the artificial neural network (ANN) classifier for decision making. Meanwhile, if the output value approaches one, it indicated a high belief that the sample belonged to that class. Figure 2 illustrates how these probabilities are extracted and fed as new features to the next stage of classification for decision making.

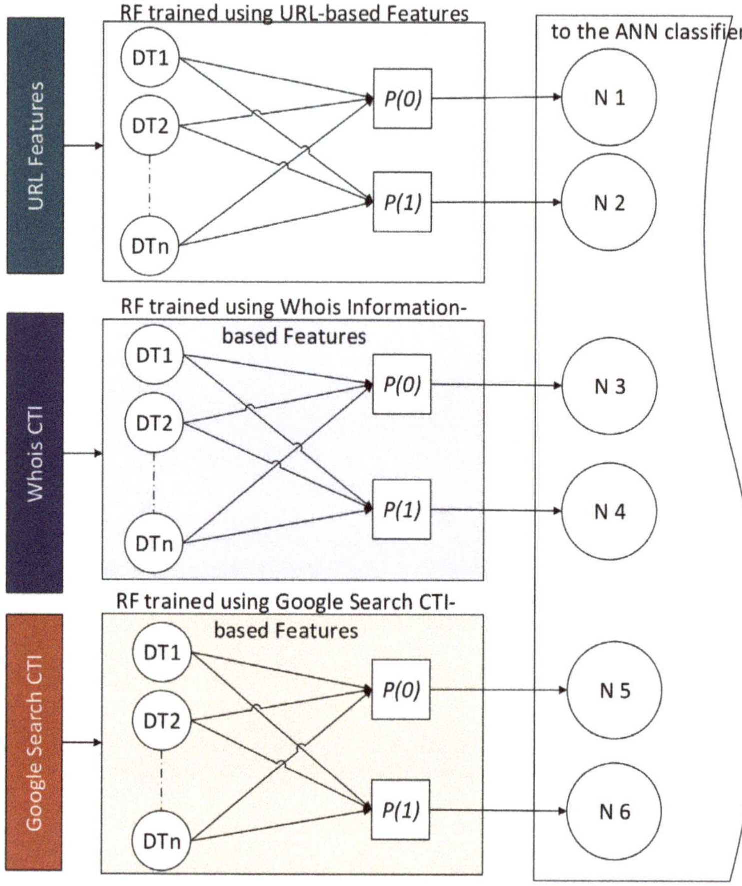

Figure 2. The new features extracted from the three ensemble models.

In Figure 2, $P(0)$ is the average probability of predicting a benign URL. In contrast, $P(1)$ is the average probability of predicting a malicious URL, DT denotes a decision tree (the weak classifier), and N1 − 6 represents the neuron node in the ANN model. These probabilistic values can be calculated as follows

$$P(0) = \frac{\sum_{i=0}^{n} p(class\_label = 0)}{n} \tag{4}$$

$$P(1) = \frac{\sum_{i=0}^{n} p(class\_label = 1)}{n} \tag{5}$$

where n denotes the total number of the estimators in each forest. These outputs are aggregated using a voting scheme in the standard RF algorithm, and the decision is based on the majority. In contrast, this study replaces the voting scheme of the three trained RF classifiers with one trained using the multilayer perceptron (MLP) algorithm for decision making. The MLP-based classifier uses the aggregated outputs of the RF classifiers as new knowledge to train the ANN classifier to learn the hidden patterns that can collectively be extracted from the outputs of these three ensemble models. A detailed description of this is explained in the next section.

3.7. Phase 7: ANN Decision Making

In this phase, a multilayer perceptron (MLP) artificial neural network (ANN) classifier was constructed for decision making. The classifier was trained using a three-layer network consisting of 6 input neurons, 6 hidden neurons, and one output neuron. ANN has better generalization and can predict the actual class even with smaller data and complex nonlinear problems. Given a set of input features $X = (x_1, x_2, x_3 \ldots x_n)$ and Y target class, the MLP learns the relationship between the X and Y. Some parameters affect the performance of the neural network such as weight initialization, biases, the activation function, the loss function, the optimizer, the number of hidden layers, and the number of neurons in each layer. The activation function provides output for the next layer by calculating the sum of the products of numerical values of input features by their weights. The loss or cost function is used to determine the classification error while the optimizer is used to reduce the error. In this study, the Broyden–Fletcher–Goldfarb–Shanno (BFGS) algorithm was used as the optimization algorithm. BFGS is a local search and gradient-based algorithm that is suitable for unconstrained nonlinear optimization problems to effectively determine the decent direction. It approximates the second derivative of the cost function (the Hessian) when the second derivative cannot be detected. The Sigmoid function in the following equation is used as an activation function:

$$Segmoid(x) = \frac{1}{1 + e^{-x}} \tag{6}$$

where x denotes the classification score and e is the natural logarithm which is approximately equal to 2.718281828.

To summarize, Figure 3 illustrates the operations of the proposed CTI-MURLD model. As can be seen in Figure 3, once the URL was intercepted (e.g., by the network sniffer of the detection system), three types of features were collected: the first types were the URL features such as the domain name, sub-domains, and types; the second types were the CTI features which were collected from a Google search; and the third types of features were collected from Whois information. These features were preprocessed, enriched using N-gram, and represented using TF-IDF techniques, as described in Sections 3.1–3.3. The important features were selected and input into the three pre-trained RF-based prediction models. Inspired by the divide and conquer principle, the RF prediction models were trained based on a single type of feature set, namely, CTI-Google, CTI-Whois, or URL features. The probabilistic aggregated outputs of the three RF prediction models (total output were 6 variables as shown in Figure 2, two values for each classifier) were used as input for the ANN-based classifier. The ANN classifier was used to learn the correlation between the RF prediction scores and the target class. It replaced the majority voting schemes used by the three RF classifiers for more accurate detection. Without this divide and conquer principle, such a correlation would not be released due to the curse of dimensionality because of the massive set of extracted multifaceted features.

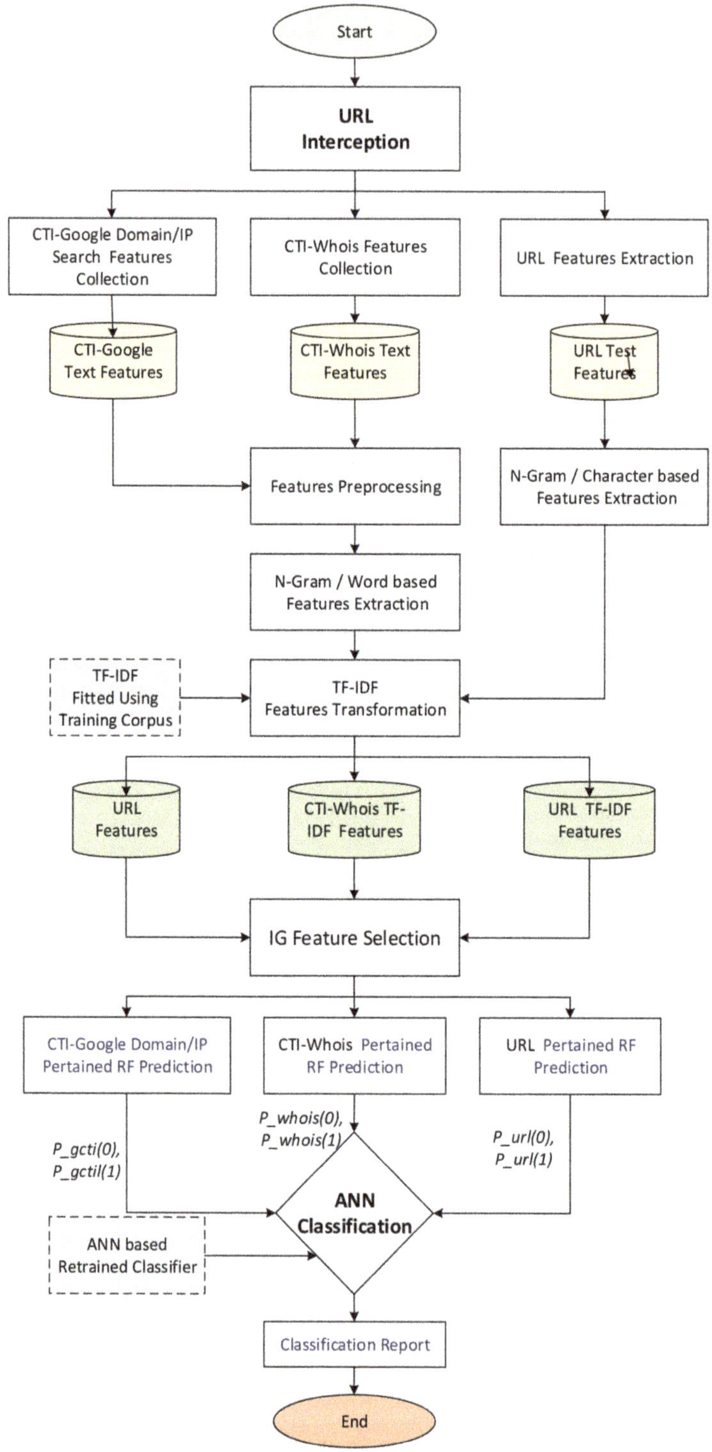

Figure 3. Flowchart of the CTI-MURLD model Operation.

4. Performance Evaluation

In this section, the used dataset, the experimental procedures, and the performance evaluation are described.

4.1. Sources and Preprocessing of Datasets

This study used a malicious URLs dataset that is publicly available on the Kaggle.com repository (available at https://www.kaggle.com/sid321axn/malicious-urls-dataset, accessed on 25 February 2022). The dataset was collected from widely-used sources by researchers of malicious URL detection domains such as Phishtank [39,40] (available at https://phishtank.org/, accessed on 25 February 2022) and URL dataset (ISCX-URL-2016) [8] (available at https://www.unb.ca/cic/datasets/url-2016.html, accessed on 25 February 2022). The URLs in the dataset were categorized into two types, malicious and benign. Malicious URLs included malware links, web defacement, spam, phishing, drive-by downloads, etc. A random sample consisting of 20,000 URLs was drawn and used in this study. Table 1 shows the number and types of URL samples in the datasets.

Table 1. Number and types of URLs used in this study.

Category	Number of Samples
Total URLs	651,191
Total Benign	428,103
Total Malicious	223,088
Malicious URLs	
Defacement	96,457
Phishing	94,111
Malware Link	32,520

4.2. Experimental Procedures

The dataset was split into training and testing sets with 70% for training and 30% for testing. The training dataset was used to train the RF and MLP/ANN classifiers. The outputs (prediction values) of RF classes were used to train the ANN-based decision-making classifier. For each RF classifier, 100 estimators were created. For the ANN prediction model, the Broyden–Fletcher–Goldfarb–Shanno (BFGS) algorithm was used as the optimization algorithm. BFGS is one of the variants of the gradient descent algorithm and has proven to have better accuracy than the plain gradient descent algorithm. The learning rate was set close to zero i.e., 10–5 for better generalizability. Meanwhile, the logistic sigmoid function was used as an activation function. The neural network prediction model consisted of three layers, the input, hidden, and output layers. The input layer consisted of 6 neurons, the hidden layer contained 6 neurons, and the output layer contained a single neuron.

4.3. Performance Evaluation

To validate the detection performance of the proposed model, five performance measures were used: the overall accuracy; the detection rate (recall); the precision; the F1 score; the false-positive rate (FPR); and the false-negative rate (FNR). These performance measures are commonly used to evaluate the accuracy of the malware detection solutions in the literature. To evaluate the proposed model, the commonly used machine learning techniques that were used to evaluate the related malicious URL detection were used. Moreover, three models were developed for the evaluation of the CTI-MURLD, Google-CTI, Whois-CTI, and lexical URL-based features as baselines [8,11,13–15,23,41]. Furthermore, two deep learning-based models were developed for the evaluation of SDL and CNN-based malicious URL-detection models. A detailed description of the results is illustrated in the following section. The following equations were used for calculating the used performance measures in this study.

$$Accuracy = \frac{TP + TN}{TP + TN + FP + FN} \quad (7)$$

$$FPR = \frac{FP}{TP + FN} \quad (8)$$

$$DR\,(Recall) = \frac{TP}{TP + FN} \quad (9)$$

$$Precision = \frac{TP}{TP + FP} \quad (10)$$

$$\text{F-measure} = \frac{2 \times Precision \times Recall}{Precision + Recall} \quad (11)$$

For comparison, the base classifier of CTI-MURLD has been trained using state-of-the-art machine learning techniques including deep learning that have been used for malicious website detection, namely naïve Bayes (NB), logistic regression (LR), decision tree (DT), random forest (RF), convolutional neural network (CNN), and sequential deep learning (SDL) models.

5. Results and Discussion

The proposed Cyber Threat Intelligence-based Malicious URL Detection (CTI-MURLD) model has been validated using the aforementioned dataset and performance measures. Additionally, it was evaluated against the commonly used feature sets including the URL-based features and Whois-based features. Different feature sets have been compared to evaluate the proposed CTI-MURLD model. Table 2 and Figures 4–9 illustrate the results obtained in terms of the detection accuracy, false-positive rate, false-negative rate, precision, recall, and F1-Measure, respectively.

Table 2. Performance of the CTI-MURLD model using Different Classifiers.

Ensemble Classifiers	Accuracy	FPR	FNR	Recall	Precession	F1
NB	75.60%	40.04%	11.27%	88.73%	68.65%	77.41%
LR	86.15%	18.85%	9.18%	90.82%	82.21%	86.30%
DT	95.70%	4.10%	4.29%	95.71%	95.69%	95.70%
RF	96.80%	3.13%	3.13%	96.88%	96.72%	96.80%
CNN	94.70%	5.27%	5.09%	94.91%	94.48%	94.69%
SDL	95.61%	4.57%	4.20%	95.80%	95.41%	95.61%

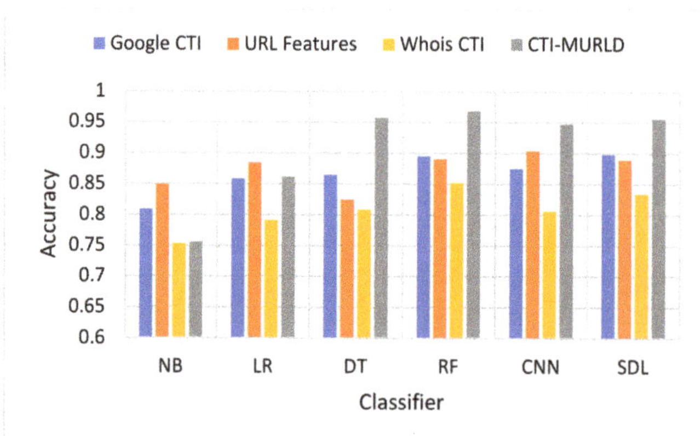

Figure 4. Comparison in terms of the detection-accuracy performance.

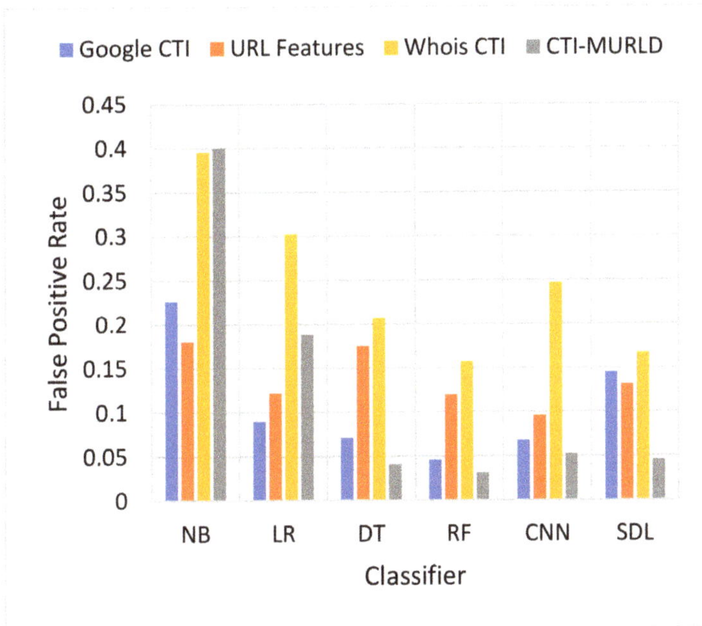

Figure 5. Comparison in terms of the FPR.

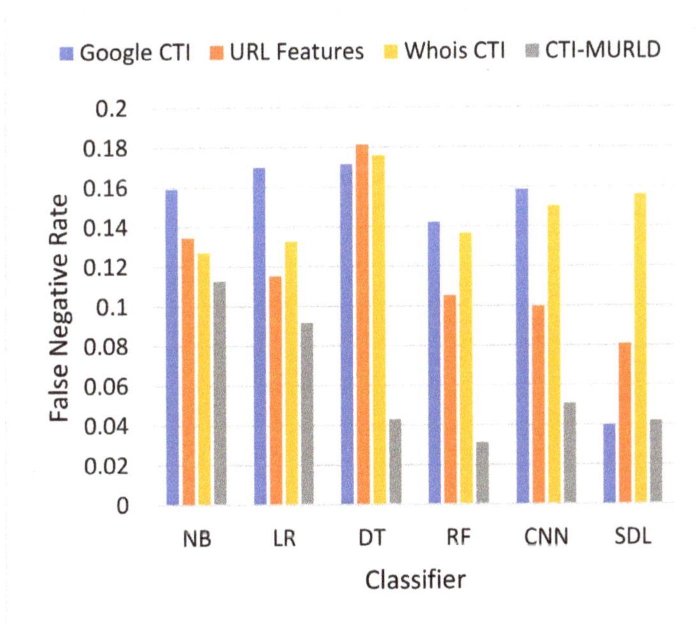

Figure 6. Comparison in terms of the FNR.

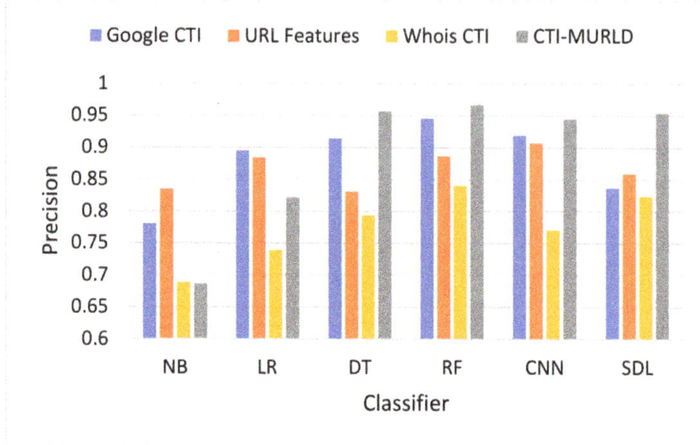

Figure 7. Comparison in terms of the precision.

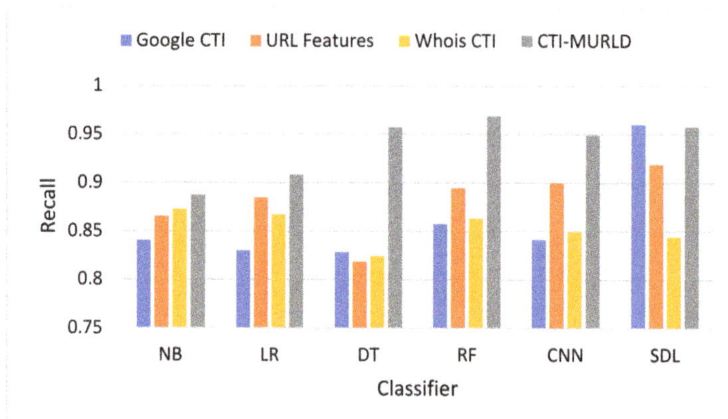

Figure 8. Comparison in terms of the recall (True Positive Rate).

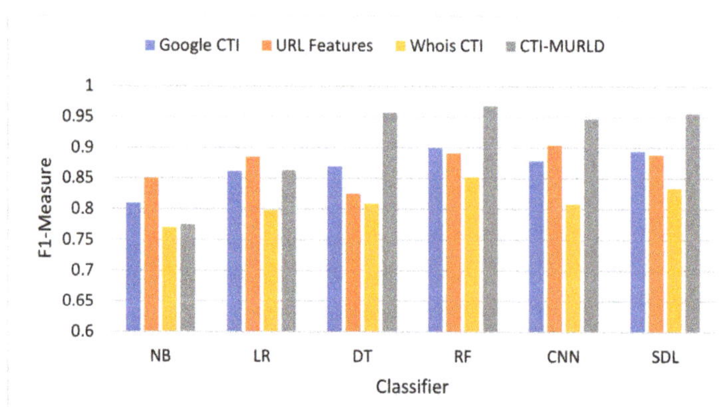

Figure 9. Comparison in terms of the F1-Measure.

Table 2 presents the performance of the CTI-MURLD model. As can be seen in Table 1, the RF technique outperformed the other machine learning algorithms. The decision tree algorithm also achieved the second-best performance while the sequential deep learning model and the CNN model performed slightly lower compared with RF and DT. However, the sequential model SDL model was better than the CNN model. It was expected that RF will perform better than DT because RF is a collection of DTs. A single DT can make a series of decisions based on the given set of features based on the information gained.

In terms of accuracy, Figure 4 depicts the accuracy performance of the proposed CTI-MURLD model compared with related works. In most cases, the CTI-based features especially the combined ones, outperformed the traditional-based features and the single set of features. For example, RF and DT DSL achieved accuracy higher than 95% with the combined CTI-based features. Meanwhile, the accuracy of the same classifiers using URL-based features was always lower than 90% except with the CNN model which achieved 90.8%. CNN was a commonly reported method that outperforms the conventional machine learning classifiers. However, it is believed that CNN performance depends on the competence of the representative features present in the image-like matrix. In the case of a huge number of features such as CTI-based features, a large portion of the dataset should be available to achieve maximum accuracy. It is worth mentioning that Whois-based features alone achieved the worst performance among the tested features while Google-based CTI achieved almost similar performance results to the URL-based features. For example, with the models DT, RF, SDL, LR, and NB, Google-based CTI features achieved the second-best accuracy performance. This indicates that the CTI-based features can complement the traditional features for detecting evasive malicious websites which try to evade detection by looking similar to benign websites.

Figures 5 and 6 present the results in terms of FPR and FNR, respectively. The models designed using the combined CTI-based features and the Google-based CTI features achieved the lowest rate of false positives while Whois-based features achieved the highest rate of false positives. However, the combined-based CTI features achieved the lowest rate of false negatives. Meanwhile, the models designed based on the Google-based CTI features suffered from a high rate of false negatives. In general, all models designed with a single feature set such as the URL, Whois, or Google CTI suffered from a considerable number of false negatives. This is because it is difficult to differentiate between some malicious websites such as spoofing websites and other benign websites. Meanwhile, the proposed combined features set with the RF classifier achieved the lowest rate of false negatives which was 3.3% followed by SDL (4.2%) and DT (4.29%).

Figure 7 depicts the performance in terms of precision. The precision measures the predictability of the positive class. The proposed CTI-MURLD models using decision trees and deep learning-based classifiers achieved the highest precision compared with the URL-based features and the single set features such as Google CTI and Whois information. The proposed CTI-MURLD model achieved 96.7% precision using the RF classifier, 95.69% using the DT, 95.41% using SDL, and 94.48% using the CNN-based classifier. The models designed using the Google-based CTI features achieved the second-best results for precision. For instance, it achieved 94.54% using the RF-based classifier and 91.43% using the DT classifier. Meanwhile, the model's design using Whois information is unprecise with all classifiers compared with the URL-based and the other CTI features.

Figure 8 shows the performance in terms of the recall or the detection rate The proposed CTI-MURLD models using all classifiers outperformed the other types of feature sets, namely, URL-based features and single-set features such as Google CTI and Whois information. It achieved a 96.88% true positive rate using the RF-based classifier, 95.8% using the SDL, 95.71% using DT, 94.91% using the CNN, 90.82%, and 88.73% using NB-based classifier. The models designed using other feature sets vary lower than 90% except with SDL-based classifier, the Google-based CTI features achieve 96% and URL-based features achieve 92%.

The harmonic means results in Figure 9 in terms of F-Measure (also called F1-Score) summarizes how well the model performs with both precision and recall. The proposed CTI-MWD model using a decision tree and deep learning-based classifiers achieved the highest F1-Score value compared with the URL-based features and the single set features such as Google CTI and Whois information. In most cases, the F1-Score of the proposed CTI-MWD model achieves a 95% or higher value. It achieves a 96.8% score using the RF classifier, 95.7% using the DT, 95.61% using SDL, and 94.69% using the CNN-based classifier. The models designed using other feature sets vary lower than 90% except with CN-based classifier and the Whois information-based features the model achieves 90% and the model designed using Google-based CTI features with RF archives 90% F-Score.

As RF implicit feature selection by applying information gain and feature importance, an experiment was conducted to evaluate the effectiveness of the feature selection used by the model before applying the RF algorithms. Given that the datasets used in this study were text data containing a massive number of features, most of these features were irrelevant and should be eliminated. The results in Table 3 (see RF without FS in Table 3) indicate that the RF with the proposed feature selection achieved better than the RF without the selection. The RF randomly selects a subset feature to train each weak classifier. The selection of the important features happened within the subset. In contrast, in this study, selecting the importance classifiers before the RF enforces the RF algorithm to select the subset features of the pool of the selected important features, which improves the accuracy of the weak classifier and thus the overall accuracy.

Table 3. Performance of the CTI-MURLD model with and without feature selection and with grid search best-found hyperparameters.

Ensemble Classifiers	Accuracy	FPR	FNR	Recall	Precession	F1
RF without FS	96.30%	3.81%	3.59%	96.20%	96.58%	96.39%
RF with FS	96.80%	3.13%	3.13%	96.88%	96.72%	96.80%
RF with GS	97.25%	2.73%	2.76%	97.26%	97.36%	97.31%

Because RF algorithms use a different range of hyperparameters, a grid search is used to search for the best parameters that improve the performance. Table 3 (see RF with GS in Table 3) shows the performance results of the grid search. Based on the grid search results, the hyperparameters that gave the best performance were: 1000 estimators, five minimum sample split, two minimum samples leaf, an unlimited number of leaf nodes, samples drawn with replacement, and an unlimited maximum number of features. As shown in Table 3, the performance was further improved, as expected.

The results obtained using the proposed CTI-MURLD model raise an interesting but fundamental question of why the ensemble-based RF and DT-based classifiers achieved the best results compared with the deep learning-based classifiers. The answer lies in the dataset itself, the number of features extracted using the Google-based CTI was huge compared to the URL or the Whois information features. This created highly noisy data with a sparse feature vector. Moreover, when the features were combined from CTI, Whois information, and URL-based features a high-dimensional problem was created. In this situation, DT and RF were suitable for high-dimensional noisy data [42]. With such high-dimensional features, deep learning models such as CNN and SDL need a larger dataset to attain their maximum performance. In addition, neural network-based classifiers create patterns by connecting neurons with each other which is difficult to generalize in high-dimensional datasets. Meanwhile, the RF classifier creates independent patterns which are suitable for high-dimensional data and small datasets. This may be an indication of why the decision tree-based classifier outperformed the deep learning model.

6. Conclusions

In this study, a malicious website detection model was designed and developed based on cyber threat intelligence extracted from Google. The first main contribution of these studies was the use of cyber threat intelligence as a new set of features with a hypothesis stating that cyber threat intelligence is an effective and safer alternative to improve the detection accuracy of malicious websites. The domain names of the websites were extracted using the Whois technique, and cyber threat intelligence was collected from Google and combined with URL-based features. Due to the diversity of attack vectors of malicious websites, high-dimensional features were created and used to train the proposed model. The second main contribution of this study is in the design of the proposed detection model. Three random forest classifiers were developed, each of which was trained based on different features, namely, URL-based features, cyber threat intelligence features based on Google, and Whois information-based features. The probabilistic outputs of the weak classifiers in each tree were aggregated and used as input features to a multilayer perceptron designed to replace the three majority voting schemes used by the trained random forest classifiers. Several types of machine learning classifiers were investigated to validate and evaluate the proposed model. Results show that the CTI-based features significantly improved the detection performance, achieving 96.80% compared with the best 90.4% achieved by the URL-based features. The false-positive rate was significantly decreased to 3.1% compared with 12% performed by the URL-based model. The main drawback of this study is that the cyber threat intelligence collected is obtained from a Google search. Such a source of data is not necessarily reliable and hence, false information is highly probable. As a result, a solution based on a trusted source could be a possible future direction for other researchers.

Author Contributions: Conceptualization, F.A.G., M.A. (Mohammed Alsaedi), F.S. and J.A.; methodology, F.A.G., M.A. (Mohammed Alsaedi), F.S. and J.A.; software, M.A., F.S. and J.A.; validation, F.A.G., M.A. (Mohammed Alsaedi), F.S. and M.A. (Mohammed Alasli); formal analysis, F.S.; investigation, J.A.; resources, F.A.G. and M.A. (Mohammed Alsaedi); data curation, M.A. (Mohammed Alasli); writing—original draft preparation, F.A.G.; writing—review and editing, F.S., M.A. (Mohammed Alsaedi) and J.A.; visualization, F.A.G.; supervision, F.S.; project administration, M.A. (Mohammed Alsaedi); funding acquisition, M.A. (Mohammed Alsaedi). All authors have read and agreed to the published version of the manuscript.

Funding: This research was funded by the Deputyship for Research & Innovation, Ministry of Education in Saudi Arabia under the project number (168/442).

Institutional Review Board Statement: Not applicable.

Informed Consent Statement: Not applicable.

Data Availability Statement: The dataset that is used in this study is publicly available on the Kaggle.com repository (https://www.kaggle.com/sid321axn/malicious-urls-dataset, last access 25 February 2022).

Acknowledgments: The authors extend their appreciation to the Deputyship for Research & Innovation, Ministry of Education in Saudi Arabia, for funding this research work; project number (168/442). Further, the authors would like to extend their appreciation to Taibah University for its supervision support.

Conflicts of Interest: The authors declare no conflict of interest.

References

1. Jang-Jaccard, J.; Nepal, S. A survey of emerging threats in cybersecurity. *J. Comput. Syst. Sci.* **2014**, *80*, 973–993. [CrossRef]
2. Khan, M.A.; Nasralla, M.; Uman, M.; Rehman, G.; Khan, S.; Choudhury, N. An Efficient Multilevel Probabilistic Model for Abnormal Traffic Detection in Wireless Sensor Networks. *Sensors* **2022**, *22*, 410. [CrossRef] [PubMed]
3. Nasralla, M.M.; García-Magariño, I.; Lloret, J. Defenses against perception-layer attacks on iot smart furniture for impaired people. *IEEE Access* **2020**, *8*, 119795–119805. [CrossRef]

4. Guo, J.; Nazir, S. Internet of Things Based Intelligent Techniques in Workable Computing: An Overview. *Sci. Program.* **2021**, *2021*, 1–15. [CrossRef]
5. Townsend, K. 18.5 Million Websites Infected With Malware at Any Time. 2018. Available online: https://www.securityweek.com/185-million-websites-infected-malware-any-time (accessed on 2 January 2022).
6. Google. Google Safe Browsing. 2022. Available online: https://transparencyreport.google.com/safe-browsing/overview?hl=en (accessed on 2 January 2022).
7. Liu, M.; Zhang, B.; Chen, W.; Zhang, X. A survey of exploitation and detection methods of XSS vulnerabilities. *IEEE Access* **2019**, *7*, 182004–182016. [CrossRef]
8. Saleem Raja, A.; Vinodini, R.; Kavitha, A. Lexical features based malicious URL detection using machine learning techniques. *Mater. Today Proc.* **2021**, *47*, 163–166. [CrossRef]
9. Subasi, A.; Balfaqih, M.; Balfagih, Z.; Alfawwaz, K. A Comparative Evaluation of Ensemble Classifiers for Malicious Webpage Detection. *Procedia Comput. Sci.* **2021**, *194*, 272–279. [CrossRef]
10. Rameem, Z.S.; Chishti, M.; Baba, A.; Wu, F. Detecting Covid-19 chaos driven phishing/malicious URL attacks by a fuzzy logic and data mining based intelligence system. *Egypt. Inform. J.* **2021**, *23*, 1–18. [CrossRef]
11. Gupta, B.B.; Yadav, K.; Psnannis, K.; Razzak, I. A novel approach for phishing URLs detection using lexical based machine learning in a real-time environment. *Comput.Commun.* **2021**, *175*, 47–57. [CrossRef]
12. Wazirali, R.; Ahmad, R.; Abu-Ein, A.A.-K. Sustaining accurate detection of phishing URLs using SDN and feature selection approaches. *Comput. Netw.* **2021**, *201*, 108591. [CrossRef]
13. Mondal, D.K.; Singh, B.; Hu, H.; Biswas, S.; Alom, Z.; Azim, M. SeizeMaliciousURL: A novel learning approach to detect malicious URLs. *J. Inf. Secur. Appl.* **2021**, *62*, 102967. [CrossRef]
14. Haynes, K.; Shirazi, H.; Ray, I. Lightweight URL-based phishing detection using natural language processing transformers for mobile devices. *Procedia Comput. Sci.* **2021**, *191*, 127–134. [CrossRef]
15. Srinivasan, S.; Vinayakumar, R.; Arunachalam, A.; Alazab, M.; Soman, K. DURLD: Malicious URL Detection Using Deep Learning-Based Character Level Representations. In *Malware Analysis Using Artificial Intelligence and Deep Learning*; Springer: Berlin/Heidelberg, Germany, 2021; pp. 535–554.
16. Chiramdasu, R.; Srivastava, G.; Bhattacharya, S.; Reddy, P.; Gadekallu, T. Malicious URL Detection using Logistic Regression. In Proceedings of the 2021 IEEE International Conference on Omni-Layer Intelligent Systems (COINS), Barcelona, Spain, 23–25 August 2021.
17. Phung, N.M.; Mimura, M. Detection of malicious javascript on an imbalanced dataset. *Internet Things* **2021**, *13*, 100357. [CrossRef]
18. Huang, Y.; Li, T.; Zhang, L.; Li, B.; Liu, X. JSContana: Malicious JavaScript detection using adaptable context analysis and key feature extraction. *Comput. Secur.* **2021**, *104*, 102218. [CrossRef]
19. McGahagan IV, J.; Bhansali, D.; Pinto, C.; Cukir, M. Discovering Features for Detecting Malicious Websites: An Empirical Study. *Comput. Secur.* **2021**, *109*, 102374. [CrossRef]
20. Samarasinghe, N.; Mannan, M. On cloaking behaviors of malicious websites. *Comput. Secur.* **2021**, *101*, 102114. [CrossRef]
21. Kim, S.; Kim, J.; Nam, S.; Jin, J. WebMon: ML-and YARA-based malicious webpage detection. *Comput. Netw.* **2018**, *137*, 119–131. [CrossRef]
22. Rakesh, R.; Muthuraijkumar, S.; Sairamesh, L.; Vijayalakmi, M.; Kannan, A. Detection of URL based attacks using reduced feature set and modified C4. 5 algorithm. *Adv. Nat. Appl.Sci.* **2015**, *9*, 304–311.
23. Kim, S.; Kim, J.; Kang, B.B. Malicious URL protection based on attackers' habitual behavioral analysis. *Comput. Secur.* **2018**, *77*, 790–806. [CrossRef]
24. He, S.; Li, B.; Peng, X.; Xin, J.; Zhang, E. An Effective Cost-Sensitive XGBoost Method for Malicious URLs Detection in Imbalanced Dataset. *IEEE Access* **2021**, *9*, 93089–93096. [CrossRef]
25. Patil, D.R.; Patil, J.B. Malicious URLs detection using decision tree classifiers and majority voting technique. *Cybern. Inf. Technol.* **2018**, *18*, 11–29. [CrossRef]
26. Li, T.; Kou, G.; Peng, Y. Improving malicious URLs detection via feature engineering: Linear and nonlinear space transformation methods. *Inf. Syst.* **2020**, *91*, 101494. [CrossRef]
27. Wang, S.; Chen, Z.; Yan, Q.; Li, K.; Peng, L.; Yang, B.; Conti, M. Deep and broad URL feature mining for android malware detection. *Inf. Sci.* **2020**, *513*, 600–613. [CrossRef]
28. Vinayakumar, R.; Soman, K.P.; Poornachandran, P. Evaluating deep learning approaches to characterize and classify malicious URL's. *J. Intell. Fuzzy Syst.* **2018**, *34*, 1333–1343. [CrossRef]
29. Kuyama, M.; Kakizaki, Y.; Sasaki, R. Method for detecting a malicious domain by using whois and dns features. In Proceedings of the Third International Conference on Digital Security and Forensics (DigitalSec2016), Kuala Lumpur, Malaysia, 6–8 September 2016.
30. Ding, C. Automatic Detection of Malicious URLs using Fine-Tuned Classification Model. In Proceedings of the 2020 5th International Conference on Information Science, Computer Technology and Transportation (ISCTT), Shenyang, China, 13–15 November 2020.
31. Cavnar, W.B.; Trenkle, J.M. N-gram-based text categorization. Proceedings of SDAIR-94, 3rd Annual Symposium on Document Analysis and Information Retrieval, Las Vegas, NV, USA, 11–14 April 1994.

32. Chauhan, T.; Palivela, H. Optimization and improvement of fake news detection using deep learning approaches for societal benefit. *J. Inf. Manag. Data Insights* **2021**, *1*, 100051. [CrossRef]
33. Goldani, M.H.; Momtazi, S.; Safabakhsh, R. Detecting fake news with capsule neural networks. *Appl. Soft Comput.* **2021**, *101*, 106991. [CrossRef]
34. Huang, Y.-F.; Chen, P.-H. Fake news detection using an ensemble learning model based on Self-Adaptive Harmony Search algorithms. *Expert Syst. Appl.* **2020**, *159*, 113584. [CrossRef]
35. Agarwal, V.; Sultana, P.; Malhotra, S.; Sarkar, A. Analysis of Classifiers for Fake News Detection. *Procedia Comput. Sci.* **2019**, *165*, 377–383. [CrossRef]
36. Ahmed, H.; Traore, I.; Saad, S. Detection of online fake news using n-gram analysis and machine learning techniques. In Proceedings of the International Conference on Intelligent, Secure, and Dependable Systems in Distributed and Cloud Environments, Vancouver, BC, Canada, 28–30 November 2017.
37. Mbona, I.; Eloff, J.H.P. Feature selection using Benford's law to support detection of malicious social media bots. *Inf. Sci.* **2022**, *582*, 369–381. [CrossRef]
38. Motiur, R.S.S.M.; Islam, T.; Jabiullah, M.I. PhishStack: Evaluation of Stacked Generalization in Phishing URLs Detection. *Procedia Comput. Sci.* **2020**, *167*, 2410–2418. [CrossRef]
39. Bell, S.; Komisarczuk, P. An analysis of phishing blacklists: Google safe browsing, openphish, and phishtank. In Proceedings of the Australasian Computer Science Week Multiconference, Melbourne, Australia, 4–6 February 2020.
40. Marchal, S.; François, J.; State, R.; Engel, T. PhishStorm: Detecting phishing with streaming analytics. *IEEE Trans. Netw. Serv. Manag.* **2014**, *11*, 458–471. [CrossRef]
41. Ranganayakulu, D.; Chellappan, C. Detecting Malicious URLs in E-mail–An Implementation. *AASRI Procedia* **2013**, *4*, 125–131. [CrossRef]
42. Islam, M.Z.; Liu, J.; Li, J.; Liu, L.; Knag, W. A semantics aware random forest for text classification. In Proceedings of the 28th ACM International Conference on Information and Knowledge Management, Beijing, China, 3–7 November 2019; pp. 1061–1070.

sensors MDPI

Article

A New Intrusion Detection System for the Internet of Things via Deep Convolutional Neural Network and Feature Engineering

Safi Ullah [1], Jawad Ahmad [2,*], Muazzam A. Khan [1,3], Eman H. Alkhammash [4], Myriam Hadjouni [5], Yazeed Yasin Ghadi [6], Faisal Saeed [7] and Nikolaos Pitropakis [2]

1 Department of Computer Science, Quaid-i-Azam University, Islamabad 44000, Pakistan; safiullah@cs.qau.edu.pk (S.U.); muazzam.khattak@qau.edu.pk (M.A.K.)
2 School of Computing, Edinburgh Napier University, Edinburgh EH10 5DT, UK; n.pitropakis@napier.ac.uk
3 Pakistan Academy of Sciences, Islamabad 44000, Pakistan
4 Department of Computer Science, College of Computers and Information Technology, Taif University, P.O. Box 11099, Taif 21944, Saudi Arabia; eman.kms@tu.edu.sa
5 Department of Computer Sciences, College of Computer and Information Science, Princess Nourah Bint Abdulrahman University, P.O. Box 84428, Riyadh 11671, Saudi Arabia; mfhaojouni@pnu.edu.sa
6 Department of Computer Science and Software Engineering, Al Ain University, Abu Dhabi 122612, United Arab Emirates; yazeed.ghadi@aau.ac.ae
7 DAAI Research Group, Department of Computing and Data Science, School of Computing and Digital Technology, Birmingham City University, Birmingham B4 7XG, UK; faisal.saeed@bcu.ac.uk
* Correspondence: j.ahmad@napier.ac.uk

Abstract: The Internet of Things (IoT) is a widely used technology in automated network systems across the world. The impact of the IoT on different industries has occurred in recent years. Many IoT nodes collect, store, and process personal data, which is an ideal target for attackers. Several researchers have worked on this problem and have presented many intrusion detection systems (IDSs). The existing system has difficulties in improving performance and identifying subcategories of cyberattacks. This paper proposes a deep-convolutional-neural-network (DCNN)-based IDS. A DCNN consists of two convolutional layers and three fully connected dense layers. The proposed model aims to improve performance and reduce computational power. Experiments were conducted utilizing the IoTID20 dataset. The performance analysis of the proposed model was carried out with several metrics, such as accuracy, precision, recall, and F1-score. A number of optimization techniques were applied to the proposed model in which Adam, AdaMax, and Nadam performance was optimum. In addition, the proposed model was compared with various advanced deep learning (DL) and traditional machine learning (ML) techniques. All experimental analysis indicates that the accuracy of the proposed approach is high and more robust than existing DL-based algorithms.

Keywords: convolution neural network; cybersecurity; deep learning; Internet of Things; intrusion detection

Citation: Ullah, S.; Ahmad, J.; Khan, M.A.; Alkhammash, E.H.; Hadjouni, M.; Ghadi, Y.Y.; Saeed, F.; Pitropakis, N. A New Intrusion Detection System for the Internet of Things via Deep Convolutional Neural Network and Feature Engineering . *Sensors* **2022**, *22*, 3607. https://doi.org/10.3390/s22103607

Academic Editor: Paolo Bellavista

Received: 7 April 2022
Accepted: 6 May 2022
Published: 10 May 2022

1. Introduction

The IoT foresees the networking of a wide range of smart things in our environment that are capable of accumulating, processing, and communicating data [1]. The IoT is a widely used technology in automated network systems across the world that has had an impact on different areas, such as the agricultural, medical, transport, and automobile industries, and water monitoring in recent years [2,3]. The use of IoT devices has increased dramatically, from 15.41 billion in 2015 to more than 35.8 billion in 2021, as homes and businesses increasingly rely on online technology [4]. The IoT is anticipated to reach 75.44 billion devices by 2025, as shown in Figure 1, which will generate 79 zettabytes (ZB) of data [5]. The IoT has been identified as a critical component of digitization for a transforming society [6].

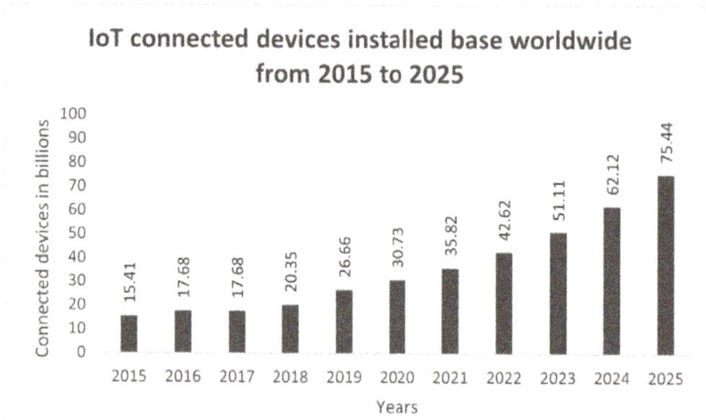

Figure 1. Growth of IoT devices from 2015 to 2025 [5].

Many IoT devices capture, store, and process personal data, making them a feasible target for assailants because of their distributed structure and openness [7]. The effective deployment of IoT networks is becoming more dependent on security [8]. An IDS is required to examine IoT network traffic for the identification of cyberattacks [9]. Several researchers have worked on IDSs in which machine learning (ML) and deep learning (DL) models play a key role [10]. ML and DL techniques are widely used in different fields, such as in agriculture [11], medical [12], and automobile industries [13,14]. DL is a branch of ML, and it is generalizable to new problems with complicated and high-dimensional data. Furthermore, DL methods allow for the training of nonlinear models on big datasets in a systematic way [15]. This is why DL performs well in detecting intrusions, as it not only handles a large amount of data but also can generalize to new types of attacks in the network [16].

The existing system has difficulties in improving performance and identifying subcategories of cyberattacks. This paper proposes a DCNN followed by a deep-neural-networks (DNN)-based IDS. The primary advantage of a DCNN is its ability to exploit the correlation between features [17]. A DCNN works on a lower number of parameters than other DL models [18]. Thus, the required computational power is decreased, and the learning process is improved. The proposed system improves the performance of existing IDSs and extends to subcategories of malicious attack detection in IoT networks. The IoT network intrusion dataset 2020 (IoTID20) was used for experiments on the proposed model. This dataset includes data for binary, multi-category, and subcategories of IoT networks.

Contributions

- We proposed a DCNN technique for malicious activity identification in IoT networks.
- We improved performance and reduced the computational power of an IDS for low-power IoT devices in the network.
- We identified the subcategory of cyberattacks in the IoT networks.
- We compared the proposed scheme with other DL and traditional ML techniques.

The remainder of the article is organized as follows. Section 2 discusses related work and presents a literature comparison. A step-by-step methodology of the proposed system is presented in Section 3. Section 4 provides a detailed analysis of the results and a comparison with state-of-the-art models. This work is concluded in Section 5.

2. Related Works

Security is an essential part of an IoT network for stability, reliability, and safe communication. Several researchers have proposed different techniques for the detection of malicious attacks in IoT networks. Basati et al. [19] presented an IDS called deep feature extraction (DFE). This model is based on a CNN. The authors mainly focused on those devices that have low processing power. They used UNSW-NB15, CICIDS2017, and KDD-Cup99 datasets for their experiments. The model was tested for both binary and multi-class classifications. Rashid et al. [20] proposed a stacking ensemble approach based on trees for intrusion detection in the IoT. Two incursion datasets, NSL-KDD and UNSW-NB15, were used to evaluate the efficacy of the proposed model. They also improved efficacy by integrating feature selection strategies to identify the most relevant features.

Fatani et al. [21] introduced a novel feature engineering technique for the IDS system while using the benefits of swarm intelligence (SI) techniques. Four popular public datasets, CIC2017, NSL-KDD, BoT-IoT, and KDD99, were utilized to test the quality of the proposed IDS technique. Alkahtani et al. [22] suggested three advanced and widely used DL models for intrusion detection. The authors conducted experiments on long short-term memory (LSTM), CNN, and a hybrid model of CNN–LSTM. They used the IoTID20 dataset for the evaluation of these DL models. Keserwani et al. [23] presented a method for extracting significant IoT network features for intrusion detection. The proposed method consists of a combination of grey wolf optimization (GWO) and particle swarm optimization (PSO). They utilized the KDDCup99, NSL-KDD, and CICIDS-2017 datasets.

A single hidden layer feedforward neural network (SLFN) method was introduced by Qaddoura et al. [24] for malicious activity detection in IoT networks. The authors used data reduction with clustering and the SMOTE oversampling technique. For the evaluation of the model, they used accuracy, precision, recall, and G-mean. Saba et al. [25] introduced a two-stage hybrid technique for the detection of malicious attacks in IoT networks. A genetic algorithm (GA) was used to choose relevant features as well as the famous ML techniques, such as support vector machine (SVM), ensemble classifier, and decision tree (DT).

The existing systems cannot identify the subcategories of multi-class attacks in the network. In addition, for binary and multi-class detection, the performance of the existing system can be improved. A comparison of the related work is given in Table 1.

Table 1. A comparison of existing work related to intrusion detection in IoT.

Authors	Year	Technique	Dataset	Multi-Class Detection	Sub-Categories Multi-Class Detection
Basati et al. [19]	2022	DFE	KDDCup99, CICIDS2017, UNSW-NB15	✓	×
Rashid et al. [20]	2022	Ensemble	NSL-KDD, UNSW-NB15	×	×
Fatani et al. [21]	2022	AQU, PSO	CIC2017, NSL-KDD, BoT-IoT, KDD99	✓	×
Alkahtani et al. [22]	2021	CNN-LSTM	IoTID20	×	×
Keserwani et al. [23]	2021	GWO–PSO–RF	KDDCup99, NSL–KDD, CICIDS-2017	✓	×
Qaddoura et al. [24]	2021	SLFN-SVM-SMOTE	IoTID20	✓	×
Saba et al. [25]	2021	GA-(SVM, Ensemble, DT)	NSL-KDD	✓	×
Propose Study	2022	CNN-DNN	IoTID20	✓	✓

3. The Proposed Framework

This section provides a detailed explanation of the utilized dataset, preprocessing approaches, the proposed deep convolutional neural network (DCNN), and evaluation metrics.

3.1. IoTID20 Dataset

The IoTID20 dataset was developed to identify cyberattacks in IoT networks. This dataset was generated through home-connected smart devices using SKT NGU and EZVIZ Wi-Fi cameras [26]. The main advantage of this dataset is that it includes modern communication data and new data on network interference detection. This dataset has 83 IoT network features and three labels [27]: binary, category, and subcategory; details are given in Table 2.

Table 2. Label details of IoTID20 dataset.

Binary	Category	Subcategory
Normal	Normal	Normal
Anomaly	DoS	DoS-Synflooding
	Mirai	Mirai-Ackflooding Mirai-HTTP Flooding Mirai-Hostbruteforceg Mirai-UDP Flooding
	MITM	MITM ARP Spoofing
	Scan	Scan Port OS Scan Hostport

3.2. Preprocessing

Data preprocessing is an essential step for ML/DL methods. Preprocessing converts data into a suitable format for any neural network. This section consists of cleaning, label encoding, feature engineering, normalization, and data splitting.

3.2.1. Dataset Cleaning

A dataset must be verified for empty and undefined instances before training a model. In this experiment, the Python built-in library (Pandas) was used to validate the dataset. The utilized IoTID20 dataset has some missing values. To clean the dataset, we removed all missing value instances.

3.2.2. Label Encoding

Label encoding is a well-known encoding approach for dealing with categorical values. It assigns a unique numeric value to each categorical value. For ML algorithms and DL neural networks to operate, the input and output values must be integers. The utilized dataset has some categorical features. Each categorical feature has several categories for which one-hot encoding requires greater memory and more time [28]. In this study, the label encoder approach was used to convert the categorical features into numeric.

3.2.3. Feature Engineering

Each dataset contains its own set of features. If a dataset contains multiple features as well as certain insignificant features that have no impact on the output label, we must eliminate those features from the dataset because they lead to overfitting and underfitting, which significantly influence the executing time and performance of the classifier. In this study, the filter approach was used. In filtering features, the extra tree classifier (ETC) technique was applied. This method calculates the impact of each feature on the output label. The utilized dataset has 83 features. We select all the features greater than 0.001 for information gain. After applying the feature filtering approach, 62 features were selected.

3.2.4. Normalization

Normalization is a method commonly used in the preprocessing of data for ML/DL algorithms. The purpose of normalization is to convert the numeric column values in a dataset to a common scale while maintaining variations in value ranges. Each feature of the IoTID20 dataset has different values. Some feature values are in the thousands, and some have negative values that reduce the model performance. To solve this problem, the data are normalized between 0 and 1 via min–max method, as represented by Equation (1).

Data are converted into an array and reshaped (number of total records, number of input features, 1) using Python's NumPy library.

$$X_{norm} = \frac{x - x_{min}}{x_{max} - x_{min}} \quad (1)$$

3.2.5. Data Splitting

Splitting the data into train and test sets is one of the common preprocessing steps used to evaluate the ML/DL models' performance. In an unbalanced dataset, random splitting of datasets can lead to an unequal split of data, which cannot evaluate the performance of the model accurately. To address this problem, we used a stratified method to split the dataset into train and test sets. A stratified sampling procedure splits the entire dataset into homogenous sets. In this work, the stratified method splits the data into 80% train and 20% test sets for each class. A detailed splitting of the cleaned dataset for binary, category, and subcategory classification is given in Table 3.

Table 3. A detailed distribution of IoTID20 dataset in train and test.

Type	Class	Instances	Train Set	Test Set
Binary	Anomaly	585,342	468,274	117,068
	Normal	40,073	32,058	8015
	Total	625,415	500,332	125,083
Category	Mirai	415,309	332,247	83,062
	Scan	75,265	60,212	15,053
	DoS	59,391	47,513	11,878
	MITM ARP Spoofing	35,377	28,302	7075
	Normal	40,073	32,058	8015
	Total	625,415	500,332	125,083
Sub-Category	Mirai-UDP Flooding	183,189	146,551	36,638
	Mirai-Hostbruteforceg	121,178	96,943	24,235
	Mirai-HTTP Flooding	55,818	44,654	11,164
	Mirai-Ackflooding	55,124	44,099	11,025
	DoS-Synflooding	59,391	47,513	11,878
	Scan Port OS	53,073	42,458	10,615
	Scan Hostport	22,192	17,754	4438
	MITM ARP Spoofing	35,377	28,302	7075
	Normal	40,073	32,058	8015
	Total	625,415	500,332	125,083

3.3. *Designing the DCNN Model*

CNN is a DL technique that consists of convolutional layers, pooling layers, and fully connected layers [29]. CNN is usually utilized for image classification and voice recognition. In this study, we used a DCNN followed by a DNN for malicious activities identification in IoT networks. The proposed approach consists of two 1D convolutional layers, two max-pooling layers, flatten, and three dense layers, as shown in Figure 2. The input shape in the first convolutional layer is (none, 62, 1). Here, "none" is the dynamic number of instances, "62" is the number of input features and "1" is the third-dimension value. The size of the kernel is three, and sixty-two filters were used in this layer, which produces output in the form of (none, 62, 62). The output of the first convolutional layer is given as an input in the max-pooling layer. In this layer, pool size four was used which produces (none, 15, 62) output. The second convolutional layer is placed here, in which the size of the kernel is three and thirty filters are used, which produce the output in the form of (none, 15, 30). The output of the second convolutional layer is given as an input in the max-pooling layer. In this layer, pool size two was used, which produces (none, 7, 30) output. The convolutional layer not only converges the most important features but also reduces noise [30]. The 1D convolutional layer is demonstrated in Equations (2) and (3).

$$x_k = b_k + \sum_{i=1}^{N}(s_i, w_{ik}) \quad (2)$$

$$y_k = f(x_k) \quad (3)$$

where x_k is the input in the 1D convolutional layer. The output of the previous layer neuron is represented by s_k, w_{ik} represents the kernel from i to k. b_k is the bias value of the neuron in the convolutional layer. The ReLU activation function is represented by $f()$. Equation (4) describes the ReLU. y_k is the output of the 1D convolutional layer. The output of the convolutional layer is the input in the pooling layer demonstrated in Equation (5). We select the maximum value from region $\Re$ which contains the output values of the convolutional layer. s_k is the output of the max-pooling layer.

$$f(x_k) = \max(0, x_k) \quad (4)$$

$$s_k =_{i\in\Re}^{\max} y_k \quad (5)$$

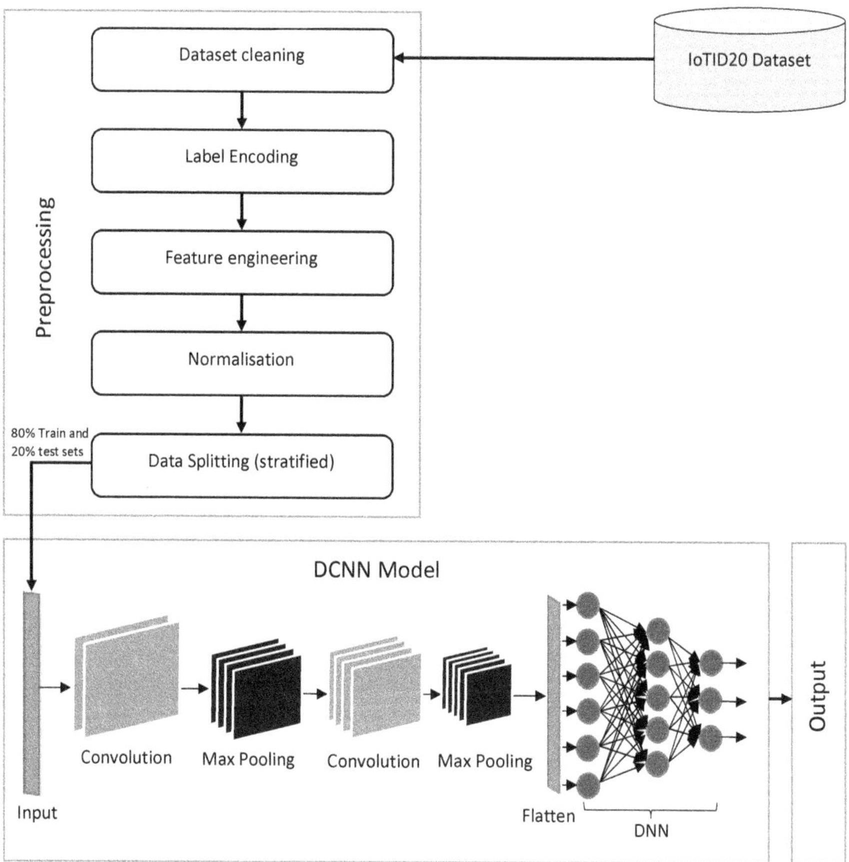

Figure 2. Architecture of the proposed DCNN model.

The flatten method is used to convert the output shape of the last pooling layer into a single-dimensional array. The output of the flatten is (none, 210) which is input in the first dense layers. The output of the first dense layer is (none, 50) which is given as input in the second dense layer. The second dense layer produces (none, 25) output which is input in the last dense layer. The ReLU activation function is used in dense layers. The last dense layer produces output results in which sigmoid function for binary classification and softmax function for multi-class classification are used, respectively. Sigmoid and softmax are demonstrated in Equations (6) and (7).

$$\sigma(x) = \frac{1}{1+e^{-x}} \tag{6}$$

$$\text{softmax}(x)_i = \frac{e^{x_i}}{\sum_{j=1}^{K} e^{x_j}} \tag{7}$$

3.4. Evaluation Metrics

The evaluation of the DCNN approach was carried out with accuracy, precision, recall, and F1-score. We start by explaining these four parameters, true positive (TP), false negative (FN), false positive (FP), and true negative (TN), which are used to compute the evaluation metrics such as accuracy, precision, recall, and the F1-score. TP refers to the number of instances that have been correctly identified as normal. The number of instances that misclassify normal data as an attack is known as the FN. FP represents the number of malicious instances that are wrongly classified as normal. TN represents the number of instances that are classified correctly as malicious. All of these evaluation metrics were calculated by using Equations (8)–(11).

$$\text{Accuracy} = \frac{\alpha+\beta}{\alpha+\beta+\gamma+\delta} \tag{8}$$

$$\text{Precision} = \frac{\alpha}{\alpha+\gamma} \tag{9}$$

$$\text{Recall} = \frac{\alpha}{\alpha+\delta} \tag{10}$$

$$\text{F1-score} = \frac{2 \times (\text{ Precision } \times \text{ Recall })}{\text{Precision } + \text{ Recall}} \tag{11}$$

where α represents TP, β represents TN, γ represents FP, and δ represents FN.

3.5. Experimental Platform

Experiments on the DCNN model were conducted with the HP ProBook G5 8th generation laptop. This laptop contains 24 GB ram and an Intel Core i5 processor. In software specifications, we used Windows 11 Pro, Python 3.8.5, Tensorflow, and Keras library.

4. Performance Analysis

This section provides a detailed evaluation of the proposed model. The proposed DCNN model was evaluated on the IoTID20 dataset. The performance of the DCNN was tested for binary, multi-class categories, and multi-class subcategories classifications. This section presents a comparison of convolutional layers followed by dense layers for multi-class categories and multi-class subcategories. The same comparison was performed for famous optimizers. The optimal solutions were selected from the comparison and compared with other ML/DL models.

4.1. Performance Evaluation of Convolutional and Dense Layers

The CNN algorithm consists of convolutional layers, pooling layers, and fully connected layers. This experiment was conducted for one and two convolutional layers, followed by fully connected dense 1–5 layers. These experiments were conducted for the multi-class category and subcategory classification. A detailed comparison is given in Tables 4 and 5. The experimental results showed that the average optimal solution is two convolutional layers and three dense layers.

Table 4. A comparison of CNN layers for multi-class category classification.

Convolutional Layers	Dense Layers	Accuracy	Precision	Recall	F1-Score
1	1	0.9465	0.92	0.9297	0.9237
1	3	0.9798	0.9712	0.9723	0.9716
2	1	0.9791	0.9756	0.9656	0.9701
2	2	0.9823	0.9744	0.9753	0.9747
2	**3**	**0.9833**	**0.9742**	**0.9788**	**0.9764**
2	4	0.9794	0.9697	0.9735	0.9713
2	5	0.9813	0.974	0.9757	0.9744

Table 5. A comparison of CNN layers for multi-class sub-category classification.

Convolutional Layers	Dense Layers	Accuracy	Precision	Recall	F1-Score
1	1	0.7232	0.7056	0.6443	0.6182
1	3	0.7633	0.7660	0.7157	0.6804
2	1	0.7690	0.7518	0.6563	0.7008
2	2	0.7731	0.7955	0.7320	0.6989
2	**3**	**0.7755**	**0.7876**	**0.7343**	**0.7600**
2	4	0.7732	0.7890	0.6790	0.6541
2	5	0.7650	0.8499	0.6527	0.6160

4.2. Performance Evaluation of Optimizers

An optimizer is a function used to update the neural network weights and learning rates. It helps to reduce the loss and improve the performance of the model [31,32]. Famous optimizers for DL algorithms are stochastic gradient descent (SGD), root mean square propagation (RMSProp), adaptive moment estimation (Adam), adaptive moment estimation maximization (AdaMax), and Nesterov-accelerated adaptive moment estimation (Nadam). The performances of these modifiers are optimal for CNN, as validated in Ref. [33]. The aforementioned five optimizers were used in this experiment. A detailed comparison of optimizers for the multi-class category and subcategory classification is shown in Tables 6 and 7, respectively. The experimental results show that Adam, Nadam, and AdaMax were the top three optimizers in this experiment.

Table 6. A detailed comparison of optimizers for multi-class category classification.

Optimizer	Accuracy	Precision	Recall	F1-Score
SGD	0.9789	0.9676	0.9706	0.9690
RMSprop	0.7630	0.7457	0.7195	0.6527
Adam	0.9801	0.9761	0.9695	0.9725
Nadam	0.9838	0.9773	0.9783	0.9777
AdaMax	0.9806	0.9726	0.9721	0.9723

Table 7. A detailed comparison of optimizers for multi-class sub-category classification.

Optimizer	Accuracy	Precision	Recall	F1-Score
SGD	0.9789	0.9676	0.9706	0.969
RMSprop	0.7630	0.7457	0.7195	0.6527
Adam	0.9801	0.9761	0.9695	0.9725
Nadam	0.9838	0.9773	0.9783	0.9777
Adamax	0.9806	0.9726	0.9721	0.9723

4.3. Performance Analysis of the Proposed DCNN

In this study, we propose a DCNN architecture for malicious activities identification in IoT networks. For DCNN, the above results show that the optimal solution for the IoTID20 dataset is two convolutional layers, followed by three dense layers. In addition, from the above results, we selected the top three optimizers (Adam, Nadam, and AdaMax) for this experiment. This section provides a detailed classification of binary-class, multi-class category, and multi-class subcategories for batch sizes 32, 64, 128, and 256.

4.3.1. DCNN Evaluation for Binary-Class Classification

The performance of the proposed approach was tested for a binary-class scenario. The DCNN model was trained with the IoTID20 dataset for 50 epochs, and the binary cross-entropy function was used to calculate the loss. In the first step, the proposed DCNN performance for the Adam optimizer is compared in the bar graphs in Figure 3. Based on the findings, the proposed model had the highest anomaly detection accuracy of 99.89% at batch size 128. For this optimizer, the other evaluation scores, namely, precision, recall, and F1-score, were 99.77%, 99.37%, and 99.57%, respectively. In the second step, all the experiments for the Nadam optimizer were rearranged with the same batch sizes. The proposed DCNN performance for the Nadam optimizer is compared in the bar graphs in Figure 4. Based on the findings, the proposed model had the highest anomaly detection accuracy of 99.91% at batch size 128. For this optimizer, the other evaluation scores, namely, precision, recall, and F1-score, are 99.87%, 99.38%, and 99.62%, respectively. In the third step, all the experiments for the AdaMax optimizer were repeated with the same batch sizes. The proposed DCNN performance for the Nadam optimizer is compared in the bar graphs in Figure 5. Based on the findings, the proposed model had the highest anomaly detection accuracy of 99.86% at batch size 128. For this optimizer, the other evaluation scores, namely, precision, recall, and F1-score, were 99.74%, 99.14%, and 99.44%, respectively.

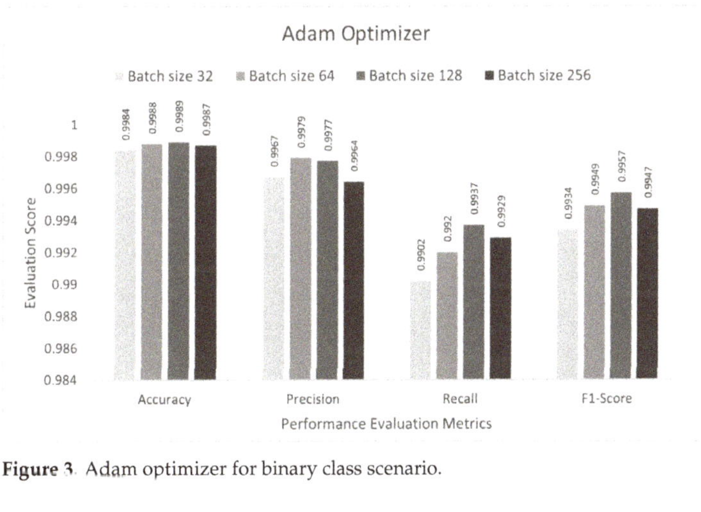

Figure 3. Adam optimizer for binary class scenario.

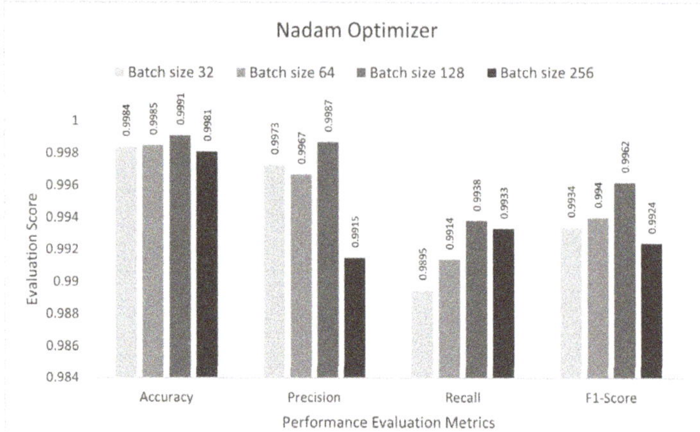

Figure 4. Nadam optimizer for binary class scenario.

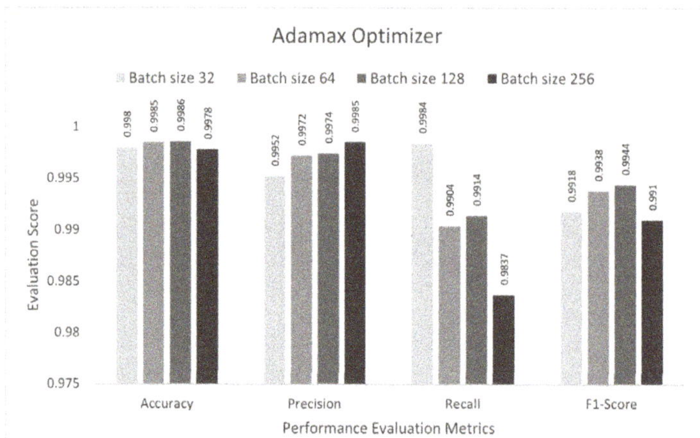

Figure 5. AdaMax optimizer for binary class scenario.

4.3.2. DCNN Evaluation for Multi-Class Category Classification

In this stage, the performance of the proposed study was evaluated for a multi-class category classification scenario. The DCNN model was trained with the IoTID20 dataset for 50 epochs, and a sparse categorical cross-entropy function was used to calculate the loss. As noted previously, for the binary-class studies, an Adam optimizer was chosen at the initial stage. The proposed DCNN performance for the Adam optimizer is compared in the bar graphs in Figure 6. Based on the analysis of the results, the proposed model had the highest anomaly detection accuracy of 98.13% at batch size 64. For this optimizer, the other performance scores, namely, precision, recall, and F1-score, were 97.40%, 97.53%, and 97.45%, respectively. In the second step, all the experiments for the Nadam optimizer were rearranged with the same batch sizes. The proposed DCNN performance for the Nadam optimizer is compared in the bar graphs in Figure 7. Based on the analysis of the results, the proposed model had the highest anomaly detection accuracy of 98.38% at batch size 32. For this optimizer, the other performance scores, namely, precision, recall, and F1-score, were 97.73%, 97.83%, and 97.77%, respectively. In the third step, all the experiments for the AdaMax optimizer were repeated with the same batch sizes. The proposed DCNN performance for the Nadam optimizer is compared in the bar graphs in Figure 8. Based on the analysis of the results, the proposed model had the highest anomaly detection accuracy of 98.06% at batch size 32. For this optimizer, the other performance scores, namely, precision, recall, and F1-score, were 97.26%, 97.21%, and 97.23%, respectively.

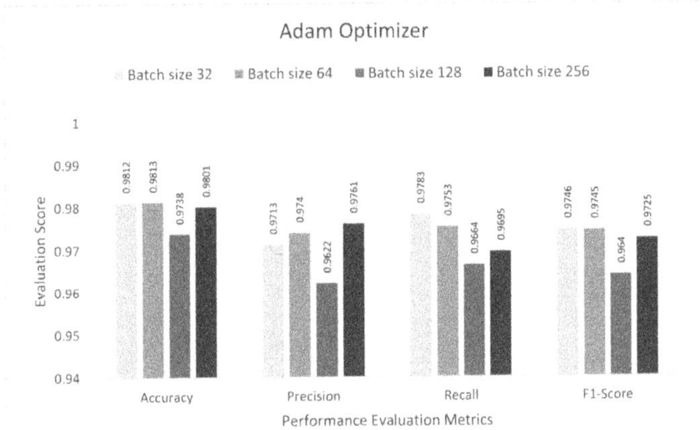

Figure 6. Adam optimizer for multi-class category classification scenario.

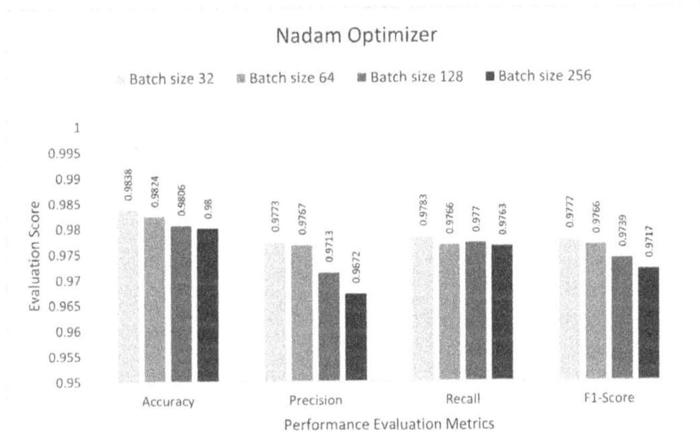

Figure 7. Nadam optimizer for multi-class category classification scenario.

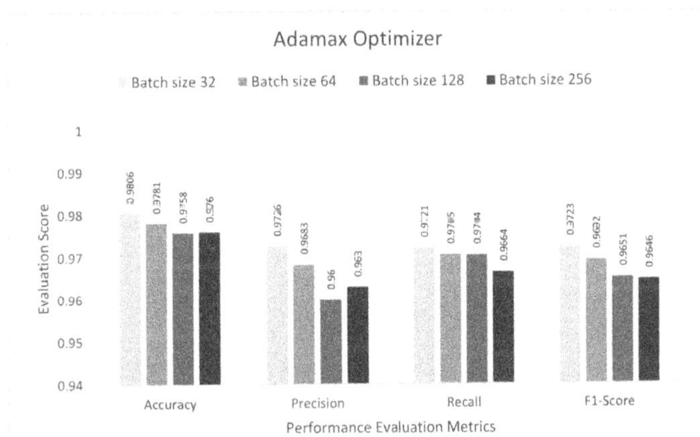

Figure 8. AdaMax optimizer for multi-class category classification scenario.

4.3.3. DCNN Evaluation for Multi-Class Subcategory Classification

In the final stage, the performance of the proposed study was evaluated for multi-class subcategory classification scenarios. The DCNN model was trained with the IoTID20 dataset for 100 epochs, and a sparse categorical cross-entropy function was used to calculate the loss. As noted previously, for the binary and multi-class category studies, an Adam optimizer was chosen at the initial stage. The proposed DCNN performance for the Adam optimizer is compared in the bar graphs in Figure 9. Based on the analysis of the results, the proposed model had the highest anomaly detection accuracy of 77.55% at batch size 32. For this optimizer, the other performance scores, namely, precision, recall, and F1-score, were 78.76%, 73.43%, and 76.00%, respectively. In the second step, all the experiments for the Nadam optimizer were rearranged with the same batch sizes. The proposed DCNN performance for the Nadam optimizer is compared in the bar graphs in Figure 10. Based on the analysis of the results, the proposed model had the highest anomaly detection accuracy of 77.44% at batch size 64. For this optimizer, the other performance scores, namely, precision, recall, and F1-score, were 86.02%, 72.58%, and 78.73%, respectively. In the third step, all the experiments for the AdaMax optimizer were repeated with the same batch sizes. The proposed DCNN performance for the Nadam optimizer is compared in the bar graphs in Figure 11. Based on the analysis of the results, the proposed model had the highest anomaly detection accuracy of 77.11% at batch size 64. For this optimizer, the other performance scores, namely, precision, recall, and F1-score, were 77.35%, 70.85%, and 73.95%, respectively.

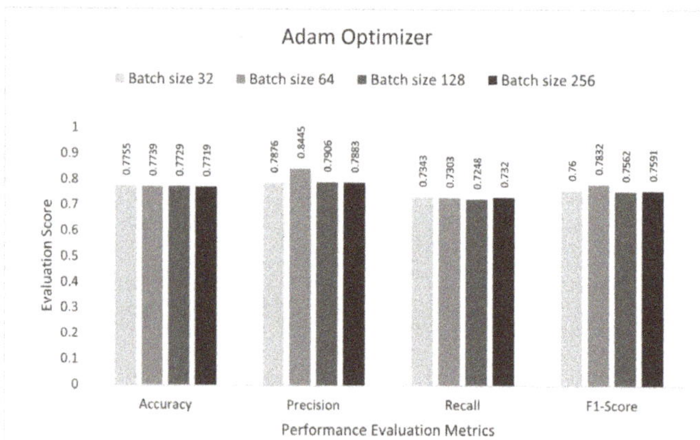

Figure 9. Adam optimizer for multi-class sub-category classification scenario.

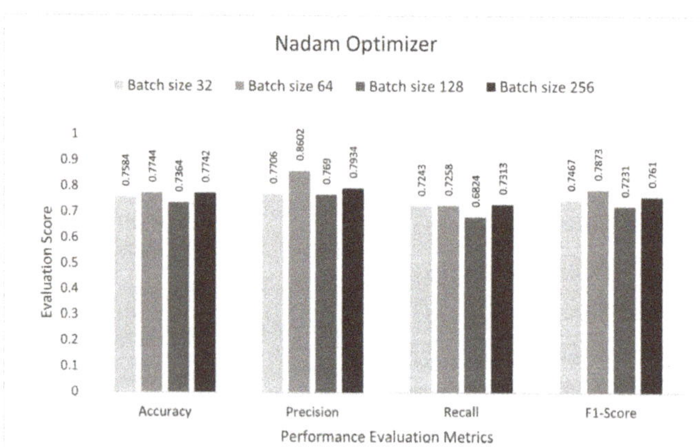

Figure 10. Nadam optimizer for multi-class sub-category classification scenario.

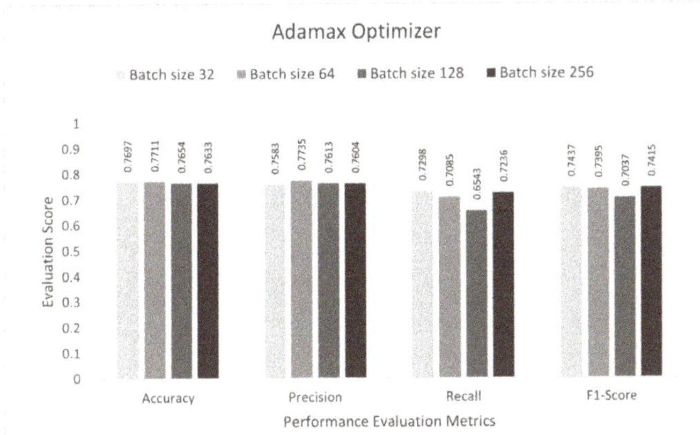

Figure 11. AdaMax optimizer for multi-class sub-category classification scenario.

4.4. Performance Discussion

The performance of the proposed DCNN was analyzed for binary, multi-class category, and multi-class subcategory classification. The results presented earlier show a comparison of optimizers and batch sizes. Based on the performance analysis of the proposed model for binary class, the Nadam optimizer with a batch size of 128 performs better than the others. Similarly, in the performance analysis of the proposed model for the multi-class category and subcategory classification, the Adam optimizer with a batch size of 32 performs better than others. For testing the performance of the proposed model, k-fold cross-validation was also used, where the "k" value is 7. The results of the k-fold cross-validation are approximately equivalent.

4.5. Performance Comparison with Other DL and Traditional ML-Based IDSs

The performance of the proposed DCNN was compared with other DL and traditional ML methods to evaluate its efficacy. LSTM, gated recurrent unit (GRU), deep neural network (DNN), deep belief network (DBN), deep autoencoder (DAE), and multilayer perceptron (MLP) are examples of DL methods. Decision tree (DT), logistic regression (LR), naive Bayes (NB), support vector machine (SVM), and k-nearest neighbors (KNN) are all examples of traditional ML methods. All of these methods were implemented in the same environment for an accurate performance comparison. The preprocessing steps were the same for all models, including the proposed model. We split the dataset into 80% train and 20% test sets. For all of the DL algorithms, we used Adam optimizer and default batch size 32. The optimal solution of each model was used for the comparison. The hidden layers used in LSTM, GRU, DNN, DBN, AE, and MLP are 3, 3, 4, 4, 6, and 10, respectively. The number of training epochs for all these models was the same as the proposed model. A detailed analysis for binary-class category, multi-class category, and subcategory classifications is shown in Tables 8–10, respectively. According to the results, the performance of the proposed DCNN model is optimal as compared to other DL models. The proposed model detection accuracy is 99.84%, 98.12%, and 77.55% for binary-class, multi-class, and subcategory classifications, respectively.

For optimal performance, each DL model requires multiple layers that maximize computational power. The proposed DCNN model improves the performance and also reduces computational power as it narrows to specific features, compared to other ML and DL models. Comparing the performance of the proposed DCNN with other ML and DL models shows the optimal results.

Table 8. A comparison of DCNN with other DL models on binary-class.

Models	Accuracy	Precision	Recall	F1-Score
LSTM	0.9952	0.9943	0.9662	0.9797
GRU	0.9959	0.9856	0.9807	0.9832
DNN	0.9981	0.9983	0.9862	0.9922
DBN	0.9969	0.9937	0.9807	0.9871
AE	0.9974	0.9895	0.9887	0.9891
MLP	0.9972	0.9938	0.9832	0.9884
DT	0.9857	0.9819	0.9861	0.9840
LR	0.9659	0.9034	0.7879	0.8345
NB	0.6504	0.5765	0.8093	0.6733
SVM	0.9744	0.9199	0.8552	0.8844
KNN	0.9983	0.9964	0.9894	0.9929
Proposed DCNN	**0.9984**	**0.9967**	**0.9902**	**0.9934**

Table 9. A comparison of DCNN with other DL models on multi-class category.

Model	Accuracy	Precision	Recall	F1-Score
LSTM	0.9584	0.9543	0.9201	0.9355
GRU	0.9681	0.9576	0.9468	0.9519
DNN	0.9547	0.9340	0.9447	0.9367
DBN	0.9589	0.9430	0.9549	0.9469
AE	0.9644	0.9515	0.9440	0.9456
MLP	0.9238	0.8933	0.8436	0.8529
DT	0.9770	0.9744	0.9737	0.9741
LR	0.8314	0.7728	0.7297	0.7311
NB	0.6772	0.6628	0.7381	0.6479
SVM	0.8557	0.8416	0.7845	0.7883
KNN	0.9793	0.9746	0.9699	0.9722
Proposed DCNN	**0.9812**	**0.9713**	**0.9783**	**0.9746**

Table 10. A comparison of DCNN with other DL models on multi-class sub-category.

Model	Accuracy	Precision	Recall	F1-Score
LSTM	0.7141	0.6993	0.5992	0.6453
GRU	0.7615	0.7571	0.6996	0.7272
DNN	0.7483	0.7244	0.6610	0.6912
DBN	0.6888	0.6916	0.6166	0.6519
AE	0.7535	0.7805	0.7016	0.7389
MLP	0.7065	0.7124	0.6263	0.6665
DT	0.7530	0.7508	0.7362	0.7413
LR	0.5481	0.4457	0.4239	0.4142
NB	0.5298	0.4878	0.5032	0.4481
SVM	0.6240	0.4888	0.4741	0.4624
KNN	0.7621	0.7634	0.7477	0.7515
Proposed DCNN	**0.7755**	**0.7876**	**0.7343**	**0.7600**

5. Conclusions

This study presents a new DCNN-based DL model and feature engineering method for malicious attack detection in IoT networks. The objective was to improve performance and reduce computational power. The proposed DCNN model successfully improves performance and reduces computational power. It is useful for low-power IoT network devices. The IoTID20 dataset was used to analyze the performance of the proposed DCNN model. The proposed model was evaluated for binary, multi-class category, and subcategory classifications. Experiments were performed for different layers of the CNN algorithm, and an optimal solution was selected. The proposed model was evaluated in-depth with Adam, Nadam, and AdaMax optimizers. The Nadam optimizer peformance was optimum for binary, multi-class category, and multi-class subcategory with 128, 32, and 64 batch sizes, respectively. The proposed model was also compared with state-of-the-art DL techniques and other traditional ML algorithms for a broader view in terms of efficacy, robustness, etc. The experimental analysis indicates that the proposed approach obtained optimum results when compared through accuracy, precision, recall, and F1-score parameters.

Author Contributions: S.U., J.A., M.A.K., E.H.A., M.H., F.S. and Y.Y.G. performed formal analysis and original draft preparation. Y.Y.G., F.S. and N.P. proposed the main ideas and validated analysis. S.U., E.H.A., M.H., F.S., N.P., Y.Y.G. and J.A. crystallized framework and also revised the manuscript. All authors have read and agreed to the published version of the manuscript.

Funding: This work is supported by Taif University Researchers Supporting Project number (TURSP-2020/292) Taif University, Taif, Saudi Arabia. This work is also supported by Princess Nourah bint Abdulrahman University Researchers Supporting Project number (PNURSP2022R193), Princess Nourah bint Abdulrahman University, Riyadh, Saudi Arabia.

Institutional Review Board Statement: Not applicable.

Informed Consent Statement: Not applicable.

Data Availability Statement: The publicly available dataset can be found at: https://sites.google.com/view/iot-network-intrusion-dataset/home (accessed on 28 January 2022).

Acknowledgments: The authors would like to acknowledge Taif University Researchers Supporting Project number (TURSP-2020/292) Taif University, Taif, Saudi Arabia. The authors would like also to acknowledge Princess Nourah bint Abdulrahman University Researchers Supporting Project number (PNURSP2022R193), Princess Nourah bint Abdulrahman University, Riyadh, Saudi Arabia.

Conflicts of Interest: The authors declare no conflict of interest.

References

1. Farooq, M.S.; Sohail, O.O.; Abid, A.; Rasheed, S. A Survey on the Role of IoT in Agriculture for the Implementation of Smart Livestock Environment. *IEEE Access* **2022**, *10*, 9483–9505. [CrossRef]
2. Ullah, I.; Mahmoud, Q.H. Design and development of a deep learning-based model for anomaly detection in IoT networks. *IEEE Access* **2021**, *9*, 103906–103926. [CrossRef]
3. Mezni, H.; Driss, M.; Boulila, W.; Atitallah, S.B.; Sellami, M.; Alharbi, N. SmartWater: A Service-Oriented and Sensor Cloud-Based Framework for Smart Monitoring of Water Environments. *Remote Sens.* **2022**, *14*, 922. [CrossRef]
4. Alam, T. A Reliable Communication Framework and Its Use in Internet of Things (IoT). *Int. J. Sci. Res. Comput. Sci. Eng. Inf. Technol.* **2018**, *3*, 450–456.
5. Al-Bahri, M.; Yankovsky, A.; Borodin, A.; Kirichek, R. Testbed for identify IoT-devices based on digital object architecture. In *Internet of Things, Smart Spaces, and Next Generation Networks and Systems*; Springer: Berlin/Heidelberg, Germany, 2018; pp. 129–137.
6. Nguyen, X.H.; Nguyen, X.D.; Huynh, H.H.; Le, K.H. Realguard: A Lightweight Network Intrusion Detection System for IoT Gateways. *Sensors* **2022**, *22*, 432. [CrossRef] [PubMed]
7. Zhang, Y.; Li, P.; Wang, X. Intrusion detection for IoT based on improved genetic algorithm and deep belief network. *IEEE Access* **2019**, *7*, 31711–31722. [CrossRef]
8. Conti, M.; Dehghantanha, A.; Franke, K.; Watson, S. Internet of Things security and forensics: Challenges and opportunities. *Future Gener. Comput. Syst.* **2018**, *78*, 544–546. [CrossRef]
9. Liu, H.; Lang, B. Machine learning and deep learning methods for intrusion detection systems: A survey. *Appl. Sci.* **2019**, *9*, 4396. [CrossRef]

10. Gao, Z.J.; Pansare, N.; Jermaine, C. Declarative parameterizations of user-defined functions for large-scale machine learning and optimization. *IEEE Trans. Knowl. Data Eng.* **2018**, *31*, 2079–2092. [CrossRef]
11. Guo, Z.; Qi, W.; Huang, Y.; Zhao, J.; Yang, H.; Koo, V.C.; Li, N. Identification of Crop Type Based on C-AENN Using Time Series Sentinel-1A SAR Data. *Remote Sens.* **2022**, *14*, 1379. [CrossRef]
12. Liu, Z.Y.C.; Chamberlin, A.J.; Tallam, K.; Jones, I.J.; Lamore, L.L.; Bauer, J.; Bresciani, M.; Wolfe, C.M.; Casagrandi, R.; Mari, L.; et al. Deep Learning Segmentation of Satellite Imagery Identifies Aquatic Vegetation Associated with Snail Intermediate Hosts of Schistosomiasis in Senegal, Africa. *Remote Sens.* **2022**, *14*, 1345. [CrossRef]
13. Salunkhe, S.S.; Pal, S.; Agrawal, A.; Rai, R.; Mole, S.; Jos, B.M. Energy optimization for CAN bus and media controls in electric vehicles using deep learning algorithms. *J. Supercomput.* **2022**, *78*, 8493–8508. [CrossRef]
14. Lin, J.; Diekmann, P.; Framing, C.E.; Zweigel, R.; Abel, D. Maritime Environment Perception Based on Deep Learning. *IEEE Trans. Intell. Transp. Syst.* **2022**. [CrossRef]
15. Heaton, J. Ian goodfellow, yoshua bengio, and aaron courville: Deep learning. *Genet. Program. Evolvable Mach.* **2018**, *19*, 305–307. [CrossRef]
16. Mighan, S.N.; Kahani, M. A novel scalable intrusion detection system based on deep learning. *Int. J. Inf. Secur.* **2021**, *20*, 387–403. [CrossRef]
17. Al-Turaiki, I.; Altwaijry, N. A convolutional neural network for improved anomaly-based network intrusion detection. *Big Data* **2021**, *9*, 233–252. [CrossRef]
18. Aldweesh, A.; Derhab, A.; Emam, A.Z. Deep learning approaches for anomaly-based intrusion detection systems: A survey, taxonomy, and open issues. *Knowl. Based Syst.* **2020**, *189*, 105124. [CrossRef]
19. Basati, A.; Faghih, M.M. DFE: Efficient IoT network intrusion detection using deep feature extraction. *Neural Comput. Appl.* **2022**, 1–21. [CrossRef]
20. Rashid, M.; Kamruzzaman, J.; Imam, T.; Wibowo, S.; Gordon, S. A tree-based stacking ensemble technique with feature selection for network intrusion detection. *Appl. Intell.* **2022**, 1–14. [CrossRef]
21. Fatani, A.; Dahou, A.; Al-Qaness, M.A.; Lu, S.; Abd Elaziz, M. Advanced Feature Extraction and Selection Approach Using Deep Learning and Aquila Optimizer for IoT Intrusion Detection System. *Sensors* **2022**, *22*, 140. [CrossRef]
22. Alkahtani, H.; Aldhyani, T.H. Intrusion detection system to advance internet of things infrastructure-based deep learning algorithms. *Complexity* **2021**, *2021*, 5579851. [CrossRef]
23. Keserwani, P.K.; Govil, M.C.; Pilli, E.S.; Govil, P. A smart anomaly-based intrusion detection system for the Internet of Things (IoT) network using GWO–PSO–RF model. *J. Reliab. Intell. Environ.* **2021**, *7*, 3–21. [CrossRef]
24. Qaddoura, R.; Al-Zoubi, A.; Almomani, I.; Faris, H. A multi-stage classification approach for iot intrusion detection based on clustering with oversampling. *Appl. Sci.* **2021**, *11*, 3022. [CrossRef]
25. Saba, T.; Sadad, T.; Rehman, A.; Mehmood, Z.; Javaid, Q. Intrusion detection system through advance machine learning for the internet of things networks. *IT Prof.* **2021**, *23*, 58–64. [CrossRef]
26. Kang, H.; Ahn, D.H.; Lee, G.M.; Yoo, J.D.; Park, K.H.; Kim, H.K. IoT Network Intrusion Dataset. 2019. Available online: https://ieee-dataport.org/open-access/iot-network-intrusion-dataset (accessed on 28 January 2022). [CrossRef]
27. Ullah, I.; Mahmoud, Q.H. A scheme for generating a dataset for anomalous activity detection in iot networks. In Proceedings of the Canadian Conference on Artificial Intelligence, Ottawa, ON, Canada, 13–15 May 2020; pp. 508–520.
28. Dahouda, M.K.; Joe, I. A Deep-Learned Embedding Technique for Categorical Features Encoding. *IEEE Access* **2021**, *9*, 114381–114391. [CrossRef]
29. Riyaz, B.; Ganapathy, S. A deep learning approach for effective intrusion detection in wireless networks using CNN. *Soft Comput.* **2020**, *24*, 17265–17278. [CrossRef]
30. Zhang, H.; Huang, L.; Wu, C.Q.; Li, Z. An effective convolutional neural network based on SMOTE and Gaussian mixture model for intrusion detection in imbalanced dataset. *Comput. Netw.* **2020**, *177*, 107315. [CrossRef]
31. Vidhya, A. A Comprehensive Guide on Deep Learning Optimizers. Available online: https://www.analyticsvidhya.com/blog/2021/10/a-comprehensive-guide-on-deep-learning-optimizers/ (accessed on 7 October 2021).
32. Ruder, S. An Overview of Gradient Descent Optimization Algorithms. Available online: https://ruder.io/optimizing-gradient-descent/ (accessed on 19 January 2016).
33. Vani, S.; Rao, T.M. An experimental approach towards the performance assessment of various optimizers on convolutional neural network. In Proceedings of the 2019 3rd International Conference on Trends in Electronics and Informatics (ICOEI), Tirunelveli, India, 23–25 April 2019; pp. 331–336.

sensors MDPI

Article

Convergence Analysis of Path Planning of Multi-UAVs Using Max-Min Ant Colony Optimization Approach

Muhammad Shafiq [1], Zain Anwar Ali [1,*], Amber Israr [1], Eman H. Alkhammash [2], Myriam Hadjouni [3] and Jari Juhani Jussila [4]

1 Electronic Engineering Department, Sir Syed University of Engineering & Technology, Karachi 75300, Pakistan; muhshafiq@ssuet.edu.pk (M.S.); aisrar@ssuet.edu.pk (A.I.)
2 Department of Computer Science, College of Computers and Information Technology, Taif University, P.O. Box 11099, Taif 21944, Saudi Arabia; eman.kms@tu.edu.sa
3 Department of Computer Sciences, College of Computer and Information Science, Princess Nourah Bint Abdulrahman University, P.O. Box 84428, Riyadh 11671, Saudi Arabia; mfhaojouni@pnu.edu.sa
4 HAMK Design Factory, Häme University of Applied Sciences, 13100 Hämeenlinna, Finland; jari.jussila@gmail.com
* Correspondence: zaali@ssuet.edu.pk

Abstract: Unmanned Aerial Vehicles (UAVs) seem to be the most efficient way of achieving the intended aerial tasks, according to recent improvements. Various researchers from across the world have studied a variety of UAV formations and path planning methodologies. However, when unexpected obstacles arise during a collective flight, path planning might get complicated. The study needs to employ hybrid algorithms of bio-inspired computations to address path planning issues with more stability and speed. In this article, two hybrid models of Ant Colony Optimization were compared with respect to convergence time, i.e., the Max-Min Ant Colony Optimization approach in conjunction with the Differential Evolution and Cauchy mutation operators. Each algorithm was run on a UAV and traveled a predetermined path to evaluate its approach. In terms of the route taken and convergence time, the simulation results suggest that the MMACO-DE technique outperforms the MMACO-CM approach.

Keywords: path planning; Max-Min Ant Colony Optimization; differential evolution; Cauchy mutation

Citation: Shafiq, M.; Ali, Z.A.; Israr, A.; Alkhammash, E.H.; Hadjouni, M.; Jussila, J.J. Convergence Analysis of Path Planning of Multi-UAVs Using Max-Min Ant Colony Optimization Approach. *Sensors* **2022**, *22*, 5395. https://doi.org/10.3390/s22145395

Academic Editor: Hyoungshick Kim

Received: 24 May 2022
Accepted: 18 July 2022
Published: 19 July 2022

1. Introduction

Motivation: Today, the applications of Unmanned Aerial Vehicles (UAVs) in the field of aeronautics are expanding day-by-day, due to their impact in every field [1]. At once, a single UAV was able to do a small task with a high operational cost. With the development of research and technology, multiple UAVs are used for complex tasks e.g., military, construction, surveying, and pattern formations [2–5]. However, one of the best reasons to use UAVs in a lethal environment is to secure humans from an ambiguous situation [6].

Biological behavior in nature is so inspiring for the real-world problem formulation of aerial robotics [7,8]. When aerial robotics becomes complex in multi-tasking applications, these biological behaviors will help find the optimal solution. In most cases related to the formation and path planning problems of UAVs, various bio-inspired algorithms become feasible [9]. However, one of the oldest optimization techniques widely used for shortest path routing problems is Ant Colony Optimization (ACO) [10–12].

Background and Related Work: The dynamics of complex aerial systems are difficult to handle especially when they are in clusters and want to achieve the same target smoothly [13]. Therefore, various controlling and optimization techniques are widely used in this area. Some famous intelligent optimizing algorithms, i.e., ACO, PSO, ABC, PIO, etc., are gaining popularity due to their problem-solving ability with the simplest structure [14–17]. Based on the food searching intelligence of real ants in nature; Ant

Colony Optimization and its variants have been used to solve the complex dynamics of a system [18]. ACO initially proposed by Dorigo et al. [19] was applied to Travelling Salesman Problem (TSP). However, the ACO algorithm resolves path planning problems of UAVs to obtain better routes and faster convergence [20].

Rapid progress in this area needs a hybrid approach based on ACO to optimize the previous attainments [21–23]. In [24], MAX–MIN Ant System (MMAS), an Ant Colony Optimization technique evolved from Ant System in which search space is bounded for better results. Later on, in [25], the author introduces efficient route planning by achieving the maximum convergence of the target. Similarly, in [26], an improved ACO is used to solve the trajectory planning of multiple UAVs. In [27], Duan et al. combined the ACO algorithm with Differential Evolution (DE) for 3D path planning of uninhabited combat air vehicles. Another author contributed in [28], for feature selection based on the combination of ACO and DE.

Contributions: This article compares the convergence rate of two state-of-the-art hybrid algorithms based on Max-Min Ant Colony Optimization techniques for the path planning of multiple UAVs. In [29], the author proposed a hybrid algorithm of Max-Min Ant Colony Optimization combined with Differential Evolution (MMACO-DE) on behalf of the path planning multiple UAVs. The author also added the dynamic environment in his research and proved with the simulation that the target achieved by his proposed algorithm has better results with a successful collision avoidance approach. The main feature of this article is that it selects only the finest ant in each cluster among all for the creation of the required path. In addition, the multiple colonies concept in the research saves the duration of target detection with fewer computations. In [30], the author used the Cauchy Mutant operator along with Max-Min Ant Colony Optimization (MMACO-CM) to improve his previous results. In this article, the same issue regarding path planning of Multiple UAVs in a dynamic environment is resolved with the new hybrid technique of MMACO-CM. This article has two important features, which include increasing the convergence speed for the avoidance falling into local optimum and achieving the shortest path for the target.

Organization: The following section is prearranged as follows. Section 2 elaborates the problem statement associated with the path planning of UAVs. Section 3 presents the preliminaries of Unmanned Aerial Vehicles. Section 4 deals with hybrid algorithm along with their mathematical modeling. Section 5 discusses the results obtained by comparison of the algorithm while Section 5 concludes the article.

2. Problem Statement

Path planning of aerial vehicles needs precise optimization techniques to obtain optimal routes despite manmade and natural threats. To achieve the target in the shortest possible time along with the shortest distance taken by each UAV, the best hybrid algorithm needs to have minimal complexions. However, the study uses the artificial obstacle theme containing mountains with different peaks and tornados in simulations. Unlike in urban environments, the mountainous area has uneven peaks and hence poses a greater challenge. Moreover, each hybrid algorithm requires a UAV to follow the same path simultaneously. Both UAVs move from the same starting point to the desired location to analyze the performance of the individual algorithm. Each UAV will travel along the specified route by implementing the hybrid algorithm and providing optimal route as well as convergence speed.

3. Preliminaries of Unmanned Aerial vehicles

3.1. Path Planning

The term "Path Planning" determines the route planned for an object for any specified mission, which includes various obstacles along the path [31]. The planned path restricts the UAV from possible crashes from obstacles or neighboring UAVs. Path planning is more feasible when operating with large quantities of robots. Path planning of aerial systems is categorized into motion-based and tracking-based approaches [32].

Motion planning: The term motion planning refers to the movement of aerial or ground robots for the specified or desired task [33]. The basics of motion planning include the shortest possible path, along with a precise turning angle. Motion planning consists of two basic configurations i.e., start and goal configurations. It uses two-dimensional (2D) or three-dimensional (3D) space configurations to show their path [34].

Table 1 shows the difficulty level of robotic motion planning in terms of information provided to the robot. The robot may calculate the size, nature, and distance of obstacles at every instant to avoid a collision [35]. There are four possible scenarios shown in Table 1. The first scenario is the simplest one with completely known information while in the third scenario, the information is partially known. In the second and fourth scenarios, the information regarding obstacles is completely and partially known for dynamic obstacles; therefore, these scenarios are considered difficult ones for robotic motion planning [36].

Table 1. Possible scenarios between obstacle types and information available.

	Static Obstacle	**Dynamic Obstacle**
Complete Information Known	1st Scenario	2nd Scenario
Partial Information Known	3rd Scenario	4th Scenario

Trajectory planning: Trajectory planning comes into existence when another motional variable encloses the path planning rather than obstacles. The velocity of robots, time taken by the path, and relative kinematics of the system are equally responsible to achieve the goal [37]. Moving from the initial point to the final point using collision avoidance is what trajectory planning is all about. In trajectory planning, both discrete and continuous methods use parametric calculations required to achieve the target. To follow a specific path with control parameters, e.g., location, rate, and acceleration, there must be the scheduling of time, and applying a control system that can accurately execute the trajectory is required [38].

When aerial vehicles fly simultaneously to achieve the desired task, they must follow the planned path to achieve that target. However, artificial intelligence and state-of-the-art technologies are still not capable of providing real-time solutions to UAVs in operating mode [39]. Therefore, nature-inspired algorithms based on the natural behavior of species are serving to provide solutions for the real time scenarios with fewer computations [40].

3.2. Collision Avoidance Protocol

Path planning of UAVs is incomplete without using a collision avoidance protocol. When UAVs form a specific shape during flight then they must follow the air collision rules to maintain a safe distance [41]. The environmental hurdles can be of many types including mountain, air disturbance, harsh weather, or any unwanted disturbance. The distance between the UAVs continues to decrease when moving towards the destination due to path planning constraints. To avoid a collision near the target, a safe distance is required by the following relationship [42].

$$\mathrm{L_{safe}(t)} = \begin{cases} \mathrm{L,\ t \le T_{d,n}} \\ \mathrm{l,\ t > T_{d,n}} \end{cases} \tag{1}$$

where L and l are the distance between UAVs and it varies according to the path planning. As a result, when planning or controlling multi-UAV formations, make sure there is no overlap between UAVs as much as possible to build air space cooperation.

Furthermore, changing the height of a UAV regularly puts the flight's safety at risk. As a result, it is best to avoid changing the altitude regularly. The fluctuating height C_h of the UAV states

$$C_h = \sqrt{\frac{1}{m_k}\sum_{k=0}^{m_k}\left((h_k - (\frac{1}{m_k+1})\sum_{l=0}^{m_k} h_k)\right)^2} \tag{2}$$

where h_k is the height of UAV, kth represents path leg and leg number is denoted by m_k. Because all UAVs are flying close together, the turning angle is a significant component of information control and path planning issues. The connection between nodes, edges, and legs is shown in Figure 1. The UAV travels between nodes, which serve as waypoints. It is limited to moving solely between nodes. When the algorithm begins, nodes are created. While the surroundings are three-dimensional, we separate it into the x, y, and z planes to make calculations easier. The nodes are therefore in 2D space. The dynamic topology of these nodes changes depending on the situational context.

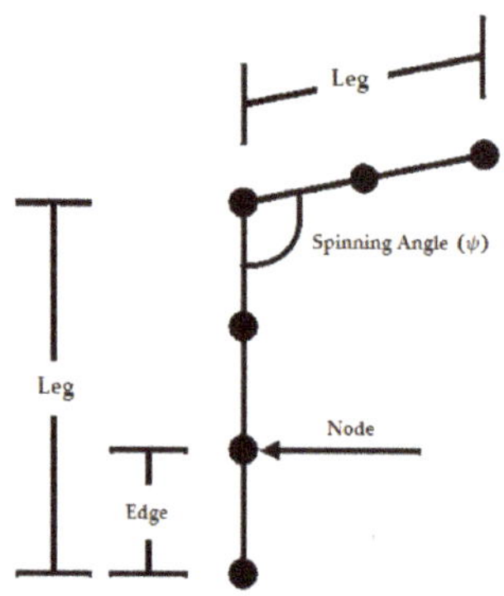

Figure 1. Correlation of Edge, Node, and Legs along with spinning angle ψ.

Each UAV has various limits in cooperatively altering its attitude due to constraints in the maneuverability of its maximum angle, which may result in an impact between UAVs. The projection of kth and $(k+1)$th leg on the parallel plan of the present location is presented by p_k and p_{k+1}. The calculated spinning angle along with its limited maneuverability is given by ψ.

$$Cos\psi_{max} \leq (p_k^T p_{k+1}/|p_k||p_{k+1}|),\ k = 1,\ 2,\ \ldots,\ n-1 \quad (3)$$

3.3. Environmental Threats

Air space contains various threats and hurdles for flying vehicles, which can produce uncertainties and delays in-flight operation [43]. In natural threats, air dynamics, weather, humidity, and temperature can disturb UAV flight operations. Similarly, artificial threats including enemies, hurdles, and obstacles can also disturb its operation. To cope with these types of threats, a 3D complex environment needs to be created in simulation [44]. This environment tests the performance of the proposed algorithm by adding uncertainties with respect to the system. In this study, there are two types of environmental threats, which contain multiple mountains with altered peaks and air disturbance in between mountains to create cyclones.

4. Hybrid Algorithm

To optimize the system more rapidly and accurately, a hybrid algorithm plays a vital role by combining two or more algorithms. It selects a suitable collection of optimization algorithms to reduce the errors in the system.

4.1. Ant Colony Optimization

In the field of optimization algorithms, the Ant Colony Optimization algorithm is extensively used in UAVs. It provides the optimal solution for a complex problem associated with air vehicles. When compared with other algorithms and techniques, ACO has an edge in its distributed computation approach and pre-mature convergence avoidance. In the ant system, real ants create the food-searching algorithm with the help of multiple tours. The most followed route is the optimal route for the rest of the ants. The pheromone trail is the basic tool in the path for indication [45].

Searching for food in insects is common in nature, where each species has their own natural procedure to follow. The most famous species in search of food are ants, due to their unmatched behavior [46–48]. The Ant Colony Optimization (ACO) algorithm gives a comprehensive food searching performance of ants [49]. In this algorithm, the social behavior of ants presents mathematically to find the optimal solution for the target. Initially, the ants distribute in different clusters and follow all possible paths to reach the food. In this journey, all ants leave pheromone (a chemical substance) along the route to help other following ants. After the first round, all the ants will try to follow the shortest route in consecutive rounds [50]. Thus, after specific rounds, a single shortest route allowed the ants to follow from the initial point to the target as shown in Figure 2.

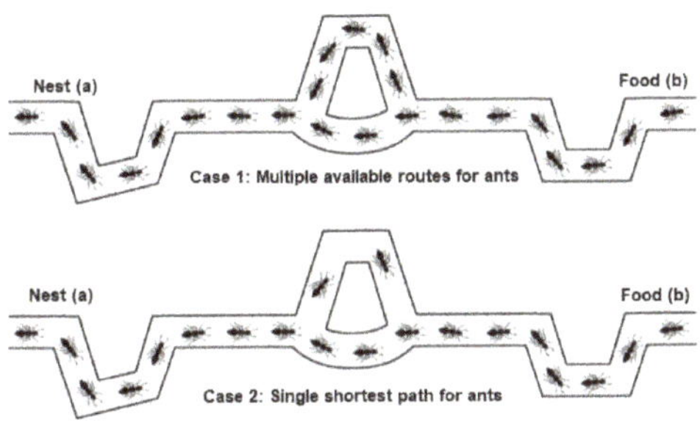

Figure 2. Food hunting procedure in Ant system, case 1 presents the multiple routes while Case 1 shows only single shortest route available for ants.

In the ACO algorithm, the pheromone value replaces with an updated one after every iteration. This process is autonomous, and all ants will follow the same instructions based on pheromone value, which can reflect the food quality and quantity on the desired path. This process continues until one of the possible paths can get more pheromone than the rest. The preferred path will now be the only path allowed for all ants to form the shortest path in their food search process [10]. However, the path with the most pheromone has a very high probability to be the best path as shown by

$$P_{i,j}^{x}(t) = \frac{\rho_{i,j}^{\alpha}(t)\upsilon_{i,j}^{\beta}(t)}{\sum_{j\in accept(i)}\rho_{i,j}^{\alpha}(t)\upsilon_{i,j}^{\beta}(t)} \tag{4}$$

where $P_{i,j}^{x}(t)$ the probability of xth ant city between i to j, $\rho_{i,j}^{\alpha}(t)$ is measured pheromone at the corner of the cities, $\upsilon_{i,j}^{\beta}(t)$ is the reciprocal of the length between two cities, α and β are the pheromone weight and distance traveled by the xth ant. Initially, the rate of pheromone is not constant at the corners of the cities but later on, the pheromone weight is greater to form a most visible line across all paths [19]. The following relation determines the rate of the pheromone.

$$\rho_{i,j}^{q}(t+1) = (1-\delta)\cdot\rho_{i,j}^{q}(t) + \Delta\rho_{i,j}^{q} \tag{5}$$

where $\delta \in (0,1)$, the amount of evaporation concerning time t to t+1

$$\rho_{i,j}^{q}(t) = \sum_{l=1}\Delta\rho_{i,j,y}^{q} \tag{6}$$

where $\Delta\rho_{m,j,y}^{q}$, the present amount of pheromone.

4.2. Maximum Minimum Ant Colony Optimization

Maximum Minimum Ant Colony Optimization (MMACO) is a technique for improving the system's search space capability and obtaining the fastest convergence time. The search space is constrained with this strategy by specifying a range, which reduces the search time and allows it to converge quickly. Updated trails in MMACO depend on decent travels of selected ants with the highest fitness value among all ants. In MMACO, consider m ants are assigned for the ith UAV for multiple routes. The mean cost μ of the route gives,

$$\mu_{i,m}(t) = \frac{1}{m}\sum_{k=1}^{m} j_{i,k}(t) \tag{7}$$

To fulfill the required condition of route cost $\mu_{i,min}(t) \geq \mu_{i,k}(t)$, Hunt stagnation alleviates a distinct solution for the finest iteration global ants for the updated trail pheromone. This form of stagnation avoids the next resolution of the pheromone trails. Although the discrepancy among pheromone trails is prohibited with the minimal effects of pheromone trails.

ACO conducts maximum and minimum pheromone trails. The following equation is responsible to update the pheromone trails in the final iteration of upgrading the trail pheromones. The entire pheromone trail is denoted by ρ_{max} and ρ_{min}. ACO carries out the maximum and minimum pheromone trails for all of the pheromone trails referred to as ρ_{max} and ρ_{min}. Now, improving the trail pheromones in the last iteration, the following equation updates the pheromone points.

$$\rho_{i,j}(t) = \begin{cases} \rho_{max};\ \rho_{i,j}(t) \geq \rho_{max} \\ \rho_{i,j}(t);\ \rho_{max} \geq \rho_{i,j}(t) \geq \rho_{min} \\ \rho_{min};\ \rho_{min}(t) > \rho_{i,j}(t) \end{cases} \tag{8}$$

4.2.1. Maximum Minimum Ant Colony Optimization with Differential Evolution

Based on the evolutionary process, a meta-heuristic search technique Differential Evolution (DE) solves the optimization issues. A population-based searching procedure improves the system performance for large search spaces. For continuous optimization problems, this adaptable optimization technique will be providing a better solution than others will. Moreover, DE has three core control factors, upon which it is based i.e., operator selection, mutated DE, and crossover DE. In this technique, a random system solution is subsequently magnified using population vectors. It creates the trail alteration first and then connects the trail mutation with the objective mutations to create an updated distinct. It will accept and replace the prior individual with an updated one who has good fitness results.

In the MMACO-DE algorithm, Zain et al. presented a hybrid algorithm that depends on Max-Min ACO and DE algorithms. This algorithm provides improved performance of path planning of multiple UAVs in 3D search space. Moreover, the multi-colonies approach carries out the shortest route to achieve the target with improved convergence speed. There are multiple sub-colonies in the ant colony system in which each sub-colony has its leader whose fitness value is best among all. This ant is responsible for forming and updating the route for other ants to follow. Similarly, the best sub-colony will lead the other colonies to obtain a robust path [24].

To deal with the issues related to path planning, the finest ant from every single colony will represent the colony, and the finest ants of the colony control the pheromone trails for the entire situation. There are three colonies, each with its colony number, which limits the total number of ants, according to the MMACO strategy model. There will be no DE operation if the value of P_{cr} is zero as shown in Equation (10). Nonetheless, one element of the mutation pheromone mechanism has now been identified which will be delivered to the newly generated pheromone matrix with confidence. However, the DE mutation method and a better trail spreading form a general equation.

$$\rho_{i,j}(t) = \begin{cases} \rho_{i,j}^{t+1}, \text{ if } t \le P_{cr} \\ \rho_{i,j}^{t}, \ t > P_{cr} \end{cases} \tag{9}$$

where $\rho_{i,j}^{t+1}$ is denoted as the number of pheromone trails between two repeated nodes i to j of the ant colony and P_{cr} is crossover probability.

4.2.2. Maximum Minimum Ant Colony Optimization with Cauchy Mutant Operator

To increase the performance of UAVs in difficult situations, one more combinatorial approach for path planning is applied. To improve the system's performance, Max-Min Ant Colony Optimization pairs with the Cauchy mutation approach. The major perceptive for combining these two algorithms is to speed up convergence and to keep track of any UAV's future failure. The Cauchy mutation distribution gives the following formula,

$$f(a : a_0, z) = 1/\pi\left[t/(a - a_0)^2 + t^2\right] \tag{10}$$

where Cauchy distribution cost is a_0, t is the thickness value related to the maximum cost of Cauchy distribution. The incremental function of the distribution function is as follows,

$$F(a : 0, 1) = \frac{1}{\pi}\tan^{-1}(a) + \frac{1}{2} \tag{11}$$

Compass and map operator: Compass operators employ conventional ant colony optimization (ACO) to identify the boundaries of the equivalent direction. Furthermore, the map operator utilizes to examine the global best objects to determine the location, velocity, and subsequent transformation of the ACO's environment to the best individuals. The compass and map operator denotes the coefficient M_1. It will not help you find the search area, but it might help you reduce the threat level. The Cauchy distribution and its mutation mass coefficient state as follows.

$$\begin{cases} \text{rand} = \frac{1}{2} + \frac{1}{\pi}\tan^{-1}(M_1) \\ M_1 = \tan(\pi(\text{rand} - 1/2)) \end{cases} \tag{12}$$

The rand function is used in the preceding equation to choose a random value of 0 or 1. The rule develops to update the position of each best ant for each iteration.

$$\hat{X}_k = X_{0k} + M_1\left(X_{0k} - X_{gbest}\right) \tag{13}$$

where $\hat{X}_k$, X_{0k} are the current locations of the kth ant with the updated one. Similarly, the global best location is denoted by X_{gbest}. Moreover, the next iteration for the specific location follows,

$$X_k = \begin{cases} f(\hat{X}_k) < \hat{X}_k, \ f(X_{0k}) \\ f(X_{0k}) < X_{0k}, \ f(\hat{X}_k) \end{cases} \tag{14}$$

When map and compass operator values of Cauchy mutation become positive, then the position updates the global optimal position. Furthermore, partial distinct entities will fail to discover the optimal and improved location. This happens due to the Cauchy mutation variations. When comparing the unadapted and current position, the superior result is obtained. This approach will not only ensure the optimization method's supremacy but will also broaden the population's range [25].

Landmark operator: Traditionally, the landmark operator in ACO reduces the population size instantly after every iteration update and a small number of ants will move towards the map's edge. In an unsuitable manner, this rapid decrease in population will induce an undeveloped convergence of the approach, which results in an unfavorable perception of the optimization at the landmark operator stage. To improve the convergence rate, we must use the Cauchy landmark function instead of the local landmark operator to update the position of each ant's colony to the best possible place.

The Cauchy landmark function states that.

$$F(a:0,1) = \frac{2}{\pi}\tan^{-1}(a) + \frac{1}{2} \tag{15}$$

The landmark operator's Cauchy mutant weight coefficient is M_2 and it follows the appropriate distribution is given by;

$$\begin{cases} \text{rand} = \frac{2}{\pi}\tan^{-1}(M_2) \\ M_2 = \tan(2/\pi * \text{rand}) \end{cases} \tag{16}$$

The above expression for the updated location is

$$X_{0k}^{N_c} = X_{0k}^{N_c-1} + M_2(X_{gbest} - X_{0k}^{N_c-1}) \tag{17}$$

where the term $X_{0k}^{N_c-1}$ is the distinct kth ant position of $N_c - 1$ iteration. During the Cauchy mutant landmark operator's operational stage, all diverse ants will gradually achieve the global ideal result. A suitable Cauchy mutant operator will efficiently move the ant colony at a proper speed and direction, ensuring the algorithm's stability and speedy convergence.

5. Results and Discussion

This section compares the effectiveness of both the algorithms and puts their results side by side to determine which algorithm performs better. The simulations were performed on a computer with an Intel Core i7-1165G7 processor, 16 GB DDR4 Ram, and Windows 11 operating system. The simulation software used was MATLAB 2021a.

The hybrid algorithms of Max-Min ACO apply to two different UAVs to obtain the best result among them in terms of the route followed and convergence rate. In UAV1, MMACO combined with DE to follow the route from the initial point to the target in the presence of a bunch of obstacles. Similarly, UAV2 specifies the effectiveness of the second algorithm i.e., MMACO with CM.

Case 1:

In case 1, the UAVs start from the same initial point and have the same target as well. Wind forces are present in this scenario. Table 2 presents the constraints of the wind force.

Table 2. Constraints of wind force.

No.	Constraints	Radius, Center Coordinates	Unit
1	Radius	1.8	km
	Center	(7,6,1)	
2	Radius	2.5	km
	Center	(6,14,1)	
3	Radius	1.3	km
	Center	(16,6,0.9)	
4	Radius	0.8	km
	Center	(15,12,1)	

Table 3 gives the initial and target points of UAVs along with the total distance traveled by these two for case 1.

Table 3. Case 1 initial and target points.

UAV	Algorithm Applied	Initial Point (x,y,z)	Target Point (x,y,z)	Distance Travelled (in KM)
UAV1	MMACO-DE	(0,2,0)	(20,20,1)	23.5
UAV2	MMACO-CM	(0,2,0)	(20,20,1)	24.1

MMACO-DE: Max-Min Ant Colony Optimization with Differential Evolution; Max-Min Ant Colony Optimization with Cauchy Mutation.

To determine the best algorithm for case 1, both UAVs start from the same location simultaneously to the same target point (*) above the ground and are allowed to choose their own shortest path according to their approaches. Figure 3a,b describes the path obtained by UAV1 and UAV2 from the initial point to the target point in a 2-dimensional system. As we can see in Figure 3a, the path constructed by the MMACO-DE is shorter than MMACO-CM and has fewer turns. While Figure 3b, presents the different view of the path followed by the compared algorithms implemented on UAVs.

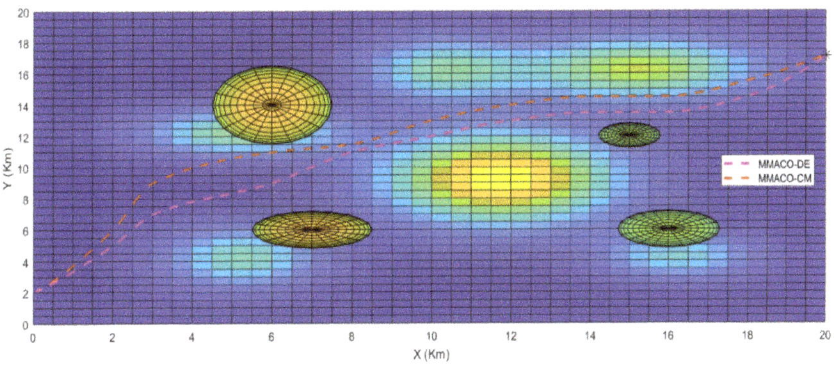

(a)

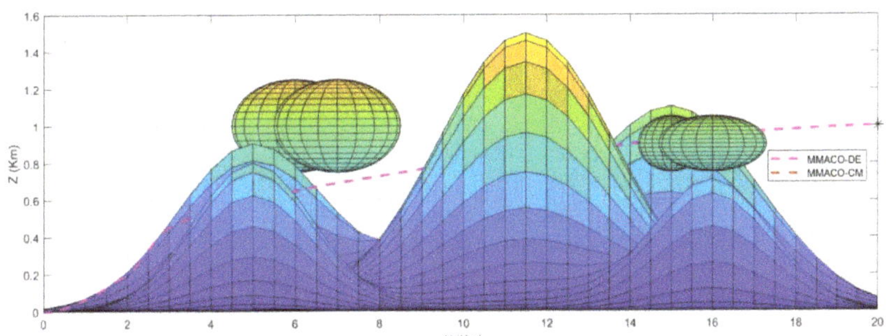

(b)

Figure 3. (**a**,**b**) Case1; 2D views of path followed by UAV1(MMACO-DE) and UAV2 (MMACO-CM) to the target (*).

The 3 dimensional view for case 1 has also been shown in Figure 4 which provides the sense of flying in a dynamic environment including obstacles such as tornados and mountains with different peaks. Again, it is clear from Figure 4 that the MMACO-DE takes fewer turns, hence saving time and minimizing the distance.

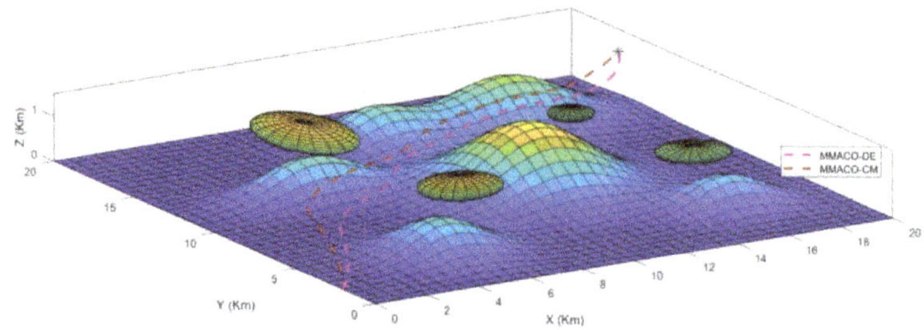

Figure 4. Case1; 3D view of path followed by UAV1 and UAV2 to the target (*).

The take-off points for both UAVs is (0,2,0) which means that both UAVs will fly simultaneously. Initially, both UAVs will fly on their route by analyzing the obstacles and target point (20,20,1) but after a 2 Km distance, MMACO-CM goes slightly off the route which increases its time of flight whereas MMACO-DE continues to follow the shortest route. At the end of the journey, UAV1 will reach the destination earlier than UAV2, and it verifies that the convergence rate of UAV1 is better than UAV2.

Case 2:

For case 2, the UAVs start from different initial points. Wind forces are present in this scenario as well. The constraints for the wind forces are the same as case 1 and given in Table 2. Table 4 gives the initial and target points of UAVs along with the total distance traveled by these two for case 2.

Table 4. Case 2 initial and target points.

UAV	Algorithm Applied	Initial Point (x,y,z)	Target Point (x,y,z)	Distance Travelled (in KM)
UAV1	MMACO-DE	(0,1,0)	(19.8,20,1)	24.221
UAV2	MMACO-CM	(0,2,0)	(19.8,19.8,1)	27.242

MMACO-DE: Max-Min Ant Colony Optimization with Differential Evolution; Max-Min Ant Colony Optimization with Cauchy Mutation.

To determine the best algorithm for case 2, both UAVs start from different locations and are allowed to choose their own shortest path according to their approaches. Figure 5 describes the path obtained by UAV1 and UAV2 from the initial point to the target point in a 2-dimensional system. As we can see in Figure 5a, the path constructed by the MMACO-DE is again shorter than the MMACO-CM and has fewer turns.

The 3-dimensional view for case 2 is shown in Figure 6 which provides the sense of flying in a dynamic environment including obstacles such as wind forces and mountains with different peaks. Again, it is clear from Figure 6 that MMACO-DE takes fewer turns, hence saving time and minimizing the distance.

Analyzing the two case studies, we can clearly see that even if the UAVs start from different locations, MMACO-DE still performs better than MMACO-CM. It picks the shortest distance while avoiding the wind forces and the uneven peaks. It also takes less turns, which in turn ensures that MMACO-DE takes less time to reach the destination than MMACO-CM.

Figure 7 presents the estimation costs of MMACO-DE and MMACO-CM to validate the work.

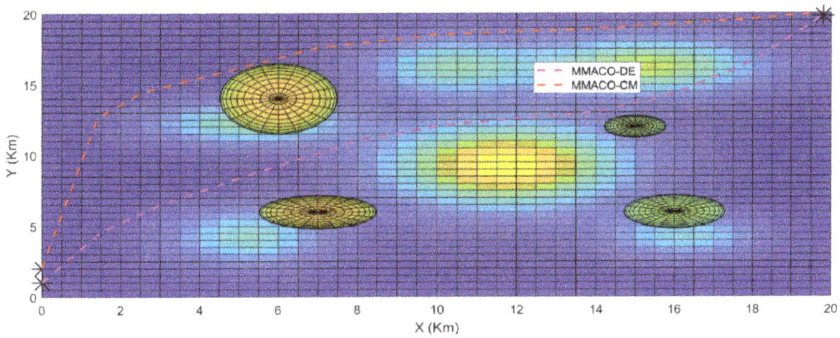

(**a**)

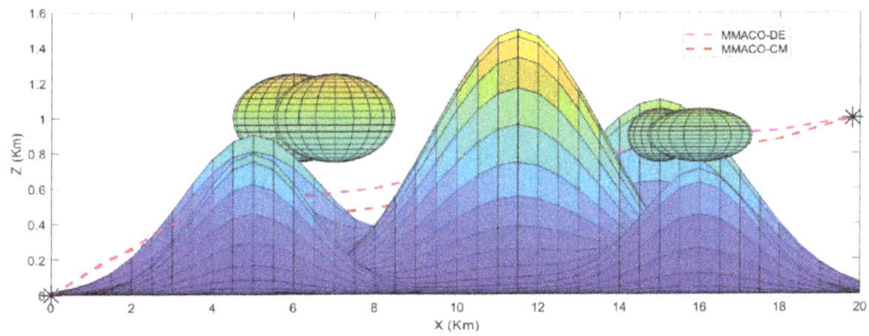

(**b**)

Figure 5. (**a**,**b**) Case 2; 2D path followed by UAV1(MMACO-DE) and UAV2 (MMACO-CM) to the target (*).

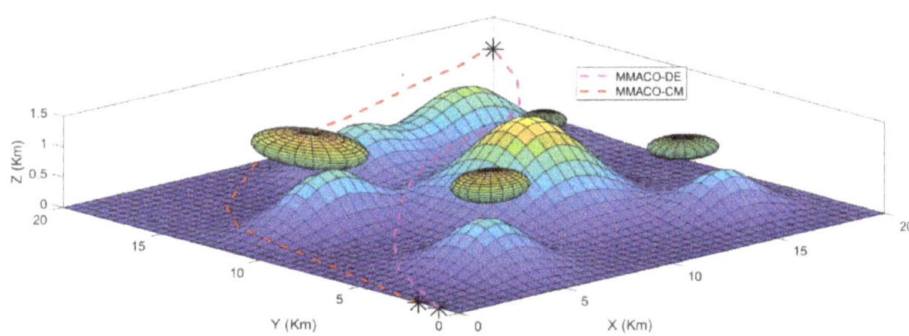

Figure 6. Case 2; 3D view path followed by UAV1 (MMACO-DE) and UAV2 (MMACO-CM) to the target (*).

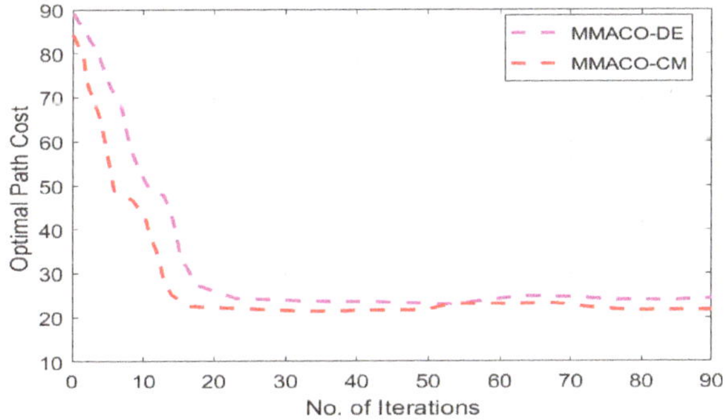

Figure 7. Estimation costs of MMACO-DE and MMACO-CM.

6. Conclusions

Nowadays, bio-inspired algorithms are becoming famous to solve issues related to path planning of unmanned aerial systems with their simplest approach. There are numerous methods in nature offered for this cause i.e., Particle Swarm Optimization, Ant Colony Optimization, Pigeon Inspired Optimization Artificial Bee Colony Optimization, etc., however, one of the most commonly used algorithms for path planning of Unmanned Aerial Vehicles (UAVs) is Ant Colony Optimization (ACO). This article compared two hybrid optimization algorithms for path planning and determines which one is more efficient. Both algorithms use a modified ACO, called Max-Min Ant Colony Optimization algorithm (MMACO), with another algorithm to enhance their effectiveness. The first hybrid algorithm is a combination of the MMACO with the Differential Evolution approach and the second hybrid algorithm is a combination of MMACO with the Cauchy Mutant approach. The MMACO algorithm has tremendous problem-solving skills, especially in complex environments. To reduce noise and disturbance in operation, along with improvement of robustness in the flying, MMACO combines with Differential Evolution and Cauchy Mutant operator. In MMACO–DE, the route followed by UAVs provides the best and shortest route than basic Ant Colony Optimization (ACO); while in MMACO-CM, a flock of UAVs when compared to basic MMACO achieves the optimal route.

Using simulations, this paper concluded that the MMACO-DE algorithm was better than MMACO-CM in achieving a shorter path with less path cost. For future work, we propose to implement the proposed algorithms on hardware and compare the experimental results with the simulations.

Author Contributions: Conceptualization, Z.A.A.; Data curation, E.H.A.; Funding acquisition, J.J.J.; Investigation, M.S.; Methodology, A.I.; Resources, M.H. and J.J.J.; Software, J.J.J.; Validation, J.J.J.; Writing—review & editing, J.J.J. All authors have read and agreed to the published version of the manuscript.

Funding: The authors would like to acknowledge Princess Nourah bint Abdulrahman University Researchers Supporting Project number (PNURSP2022R193), Princess Nourah bint Abdulrahman University, Riyadh, Saudi Arabia, and the European Regional Development Project Green Smart Services in Developing Circular Economy SMEs (A77472).

Data Availability Statement: All data used to support the findings of this study are included within the article.

Acknowledgments: The authors would like to acknowledge Princess Nourah bint Abdulrahman University Researchers Supporting Project number (PNURSP2022R193), Princess Nourah bint Abdulrahman University, Riyadh, Saudi Arabia and the European Regional Development Project Green Smart Services in Developing Circular Economy SMEs (A77472).

Conflicts of Interest: The authors declare that they have no conflict of interest.

References

1. Otto, A.; Agatz, N.; Campbell, J.; Golden, B.; Pesch, E. Optimization approaches for civil applications of unmanned aerial vehicles (UAVs) or aerial drones: A survey. *Networks* **2018**, *72*, 411–458. [CrossRef]
2. Shima, T.; Rasmussen, S. (Eds.) *UAV Cooperative Decision and Control: Challenges and Practical Approaches*; Society for Industrial and Applied Mathematics: Philadelphia, PE, USA, 2009.
3. Lemaire, T.; Rachid, A.; Lacroix, S. A distributed tasks allocation scheme in multi-UAV context. In Proceedings of the IEEE International Conference on Robotics and Automation, New Orleans, LA, USA, 26 April–1 May 2004; Volume 4, pp. 3622–3627.
4. Roldán, J.J.; del Cerro, J.; Barrientos, A. A proposal of methodology for multi-UAV mission modeling. In Proceedings of the 23rd Mediterranean Conference on Control and Automation (MED), Torremolinos, Spain, 16–19 June 2015; pp. 1–7.
5. Samad; Manan, A.; Kamarulzaman, N.; Hamdani, M.A.; Mastor, T.A.; Hashim, K.A. The potential of Unmanned Aerial Vehicle (UAV) for civilian and mapping application. In Proceedings of the IEEE 3rd International Conference on System Engineering and Technology, Shah Alam, Malaysia, 19–20 August 2013; pp. 313–318.
6. Glade, D. *Unmanned Aerial Vehicles: Implications for Military Operations*; Center for Strategy and Technology, Air War College, Air University: Montgomery, AL, USA, 2000.
7. Iida, F. Biologically inspired visual odometer for navigation of a flying robot. *Robot. Auton. Syst.* **2003**, *44*, 201–208. [CrossRef]

8. Yakıcı, E. Solving location and routing problem for UAVs. *Comput. Ind. Eng.* **2016**, *102*, 294–301. [CrossRef]
9. Chen, Y.B.; Yu, J.Q.; Su, X.L.; Luo, G.C. Path planning for multi-UAV formation. *J. Intell. Robot. Syst.* **2015**, *77*, 229–246. [CrossRef]
10. Dorigo, M.; Birattari, M.; Stutzle, T. Ant colony optimization. *IEEE Comput. Intell. Mag.* **2006**, *1*, 28–39. [CrossRef]
11. Blum, C. Ant colony optimization: Introduction and recent trends. *Phys. Life Rev.* **2005**, *2*, 353–373. [CrossRef]
12. Dorigo, M.; Blum, C. Ant colony optimization theory: A survey. *Theor. Comput. Sci.* **2005**, *344*, 243–278. [CrossRef]
13. Hajiyev, C.; Soken, H.E. Robust adaptive Kalman filter for estimation of UAV dynamics in the presence of sensor/actuator faults. *Aerosp. Sci. Technol.* **2013**, *28*, 376–383. [CrossRef]
14. Duan, H.; Luo, Q. New progresses in swarm intelligence–based computation. *Int. J. Bio-Inspired Comput.* **2015**, *7*, 26–35. [CrossRef]
15. Li, W.; Wang, G.-G.; Gandomi, A.H. A survey of learning-based intelligent optimization algorithms. *Arch. Comput. Methods Eng.* **2021**, *28*, 3781–3799. [CrossRef]
16. Duan, H.; Qiao, P. Pigeon-inspired optimization: A new swarm intelligence optimizer for air robot path planning. *Int. J. Intell. Comput. Cybern.* **2014**, *7*, 24–37. [CrossRef]
17. Kennedy, J.; Eberhart, R. Particle swarm optimization. In Proceedings of the ICNN'95-International Conference on Neural Networks, Perth, WA, Australia, 27 November–1 December 1995; Volume 4, pp. 1942–1948.
18. Wang, L.; Zhang, Q.; Wan, N. Simulation study on searching for food by cooperation of multi-robots with swarm intelligence. In Proceedings of the IEEE International Conference on Automation and Logistics, Washington, DC, USA, 18–21 August 2007; pp. 2296–2300.
19. Dorigo, M.; Gambardella, L.M. Ant colonies for the travelling salesman problem. *Biosystems* **1997**, *43*, 73–81. [CrossRef]
20. Aggarwal, S.; Kumar, N. Path planning techniques for unmanned aerial vehicles: A review, solutions, and challenges. *Comput. Commun.* **2020**, *149*, 270–299. [CrossRef]
21. Jiang, H.; Zhang, J.; Xuan, J.; Ren, Z.; Hu, Y. A hybrid ACO algorithm for the next release problem. In Proceedings of the 2nd International Conference on Software Engineering and Data Mining, Chengdu, China, 23–25 June 2010; pp. 166–171.
22. Rivas, A.E.L.; Gallego Pareja, L.A.; Abrão, T. Coordination of distance and directional overcurrent relays using an extended continuous domain ACO algorithm and an hybrid ACO algorithm. *Electr. Power Syst. Res.* **2019**, *170*, 259–272. [CrossRef]
23. Shuang, B.; Chen, J.; Li, Z. Study on hybrid PS-ACO algorithm. *Appl. Intell.* **2011**, *34*, 64–73. [CrossRef]
24. Stützle, T.; Hoos, H.H. MAX–MIN ant system. *Future Gener. Comput. Syst.* **2000**, *16*, 889–914. [CrossRef]
25. Karakaya, M. UAV route planning for maximum target coverage. *arXiv* **2014**, arXiv:1403.2906. [CrossRef]
26. Li, B.; Qi, X.; Yu, B.; Liu, L. Trajectory planning for UAV based on improved ACO algorithm. *IEEE Access* **2019**, *8*, 2995–3006. [CrossRef]
27. Duan, H.; Yu, Y.; Zhang, X.; Shao, S. Three-dimension path planning for UCAV using hybrid meta-heuristic ACO-DE algorithm. *Simul. Model. Pract. Theory* **2010**, *18*, 1104–1115. [CrossRef]
28. Khushaba, R.N.; Al-Ani, A.; AlSukker, A.; Al-Jumaily, A. A combined ant colony and differential evolution feature selection algorithm. In *Ant Colony Optimization and Swarm Intelligence, Proceedings of the 6th International Conference on Ant Colony Optimization and Swarm Intelligence, Brussels, Belgium, 22–24 September 2008*; Springer: Berlin/Heidelberg, Germany, 2008; pp. 1–12.
29. Ali, Z.A.; Zhangang, H.; Zhengru, D. Path planning of multiple UAVs using MMACO and DE algorithm in dynamic environment. *Meas. Control* **2020**, *53*, 0020294020915727. [CrossRef]
30. Ali, Z.A.; Zhangang, H.; Hang, W.B. Cooperative path planning of multiple UAVs by using max–min ant colony optimization along with Cauchy mutant operator. *Fluct. Noise Lett.* **2021**, *20*, 2150002. [CrossRef]
31. Stentz, A. Optimal and efficient path planning for partially known environments. In *Intelligent Unmanned Ground Vehicles*; Springer: Boston, MA, USA, 1997; pp. 203–220.
32. Forsmo, E.J. Optimal Path Planning for Unmanned Aerial Systems. Master's Thesis, Institutt for Teknisk Kybernetikk, Trondheim, Norway, 2012.
33. Latombe, J.-C. *Robot Motion Planning*; Springer Science & Business Media: Berlin/Heidelberg, Germany, 2012; Volume 124.
34. Kavraki, L.E.; Svestka, P.; Latombe, J.-C.; Overmars, M.H. Probabilistic roadmaps for path planning in high-dimensional configuration spaces. *IEEE Trans. Robot. Autom.* **1996**, *12*, 566–580. [CrossRef]
35. Lumelsky, V.J.; Harinarayan, K.R. Decentralized motion planning for multiple mobile robots: The cocktail party model. *Auton. Robot.* **1997**, *4*, 121–135. [CrossRef]
36. Saunders, J.; Call, B.; Curtis, A.; Beard, R.; McLain, T. Static and dynamic obstacle avoidance in miniature air vehicles. In Proceedings of the Infotech@ Aerospace, Arlington, VA, USA, 26–29 September 2005; p. 6950.
37. Gasparetto, A.; Zanotto, V. A new method for smooth trajectory planning of robot manipulators. *Mech. Mach. Theory* **2007**, *42*, 455–471. [CrossRef]
38. Najm, A.A.; Ibraheem, I.K. Nonlinear PID controller design for a 6-DOF UAV quadrotor system. *Eng. Sci. Technol. Int. J.* **2019**, *22*, 1087–1097. [CrossRef]
39. Kim, J.-H.; Sukkarieh, S.; Wishart, S. Real-time Navigation, Guidance, and Control of a UAV using Low-cost Sensors. In *Field and Service Robotics*; Springer: Berlin/Heidelberg, Germany, 2003; pp. 299–309.
40. Zhang, C.; Zhen, Z.; Wang, D.; Li, M. UAV path planning method based on ant colony optimization. In Proceedings of the Chinese Control and Decision Conference, Xuzhou, China, 26–28 May 2010; pp. 3790–3792.
41. Park, J.-W.; Oh, H.-D.; Tahk, M.-J. UAV collision avoidance based on geometric approach. In Proceedings of the SICE Annual Conference, Tokyo, Japan, 20–22 August 2008; pp. 2122–2126.

42. Zhang, W.; Ning, Y.; Suo, C. A method based on multi-sensor data fusion for UAV safety distance diagnosis. *Electronics* **2019**, *8*, 1467. [CrossRef]
43. Jackson, B.A.; Frelinger, D.; Lostumbo, M.; Button, R.W. *Evaluating Novel Threats to the Homeland: Unmanned Aerial Vehicles and Cruise Missiles*; Rand Corporation: Santa Monica, CA, USA, 2008; Volume 626.
44. Yan, F.; Liu, Y.-S.; Xiao, J.-Z. Path planning in complex 3D environments using a probabilistic roadmap method. *Int. J. Autom. Comput.* **2013**, *10*, 525–533. [CrossRef]
45. Al Salami, N.M.A. Ant colony optimization algorithm. *UbiCC J.* **2009**, *4*, 823–826.
46. Bell, W.J. Searching behavior patterns in insects. *Annu. Rev. Entomol.* **1990**, *35*, 447–467. [CrossRef]
47. Hassell, M.P.; Southwood, T.R.E. Foraging strategies of insects. *Annu. Rev. Ecol. Syst.* **1978**, *9*, 75–98. [CrossRef]
48. Cain, M.L. Random search by herbivorous insects: A simulation model. *Ecology* **1985**, *66*, 876–888. [CrossRef]
49. Dorigo, M.; Di Caro, G. Ant colony optimization: A new meta-heuristic. In Proceedings of the 1999 Congress on Evolutionary Computation-CEC99 (Cat. No. 99TH8406), Washington, DC, USA, 6–9 July 1999; Volume 2, pp. 1470–1477.
50. Vander, M.; Robert, K.; Leeanne, E. Alonso. Pheromone directed behavior in ants. In *Pheromone Communication in Social Insects*; CRC Press: Boca Raton, FL, USA, 1998; pp. 159–192.

sensors

MDPI

Article

Cyber-Threat Detection System Using a Hybrid Approach of Transfer Learning and Multi-Model Image Representation

Farhan Ullah [1,*], Shamsher Ullah [1], Muhammad Rashid Naeem [2], Leonardo Mostarda [3], Seungmin Rho [4] and Xiaochun Cheng [5]

1 School of Software, Northwestern Polytechnical University, 127 West Youyi Road, Beilin District, Xi'an 710072, China
2 School of Electronic Information and Artificial Intelligence, Leshan Normal University, Leshan 614000, China
3 Computer Science Department, Camerino University, 62032 Camerino, Italy
4 Department of Industrial Security, Chung-Ang University, Seoul 06974, Korea
5 Department of Computer Science, Middlesex University, London NW4 4BT, UK
* Correspondence: farhankhan.cs@yahoo.com

Abstract: Currently, Android apps are easily targeted by malicious network traffic because of their constant network access. These threats have the potential to steal vital information and disrupt the commerce, social system, and banking markets. In this paper, we present a malware detection system based on word2vec-based transfer learning and multi-model image representation. The proposed method combines the textual and texture features of network traffic to leverage the advantages of both types. Initially, the transfer learning method is used to extract trained vocab from network traffic. Then, the malware-to-image algorithm visualizes network bytes for visual analysis of data traffic. Next, the texture features are extracted from malware images using a combination of scale-invariant feature transforms (SIFTs) and oriented fast and rotated brief transforms (ORBs). Moreover, a convolutional neural network (CNN) is designed to extract deep features from a set of trained vocab and texture features. Finally, an ensemble model is designed to classify and detect malware based on the combination of textual and texture features. The proposed method is tested using two standard datasets, CIC-AAGM2017 and CICMalDroid 2020, which comprise a total of 10.2K malware and 3.2K benign samples. Furthermore, an explainable AI experiment is performed to interpret the proposed approach.

Keywords: malware detection; malware visualization; transfer learning; network traffic; explainable AI; cyber security

Citation: Ullah, F.; Ullah, S.; Naeem, M.R.; Mostarda, L.; Rho, S.; Cheng, X. Cyber-Threat Detection System Using a Hybrid Approach of Transfer Learning and Multi-Model Image Representation. *Sensors* **2022**, 22, 5883. https://doi.org/10.3390/s22155883

Academic Editor: Arshad Arshad

Received: 14 June 2022
Accepted: 4 August 2022
Published: 6 August 2022

1. Introduction

We have entered the "mobile era" with the advent of sophisticated technologies and smartphones becoming increasingly common. Traditional cognitive platforms that power desktop computers are being displaced by smartphones and tablets with massive computational capability. Apps that were previously only available on high-end desktop computers are now available on a variety of mobile platforms. Mobile phones have evolved into devices that allow users to conduct online transactions, communicate with friends, and play games [1]. The number of apps accessible for download on the Google Play Store expanded between 2009 and 2017. The Google Play Store (https://www.statista.com/statistics/266210/number-of-available-applications-in-thegoogle-play-store (accessed on 20 February 2022)) had more than 3.5 million apps as of December 2017, an increase from slightly more than 1 million in July 2013. Furthermore, mobile network data are rapidly growing, and cloud services are hastening this process. Android has the largest market share in terms of mobile operating systems. The rapid expansion of Android has spawned a thriving developer community. Hundreds of millions of apps can be downloaded in seconds from various Android marketplaces. As smartphones and tablets become more popular, the number of mobile malware threats targeting them grows [2]. The number of ransomware attacks nearly doubled in 2021, as reported by the National Computing Centre (NCC) (https://www.nccgroup.com/uk/ (accessed on 3 July 2022)) Group. For instance, the number of reported ransomware attacks increased by 92.7% between 2020 and 2021, from 1389 to 2690. Network-based malware is becoming more sophisticated and difficult to combat. This means that we must now deal with everything from network-based malware to internet services that are protected by mobile devices. Furthermore, adversaries are becoming increasingly capable of creating malware that can avoid traditional sandboxing [3]. It is critical to establish a strong network-based malware classification and detection mechanism.

Mobile malware detection solutions can be classified as static, dynamic, or traffic-based [4]. Several previous studies used a static technique to detect vulnerabilities and malware in Android apps. The complexity and diversity of the required codes make this method challenging. Many dynamic approaches attempt to change the operating system of the phone to monitor and recover sensitive information. These strategies are effective, but they necessitate a significant amount of computing power to investigate all possible app patterns [5]. Several malware detection algorithms focus on network traffic generated by Android apps. Malware can be identified by abnormal network behavior patterns. This type of malware detection technology is very useful because the vast majority of Android malware performs harmful behaviors via network activity. To perform malicious acts, malware must communicate with a host system via the network. These traces allow different types of malwares to be tracked and identified. Furthermore, compared to previous methods, developing a network-based malware detection system is less difficult. For instance, such a method can be used at an entry point or gateway without overburdening the mobile device. These solutions are solely based on data generated by consumers, ensuring that users have access to desired mobile apps. Furthermore, other than granting rights to the identifying service, these solutions require no user engagement [6,7]. The goal of network traffic-based approaches is to discover distinctive features of malware that may be used to classify it accurately.

1.1. Problem Statement

Network traffic malware may employ several malicious URL scripts to affect a target Android app. Text-based feature analysis can identify potentially harmful scripts in terms of behavioral segmentation. Figure 1 depicts the malicious activities of adware and riskware. Riskware can embed malicious bytes required for remote code execution. For instance, "application/x-javascript" is incorporated in network traffic to be executed on a remote device to prevent normal access. Similarly, Adware is a type of malware that hides on the target system and displays advertisements. Some adware also monitors internet activity to serve relevant advertisements. Such behavior cannot be achieved solely through image visualization. However, text-based analysis is associated with several issues, such as code obfuscation, insertion, reordering, etc. Image-based malware classification is widely used because it can collect all types of structural information such as memory, process, header, etc. As a result, visual images can be used to retrieve any type of dynamic or obfuscated data. However, it can alter the overall structure of network traffic files, rendering it impossible to target a specific script, such as a malicious script, URL, etc. Furthermore, this method is completely reliant on image attributes. For instance, a hacker can attack a malware image, affecting overall classification performance. As a result, we combined text-based features to detect potential malicious scripts and textural image features to detect other dangerous behaviors, such as memory or resource utilization. A hybrid approach can efficiently use and classify malware and benign files.

Packet	Hostname	Content Type	Size	Filename
48	api.fanjie.com:8001	text/html	19 bytes	getAds.php?pn=com.elm&ver_p=3&imei=352584066095073&appkey=17058
63	api.fanjie.com:8001	text/html	19 bytes	getAds.php?pn=com.elm&ver_p=3&imei=352584066095073&appkey=17058
82	api.fanjie.com:8001	text/html	19 bytes	getAds.php?pn=com.elm&ver_p=3&imei=352584066095073&appkey=17058
111	api.fanjie.com:8001	text/html	19 bytes	getAds.php?pn=com.elm&ver_p=3&imei=352584066095073&appkey=17058
128	api.fanjie.com:8001	text/html	19 bytes	getAds.php?pn=com.elm&ver_p=3&imei=352584066095073&appkey=17058
156	api.fanjie.com:8001	text/html	19 bytes	getAds.php?pn=com.elm&ver_p=3&imei=352584066095073&appkey=17058
176	api.fanjie.com:8001	text/html	19 bytes	getAds.php?pn=com.elm&ver_p=3&imei=352584066095073&appkey=17058
189	api.fanjie.com:8001	text/html	19 bytes	getAds.php?pn=com.elm&ver_p=3&imei=352584066095073&appkey=17058
202	api.fanjie.com:8001	text/html	19 bytes	getAds.php?pn=com.elm&ver_p=3&imei=352584066095073&appkey=17058
219	api.fanjie.com:8001	text/html	19 bytes	getAds.php?pn=com.elm&ver_p=3&imei=352584066095073&appkey=17058
261	api.fanjie.com:8001	text/html	19 bytes	getAds.php?pn=com.elm&ver_p=3&imei=352584066095073&appkey=17058
280	api.fanjie.com:8001	text/html	19 bytes	getAds.php?pn=com.elm&ver_p=3&imei=352584066095073&appkey=17058
293	api.fanjie.com:8001	text/html	19 bytes	getAds.php?pn=com.elm&ver_p=3&imei=352584066095073&appkey=17058
306	api.fanjie.com:8001	text/html	19 bytes	getAds.php?pn=com.elm&ver_p=3&imei=352584066095073&appkey=17058

(a)

Packet	Hostname	Content Type	Size	Filename
6057	xtra2.gpsonextra.net	application/octet-stream	57kB	xtra2.bin
8619	s3.qhmsg.com	application/x-javascript	6511 bytes	carousel.1.0.4.js
8623	s1.qhmsg.com	application/x-javascript	7330 bytes	widget.1.0.2.js
8636	s0.qhmsg.com	application/x-javascript	29kB	zepto_touch_fx.112.js
8661	s8.qhmsg.com	application/x-javascript	5364 bytes	monitor131227.js

(b)

Figure 1. Malicious behaviors of adware and riskware network traffic. (**a**) Adware, (**b**) riskware.

1.2. Research Contributions

In this paper, we propose a novel method for analyzing and characterizing network-based malware. The HTTP and TCP flows are filtered from encrypted communications for broad analysis. Then, word2vec is utilized to capture the trained vocab features. Then, the network-based byte stream is converted to an image. The text-based and visual features are combined for effective malware classification. We observed that these two sorts of features complement one other and that combining them can increase the detection rate of malware. The main contributions of the paper are as follow:

- A malware classification and detection system is proposed using a hybrid approach of transfer learning and texture features. The proposed method adopts the benefits of both methods, i.e., textual and visual analysis.
- An explainable AI experiment is designed to interpret and validate the proposed approach.

The remainder of this paper is organized as follows. In Section 2, we describe the related work, and in Section 3, we describe the proposed method. In Section 4, we thoroughly discuss the experiments, and in Section 5 we present our conclusions.

2. Related Work

Several studies [8,9] had demonstrated how the Android platform protects infected target devices using a variety of security measures, including permission processes. However, individuals have to be adequately qualified with respect to security concerns to benefit from admin privilege protection. These limits imposed by excessive reliance on the customer enable Android malware to infiltrate and proliferate via portable devices. The majority of such analyzers examine aspects such as permissions and potentially unwanted programs to determine whether an application is suspicious or not. Antivirus apps protect computers against malware threats. However, malicious software is always evolving and expanding. As a consequence, malware detection methods need improvement. Several malware detection systems can currently decipher malicious activity in APK files without executing them.

Sanz et al. [10] developed a static approach that accurately classifies infections by capturing an app's uses-permission and uses-feature details, as well as the user's permission information for log files. The proposed method achieved 86.41% classification accuracy. Puerta et al. [11] used the same approach to detect malware using the Drebin dataset and achieved 96.05% accuracy. Liu et al. [12] proposed a two-phase malware detection method. The first phase involves analyzing the app's Manifest.xml document, which provides requested permissions. The second phase is to preprocess the APK file using APK tools to obtain the smali code. The smali code may contain details about asserted permissions, including API calls, which may be used to detect malicious acts. The suggested technique has a detection performance of 98.6%. Shanshan et al. [13] proposed an HTTP- and TCP-based malware detection system for abnormal network assessment. The network device replicates the portable app's data flow. All information retrieval and malware identification take place on the web, utilizing the fewest resources possible. Network-based characteristics and neural network models are coupled to identify mobile malware with an accuracy of 97.89%. Aresu et al. [14] investigated HTTP-based datagrams produced by Android apps when they interact with distant malicious servers. It also applies a grouping method of producing profiles from several malware variants. These markers are then employed to determine unusual operations. Wang et al. [15] developed the TextDroid methodology, which divides an HTTP content flow into special symbols and then generates n-gram sequences to study the layout of the resulting attributes. TextDroid also collects sequential information to feed into a learning algorithm for malware identification. This text-based technique achieved a classification score of 76.99%. Shanshan et al. [16] presented data traffic as a concept for detecting mobile malware. Natural language processing (NLP) tools are used to exploit an HTTP text file for knowledge representation. The next step is to detect malware by inspecting the linguistic characteristics of network data. The presented scheme has a classification performance of 95%. Data from TCP and HTTP traffic features are extracted by TrafficAV and compared to each other using a C4.5 decision tree for accuracy comparison. However, this method does not integrate TCP and HTTP network traces for the machine learning model. It provides a malware detection rate of 98.16% based on HTTP flows [17]. Johann et al. [18] proposed a WebEye framework that generates feasible HTTP traffic on its own, enriches captured traffic with detailed information, and classifies records as malicious or benign using various classifiers, with an accuracy rate of 89.52%.

Numerous studies [19,20] using deep learning to classify malware have produced promising results. A perceptron called the multi-layer perceptron (MLP) [21] works with other perceptrons stacked in multiple layers to categorize malware. A CNN [22] is primarily used to deal with texture features from malware images in order to classify malware. Gradient boosting [23] uses an ensemble of weak prediction models, usually decision trees, to classify malware. A temporal convolutional network (TCN) [24] is influenced by convolutional architectures, which combine easiness, vector autoregression prediction, and enormously long memory for malware classification. A general meta-approach to machine learning called ensemble learning combines the predictions from various models to improve malware classification performance [25]. Chen et al. [26] proposed a CNN model for categorizing mobile apps that relies on HTTP logs. The use of CNN speeds up the selection of features, resulting in more precise traffic detection outputs. The presented method achieved an identification rate of 98%. David et al. [27] introduced the DeepSign method, which is based on deep belief networks. It is capable of producing immutable, concise definitions of malware activities, which can enable it to effectively differentiate nearly all current malware variants with an accuracy of 98.6%. Shanshan et al. [28] introduced an HTTP-based malware classification method. A multi-view neural network is used to detect destructive behavior with varying levels of penetration. This method can be used to focus on certain attributes of input parameters by allocating continuous attention to features. The highest and lowest accuracy rates are 98.81% and 89.33%, respectively.

3. Proposed Method

Figure 2 explains the architectural framework of the proposed method. Android network traffic is monitored and extracts encrypted communication in the form of packet capture files. The network traffic in two ways, i.e., via textual or visual features.

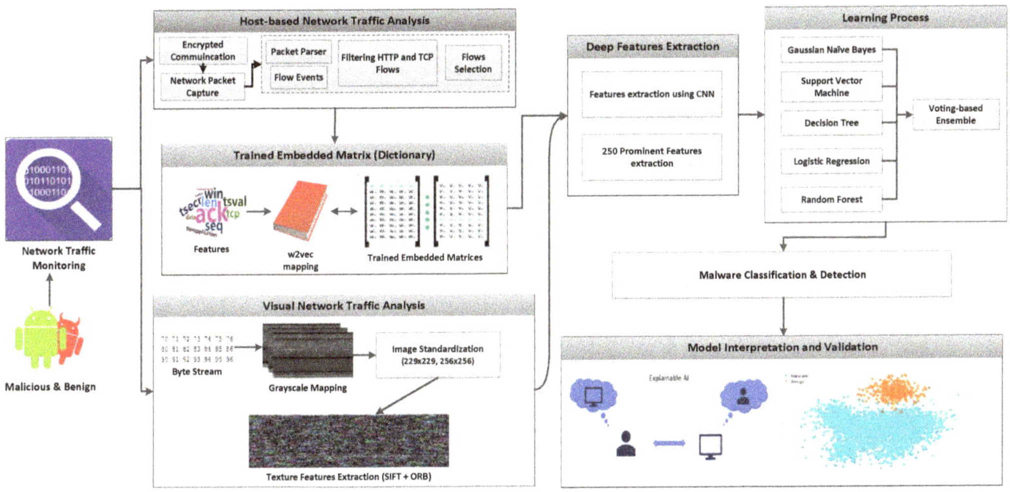

Figure 2. Cyber-threat detection system using a hybrid approach of word2vec-based transfer learning and visual representation.

3.1. Network Trace Collection

3.1.1. Network Data Preprocessing

HTTP traffic is used because it is the most widely used protocol for global communication. HTTP headers contain data that can be used to detect malicious behavior. However, because mobile apps communicate via HTTP, critical information cannot be obtained. To address this issue, we analyze TCP streams with HTTP traces from packet capturing (PCAP) files. PCAP files are source documents generated by network communication. Such files contain network traffic information and are used to assess the underlying information exchange between malicious nodes. Furthermore, they make network traffic management and network activity detection easier. A packet parsing method that filters secure communication and extracts HTTP and TCP flows is developed. The packet parser algorithm is used to filter the PCAP file, as shown in Algorithm 1.

Algorithm 1: Packet Parser Algorithm

Input: Packet Capturing Files (PCAP)
Output: TCP, HTTP as output files
Step 1: Set $P = \{p_1, p_2, \ldots, p_n\}$, where P is a packet
Step 2: $Filter\ (P) = P'$
Step 3: Compute PCAP from P', where $P' = (\text{IP}, \text{TCP}, \text{HTTP}, \ldots, n)$
Step 4: Select NF from PCAP, where NF is the required network flows
Step 5: Display/select HTTP + TCP

HTTP traces include source IP, destination IP, port, host address, source info, bytes, packet length, frame length, and TTL. The source information section includes GET, POST, and URLs, such as "www.yahoo.com" (accessed on 5 December 2021). TCP flows provide three-way handshake information, including uploaded and downloaded bytes and total packet numbers during different sessions. Such information can be filtered to capture meaningful information, preserving the actual semantics. We developed a semantic tokenizer that can filter such information. The main steps taken during data preprocessing are as follow:

- Remove consecutively identical features from input sequences to avoid duplicated data.
- Short sequences may not include enough information to identify the relevant network traffic and are eliminated from the dataset.

- Because different sequence lengths confuse neural network models, unifying sequence length is critical for malware classification. This approach uses a preset sequence length (L) to balance the lengths. Sequences greater than L keep their first L names, but those shorter than L are unified through zero padding.

3.1.2. Transfer Learning with Word2vec

The neural network operates through the use of vectors. Network traffic is represented by a fixed-size vector (L), and a one-hot vector can be employed. However, its scope is limited by the variety of features. This method is unsuitable for learning large datasets. Therefore, a reduced and meaningful vector is required. Word2vec [29] satisfies these criteria. Our goal is to construct a dense vector for each network element that records its contexts in a big dataset. Geometric techniques can be used on network vectors to detect their logical similarities, i.e., intruders use the same web address or TCP conversation for the same victim. Figure 3 demonstrates word2vec with TensorFlow embedding. In our situation, word2vec is used to mine trained vocab features from legitimate and malignant apps. The embedding word model output is a matrix, K $\times$ A, where K is the embedding vector size, and A is the number of unique network features. The encoded-word vector can be trained independently for malware classification [30]. The embed vectors are trained with 8-dimensionally for small datasets and with 1024-dimensionally for large datasets. We selected 300 dimensions for HTTP and TCP. Higher-dimensional embeddings require more data for finer word correlations. The trained vocab features are extracted from word2vec using dynamic fine tuning. Using this procedure, each feature is transferred to a large number of vectors with the same meaning. As a result, this mapping function allows for multiple interpretations of the same feature, which may change over time. Algorithm 2 shows trained feature extraction process from network flows.

Algorithm 2: Trained Feature Mining

Input: HTTP and TCF flows
Output: Trained features
Step 1: Select HTTP and TCP flows
Step 2: Tokenize and filter HTTP and TCP flows = clean features
Step 3: Apply fine-tune embedding
- Dynamic word2vec = train feature

Step 4: Extraction = train feature
Step 5: Compute trained files = mining trained files
Step 6: Finish

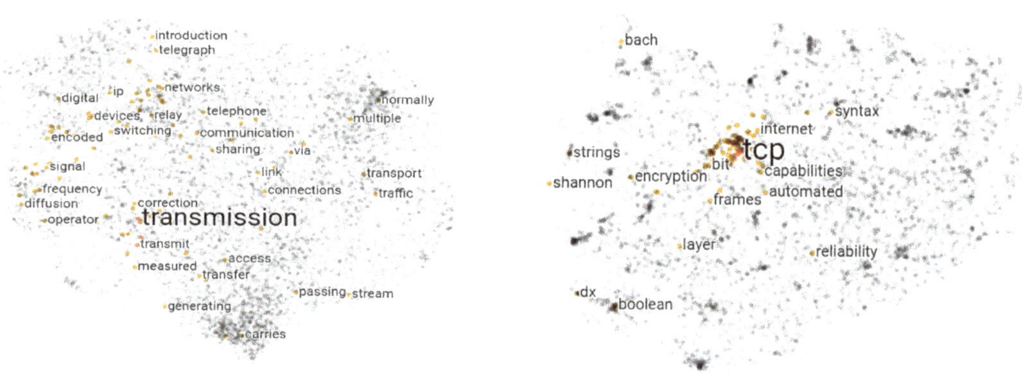

Figure 3. Visualization of trained features (transmission, tcp) using word2vec and TensorFlow.

3.2. Texture Feature Collection

Considering that malware is frequently changed to circumvent static and dynamic identification, we analyzed a malware detection system based on texture properties. This technique detects the malware as a whole by turning the malware into an image and obtaining the textural features. It is not necessary to collect malware fingerprints or use reverse engineering tools. This strategy is effective against antidetection technologies, such as signature modification and dynamic feature detection evasion. We developed a malware-to-image conversion algorithm capable of retrieving images from PCAP files. The eight-bit vectors are retrieved from network traffic first and then processed to produce grayscale malware images. The image sizes are then standardized to 229 × 229 and 256 × 256. Figure 4 depicts a collection of malware images for adware (229 × 229), banking (229 × 229), adware (256 × 256), and SMS (256 × 256). A large PCAP size is transformed to a smaller image size. For instance, the PCAP is converted from megabytes to kilobytes in the image. As a result, it may be possible to reduce computation power. The extraction of texture features is illustrated in Algorithm 3. The extracted network bytes from PCAP files are utilized to mine texture features. These network bytes are represented as images. The texture features are then extracted from these images by combining SIFT and ORB descriptors. SIFT identifies key points or local features within a texture. These steady characteristics can be used for image comparison, object tracking, and scene recognition, among other applications. SIFT consistently outperforms ORB, although ORB is the fastest method. When the angle of rotation is 90 degrees, ORB and SIFT exhibit similar behavior [31]. In order to take advantage of both techniques, we combined SIFT and ORB descriptors to obtain pixel values representing texture features.

Algorithm 3: Texture Feature Mining

Input: Network traffic (Bytes)
Output: Texture features
Step 1: Compute $B = \{B_1, B_2, \ldots, B_n\}$, where B is for Bytes
Step 2: Compute I, where I is image
Step 3: Decompose I in SS_1 & SS_2, where $SS_1 = 229 \times 229$ and $SS_2 = 256 \times 256$
Step 4: Apply SIFT and ORB on SS_1
Step 5: Apply SIFT and ORB on SS_2
Step 6: Generate texture features from the combination of SIFT and ORB
Step 7: Get texture features
Step 8: Finish

Adware (229 × 229) (5.9KB) Banking (229 × 229) (5.02KB) Adware (256 × 256) (8KB) SMS (256 × 256) (8KB)

Figure 4. A chunk of malware images (229 × 229, 256 × 256) extracted from network traffic.

3.3. Deep and Prominent Feature Selection Using CNN

A CNN network is designed to mine a large number of features and extract deep and prominent characteristics that can lessen the load and processing power on the classification model. To achieve this, the pretrained dictionary and visually based texture features are combined and fed into the CNN. Several studies [32,33] have used CNN to categorize malware. The CNN model performs better with a variety of information, including text, images, and video files. We use a one-dimensional CNN network containing convolutional layers, pooling layers, dropout layers, and a fully connected layer. Convolution acts as a filter, repeatedly cycling through the combined features and obtaining the best feature representations. Each filter generates a new set of features, called a feature map. The optimal number of filters is determined by adjusting the hyperparameters. We used three convolution layers with 32, 64, and 128 filters, respectively. Max pooling reduces the size of the feature space, the range of features, and the computational cost. This layer also generates a feature map with the most important features from the preceding set. Furthermore, we combine the Keras batch normalization layer with the CNN network. Batch normalization keeps the resultant mean close to zero and the standard deviation close to one. Notably, it operates differently throughout training and testing. This stabilizes the learning process and reduces the number of training epochs deep networks need. In the proposed CNN network, softmax and dropout layers address overfitting. Equation (1) represents the CNN network's output.

$$o_k^l = f(c_k^l + \sum_{i=1}^{N_{l-1}} \mathrm{Con1D}(X_{ik}^{l-1}, t_i^{l-1})) \tag{1}$$

where c_k^l is the parameter bias of the kth neuron in the first layer, t_i^{l-1} is the outcome of the ith neuron in layer $l-1$, X_{ik}^{l-1} is the kernel strength from the ith neuron in layer $l-1$ to the kth neuron in layer l, and "f ()" is the activation function. After analyzing the deep features, we chose the top 250 prominent features for accurate malware classification.

3.4. Ensemble Model for Malware Classification

The deep and prominent features are fed into the voting-based ensemble model for malware classification and detection.

3.4.1. Naive Bayes (SVM)

To perform classification tasks, the NB algorithm, commonly known as the probabilistic algorithm, is utilized. It is a simple algorithm that works well in a variety of circumstances. The Bayes theorem is utilized to construct the classifier in Equation (2).

$$P(y|X) = \frac{P(X|y)P(y)}{P(X)} \tag{2}$$

where y indicates the class variable, whereas X indicates the characteristics or attributes. Here, X is defined as $(x_1, x_2, \ldots, x_n)$. Gaussian naive Bayes (GNB) conditional probability arises from normal distribution, as shown in Equation (3).

$$P(x_1|y) = \frac{1}{\sigma_y\sqrt{2\pi}} e^{-(x_i-\mu_y)^2}/2\sigma_y{}^2 \tag{3}$$

3.4.2. Support Vector Machine (SVM)

SVM is a supervised learning approach for classification and regression. It classifies by finding the most distinct hyperplane. It locates the hyperplane by widening the distance. Using the kernel function, the kernel trick converts a non-separable job into a separable solution. It is especially useful when dealing with non-linear discrete problems. We used sigmoid as a kernel function. The soft margin of an SVM classifier is calculated by reducing an expression of the kind given in Equation (4).

$$\left[\frac{1}{n}\sum_{i=1}^{n} \max(0.1 - y_i(w^T x_i - b))\right] + \lambda\|w\|^2 \tag{4}$$

3.4.3. Decision Tree (DT)

Each leaf node in a decision tree represents the outcome, a branch represents a decision rule, and an internal node represents a task. The top node is the root node. It usually segments based on the level of an attribute. A tree is partitioned using iterative segmentation. This flow design could help make better decisions. It uses loss functions to assess the integrity of produced nodes. We employed entropy to estimate the decision node's impurity, as illustrated in Equation (5).

$$Entropy = -\sum_{i=1}^{K} p_i \times log_2 p_i \quad (5)$$

The entropy value varies between 0 and 1. The lower the entropy, the higher the purity of the node. Using entropy as a loss function allows for division only if the new nodes tend to have lower entropy than the parent node.

3.4.4. Logistic Regression (LR)

LR accurately predicts binary outcomes (y = 0 or 1). LR is better than linear regression for forecasting classification. Equation (6) shows the logistic function.

$$fx = \frac{1}{1 + e^{-x}} \quad (6)$$

3.4.5. Random Forest (RF)

RF is an estimator that uses DT models to improve the detection rate and reduce overfitting. DTs are often trained by "bagging", which creates a "forest" of trees. The bagging technique claims that integrating many DT models will yield excellent performance. During training, it may handle the growth of numerous DTs and extract information, aggregating the results of each DT [34].

3.4.6. Voting-Based Ensemble Learning

Ensemble is a robust model created by systematically combining base technologies. Unlike individual models, the ensemble model is able to solve classification and regression problems. The proposed investigation employs the soft polling ensemble approach. To begin, we used training data to build basic GNB, SVM, DT, LR, and RF models. The efficiency of the base models is then validated using test data, with each model producing a unique classification. To obtain the final classification performance, ensemble learning employs the estimations of several approaches as supplementary information [35]. The trained and texture features are combined for malware classification, as shown in Algorithm 4.

Algorithm 4: Malware Classification

Input: Trained and texture features
Output: Malware classification
Step 1: Insert T and I
Step 2: $T' = \text{CNN}(T)$ to apply the CNN technique of trained features
Step 3: $I' = \text{CNN}(I)$ to apply CNN the technique of texture features
Step 4: Calculate deep PF as a prominent features as a prominent features
Step 5: Apply voting-based ensemble learning on deep PF
Step 6: App classification as malware or benign
Step 7: Finish

Computational complexity is concerned with categorizing computational issues based on their resource utilization and relating these classes to one another. We analyzed the computational complexity for each algorithm presented in Table 1. The complexity is based on the space required for the proposed approach.

Table 1. Computational cost analysis.

Algorithm	Computational Costs		
	P/T	D/E/D	C/S
Algorithm 1	$\lvert P\rvert = \lvert p_1\rvert + \lvert p_2\rvert + \ldots + \lvert p_n\rvert$	$\lvert P \rvert = \lvert n \rvert$	$\lvert P' \rvert = \lvert n \rvert$
Algorithm 2	$\lvert HTTP \rvert + \lvert TCP \rvert$	$\lvert TF \rvert$	$\lvert MTF \rvert$
Algorithm 3	$\lvert B\rvert = \lvert B_1\rvert + \lvert B_2\rvert + \ldots + \lvert B_n \rvert \lvert I\rvert + \lvert SS_1\rvert + \lvert SS_2\rvert$		$\lvert SS_1\rvert + \lvert SS_2\rvert$
Algorithm 4	$\lvert T \rvert + \lvert I \rvert = \lvert n \rvert$	$\lvert T'\rvert + \lvert I'\rvert = \lvert n \rvert$	$\lvert PF \rvert$

P/T, packets/tokenize; D/E/D, decryption/extraction/decomposed, TF, trained feature; MTF, mining trained files; C/S, computes/shifting.

4. Results and Discussions

4.1. Dataset Preparation

The proposed method is thoroughly examined using two datasets obtained from the Canadian Institute for Cybersecurity (https://www.unb.ca/cic/datasets/index.html (accessed on 6 September 2021)). The first dataset, the Canadian Institute of Cybersecurity Android Adware and General Malware (CICAAGM2017) dataset [36] is gathered semi-automatically by installing Android apps on authorized mobile devices. The dataset is generated using 1900 apps and is separated into three classes: adware, general malware, and benign. The adware contains 250 malicious apps, including Airpush, Dowgin, kemoge, mobidash, and shuanet. The general malware consists of 150 malicious apps, including AVpass, fakeAV, fakeflash, GGtracker, and penetho. A total of 1500 apps are included in the benign set. Table 2 contains a detailed description of the dataset. The second dataset, CICMalDroid 2020 [25,37], collected over 17,341 Android samples from different sources, including the VirusTota l service, the Contagio security blog, AMD, and MalDozer between December 2017 and December 2018. The classification of Android apps as malware is critical for cybersecurity investigators to implement effective classification and detection systems. As a result, this dataset contains adware, banking, riskware, and SMS as malware, as well as benign apps. The number of adware, banking, riskware, SMS, and benign apps is 1253, 2100, 2546, 3904, and 1795, respectively. A detailed description of each app is presented in Table 3.

Table 2. Android Adware and General Malware Dataset (CIC-AAGM2017) (dataset 1).

App	No. of Apps	Family	Description
Adware	250	Airpush	Distributes intrusive adverts to bypass security
		Dowgin	Ad package that collects data
		Kemoge	Takes over the user's Android phone
		Mobidash	Created to broadcast ads and illegal access
		Shuanet	Takes over the user's device
General Malware	150	AVpass	A utility software masquerading as a clock
		FakeAV	Phishing scam to obtain full-version apps
		FakeFlash	Fake Flash software that redirects viewers to a fake website
		Ggtracker	Employed to obtain data via SMS fraud
		Penetho	Fake tool to recover WiFi passwords
Benign	1500	Benign	Clean apps (not malicious)

Table 3. CICMalDroid 2020 dataset (dataset 2).

App	Family	No. of Apps	Description
Malware	Adware	1253	Ads can be hidden within malware-infected programs
	Banking	2100	Connects directly to the user's online payments
	Riskware	2546	Any legitimate program can be abused to inflict harm
	SMS	3904	Attacks via SMS
Benign	Benign	1795	Clean apps (not malicious)

4.2. Result Analysis and Performance Comparison

The trained textual features are combined with visual texture features before being fed into the designed model. We generated texture features with 229 × 229 and 256 × 256 and then combined them with textual features to analyze the impact. Figure 5 shows the training and testing curves for malware classification and detection using dataset 1. We utilized two standard image sizes: 229 × 229 and 256 × 256. In terms of model accuracy, the blue and red curves represent the training and testing data points, respectively. In terms of model loss, the yellow and green curves represent the training and testing points, respectively. (a–d) demonstrate classification and detection for 229 × 229 images, whereas (e–h) demonstrate classification and detection for 256 × 256 images. These curves represent the dynamic behavior of the specified model during the training phase. Using 229 × 229 texture features, the model accuracy curves range from 40% to 98% for classification and 40% to 99% for detection. The model accuracy curves for 256 × 256 texture features result in 35% to 98.1% classification and 30%to 99.16% detection accuracy. As a result, the combined features with 256 × 256 texture features outperform. The model loss is inversely proportional to the model accuracy. Figure 6 depicts the training and testing curves for model accuracy and loss using dataset 2. The model accuracy curves achieve between 50% and 98.1% accuracy for classification and between 40% and 99.1% for detection using dataset 1. Similarly, the same curves provide performance accuracy ranging from 30% to 98.11% for classification and from 40% to 99% for detection. It is clear that textual features with 256 × 256 work better for malware detection.

The confusion matrices for malware detection are obtained to examine misclassification errors for each class, such as malware and benign. Figure 7 depicts the confusion matrices for the individual approaches and the ensemble model, allowing for detailed comparison. The ensemble model outperforms RF in terms of classification. For instance, both approaches had 99% classification and 12% misclassification accuracy for malware and 90% and 10% for benign, respectively. The LR model behaves similarly to ensemble learning but with different results. For example, LR has a 100% classification accuracy and 0% misclassification for malware and 91% classification and 9% misclassification for benign. Figure 8 depicts the confusion matrices for malware classification using 256 × 256 dataset 2. Ensemble and RF models outperform other methods. For instance, they provide classification and misclassification rates of 99% and 1%, respectively, for each class, such as adware, banking, riskware, and SMS.

Table 4 shows the precision, recall, f1-score, and accuracy measures for both datasets using 229 × 229. Performance matrices are provided for each approach, as well as for the ensemble. The ensemble model outperforms the other models in terms of malware classification and detection when utilizing dataset 1. For malware classification, the precision, recall, f1-score, and accuracy measures are 98%, 97, 98%, and 98.18%, respectively. The same performance measures achieve 99%, 99%, 99%, and 99.02% accuracy for malware and detection, respectively. Using dataset 2, the ensemble approach performs better for malware classification; however, the RF approach works better for malware detection. Malware categorization performance measures are 98, 98%, 98%, and 98.1%, respectively. Similarly, the performance measures for malware detection are 99%, 99%, 99%, and 99.04%, respectively. Table 5 shows the performance measures for malware classification and detection using both 256 × 256 datasets. The proposed approach achieves the best classification results using both datasets with 256 × 256 dimensions. Table 6 shows the malware classification performance measures for each class label using dataset 1. Table 7 shows the performance measures for each class label using dataset 2. The methods with a bold style demonstrate that they outperform others for the designed experiment.

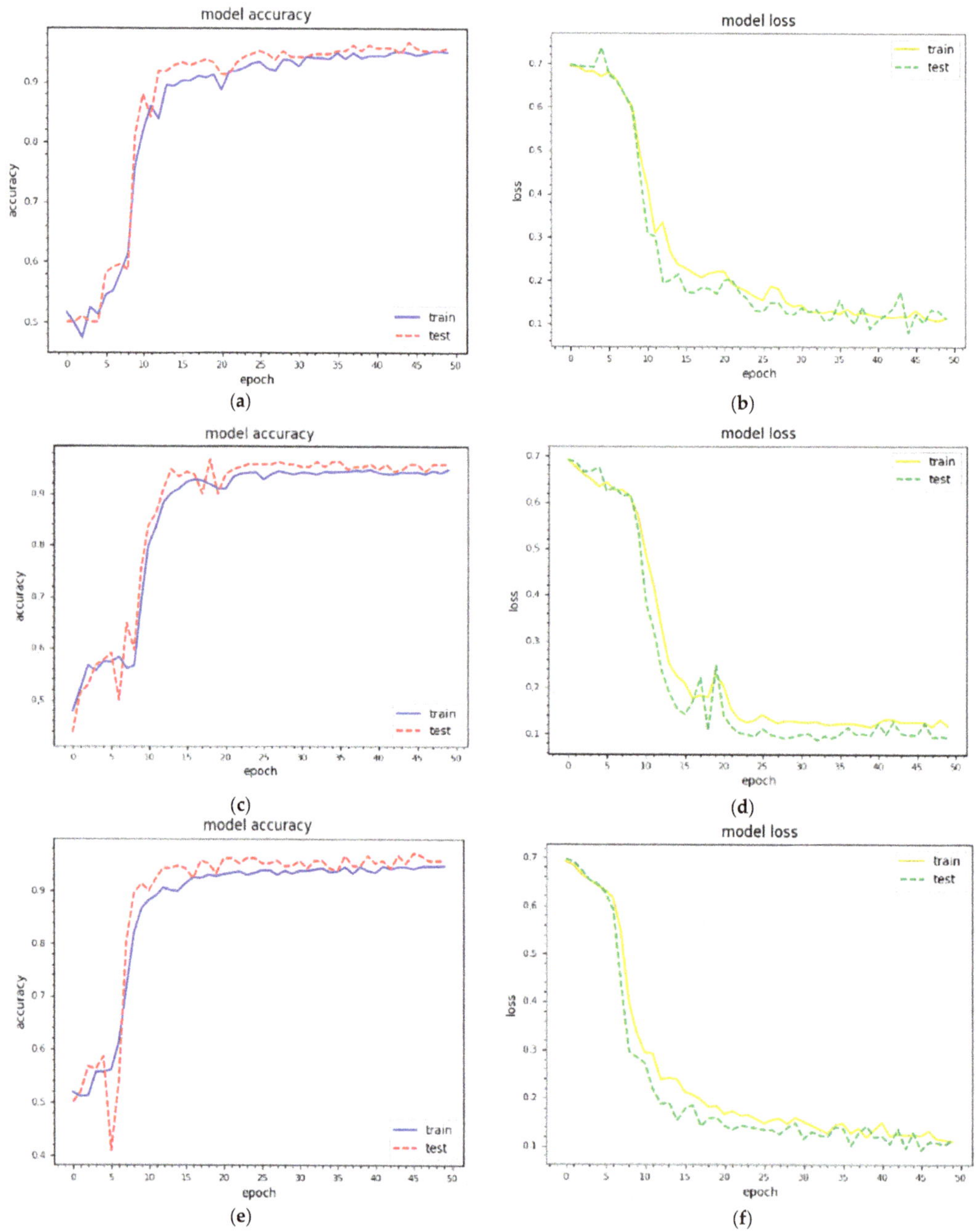

Figure 5. *Cont.*

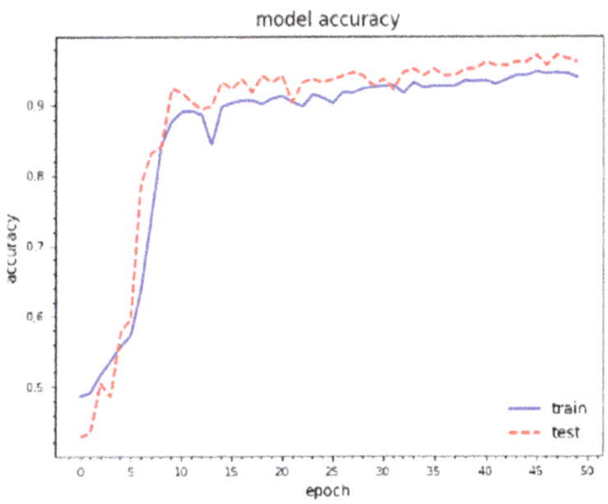

(**g**)

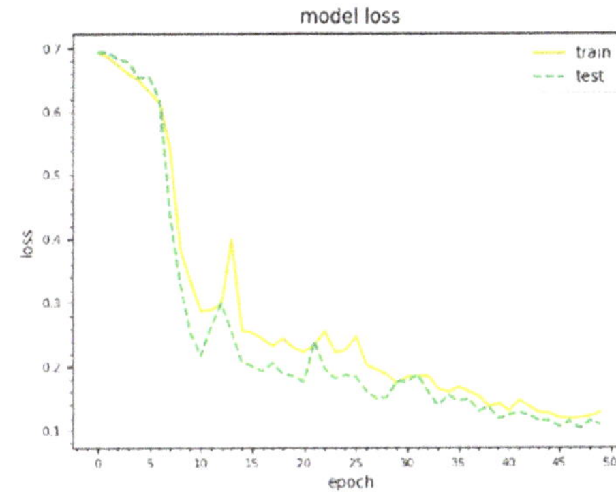

(**h**)

Figure 5. Epoch curves for training and testing data points using different visual representations for **dataset 1**, i.e., 229 × 229, 256 × 256 (training accuracy, training loss; testing accuracy, testing loss). (**a**) 229 × 229 (classification); (**b**) 229 × 229 (classification); (**c**) 229 × 229 (detection); (**d**) 229 × 229 (detection); (**e**) 256 × 256 (classification); (**f**) 256 × 56 (classification); (**g**) 256 × 256 (detection); (**h**) 256 × 256 (detection).

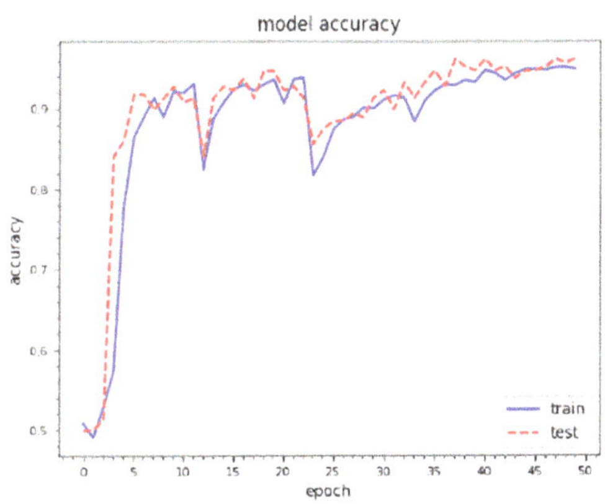

(**a**)

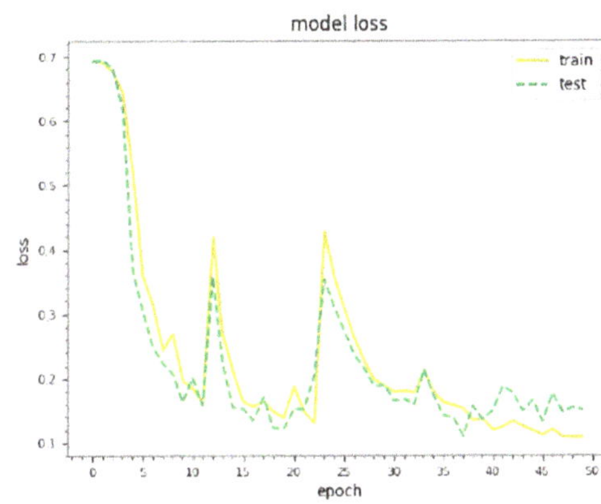

(**b**)

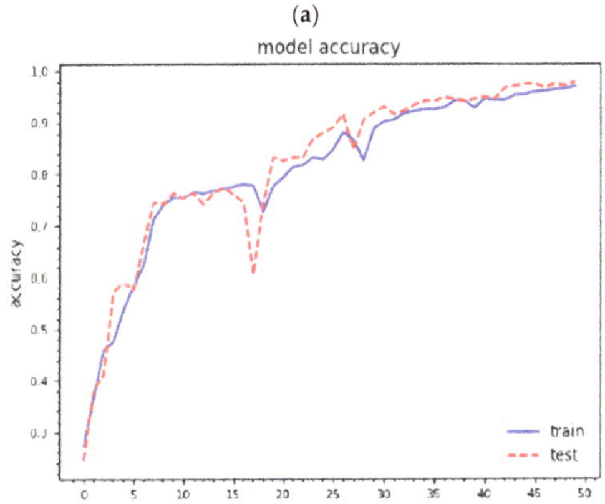

(**c**)

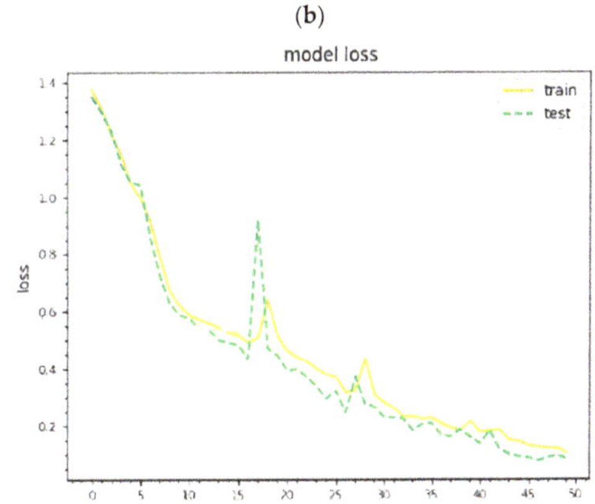

(**d**)

Figure 6. *Cont.*

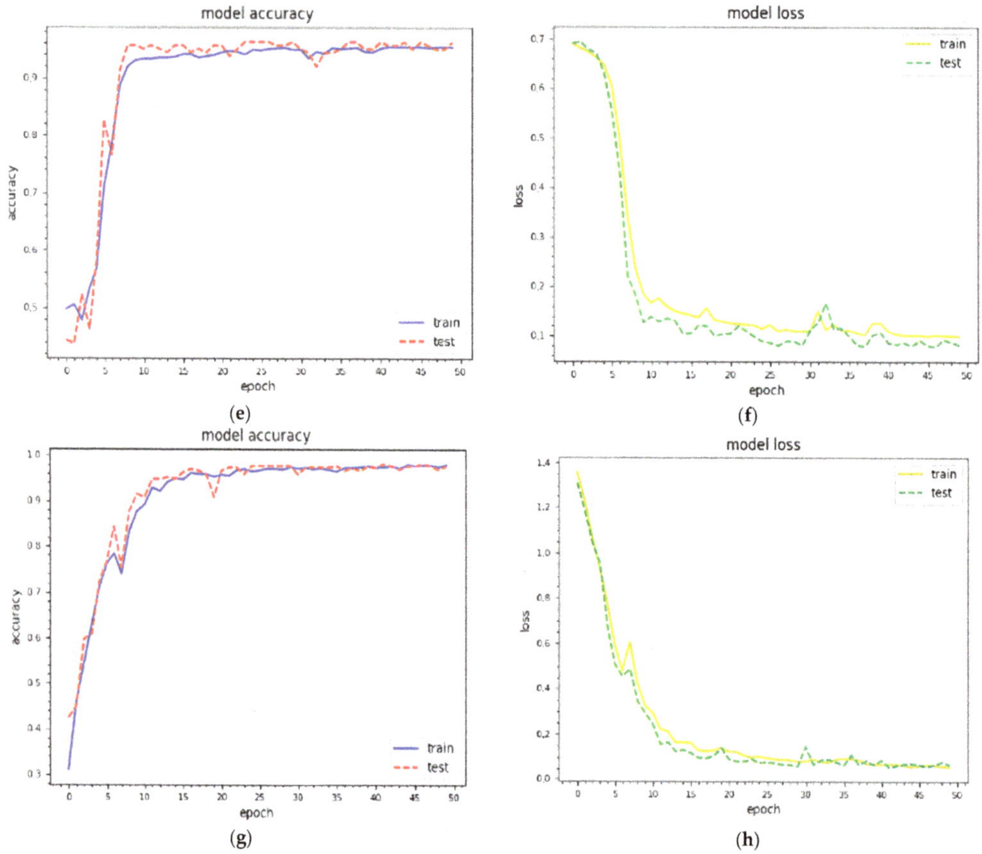

Figure 6. Epoch curves for training and testing data points using different visual representations for **dataset 2**, i.e., 229 × 229, 256 × 256. (training accuracy, training loss; testing accuracy, testing loss). (**a**) 229 × 229 (classification); (**b**) 229 × 229 (classification); (**c**) 229 × 229 (detection); (**d**) 229 × 229 (detection); (**e**) 256 × 256 (classification); (**f**) 256 × 56 (classification); (**g**) 256 × 256 (detection); (**h**) 256 × 256 (detection).

Table 8 depicts the analysis of the optimum features used to determine the best feature selection. The proposed method is tested with a variety of feature counts, such as 100, 150, 200, 250, etc., corresponding to classification accuracy. Dataset 1 is used to examine feature selection with various feature counts. The NB, SVM, DT, LR, RF, and ensemble models provide the highest classification accuracy for 250 features. The classification accuracy increases from 100 to 200 features but decreases after 250. With 400 classification features, classification accuracy increases slightly but then decreases. According to this analysis, 250 is the optimal number of features for the proposed approach.

Generally, classification models produce different results after each execution. To evaluate performance, the datasets are randomly divided into train and test models. As a result, each execution produces unique results for each classification model. We used the same random seed on all classification models with 10 executions to test the scalability and reliability of the proposed ensemble model. Table 9 shows the classification model performance using the same random seeds. On 8 of 10 random seeds, the ensemble model outperforms other classification models, demonstrating that the ensemble model configuration is more reliable than a single classification model. At execution times 2 and 10, the RF slightly outperforms other models relative to the ensemble. Surprisingly, the average performance of 10 executions demonstrates that the ensemble model is more scalable and reliable than the random forest, and it is adopted as the best solution for malware detection and classification. Furthermore, the ensemble model has an accuracy range of 98.98% to 99.02%, whereas the RF has an accuracy range of 98.86% to 99.02%.

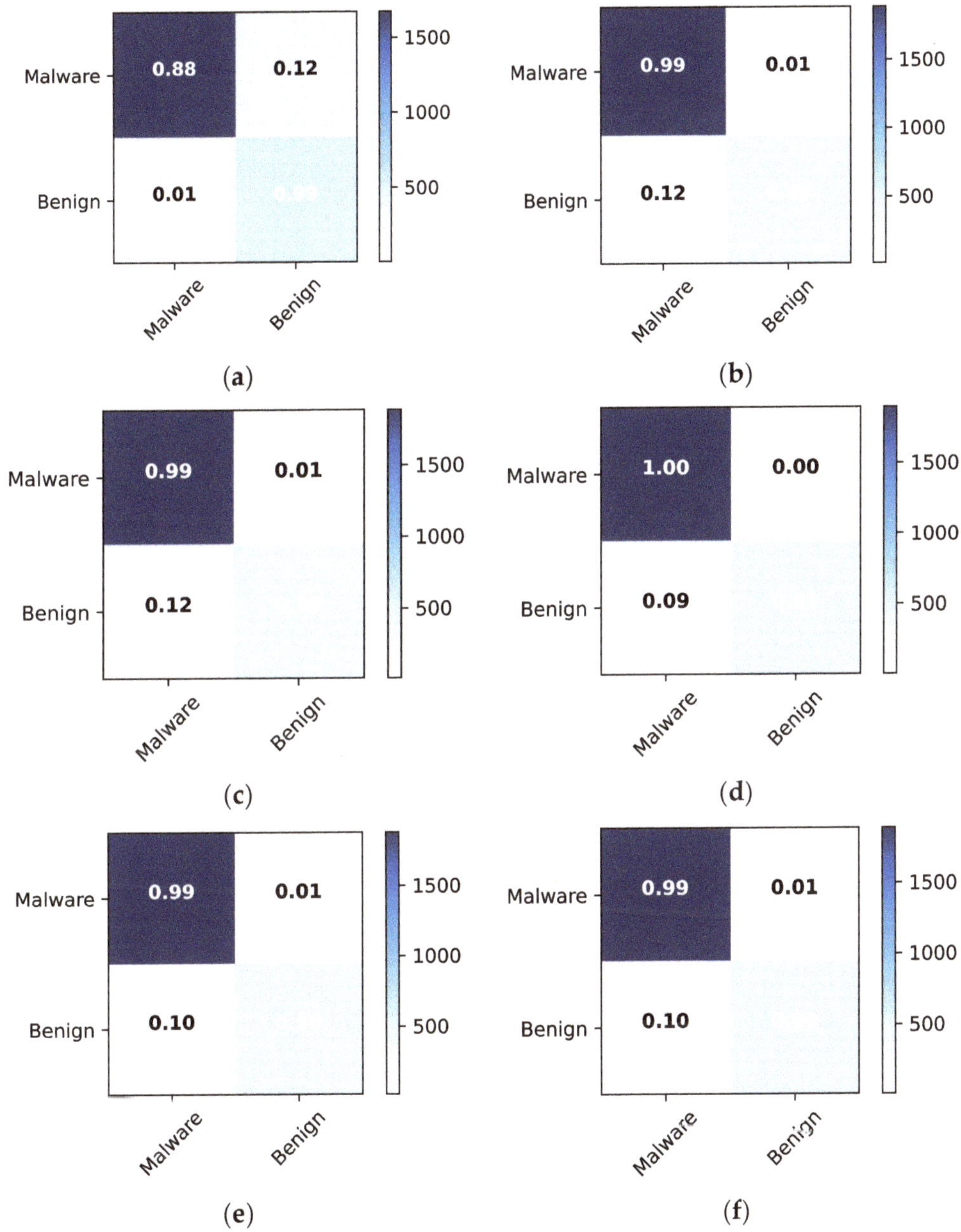

Figure 7. Confusion matrices for malware detection using dataset 2 with 256 × 256. (**a**) GNB; (**b**) SVM; (**c**) DT; (**d**) LR; (**e**) RF; (**f**) ensemble.

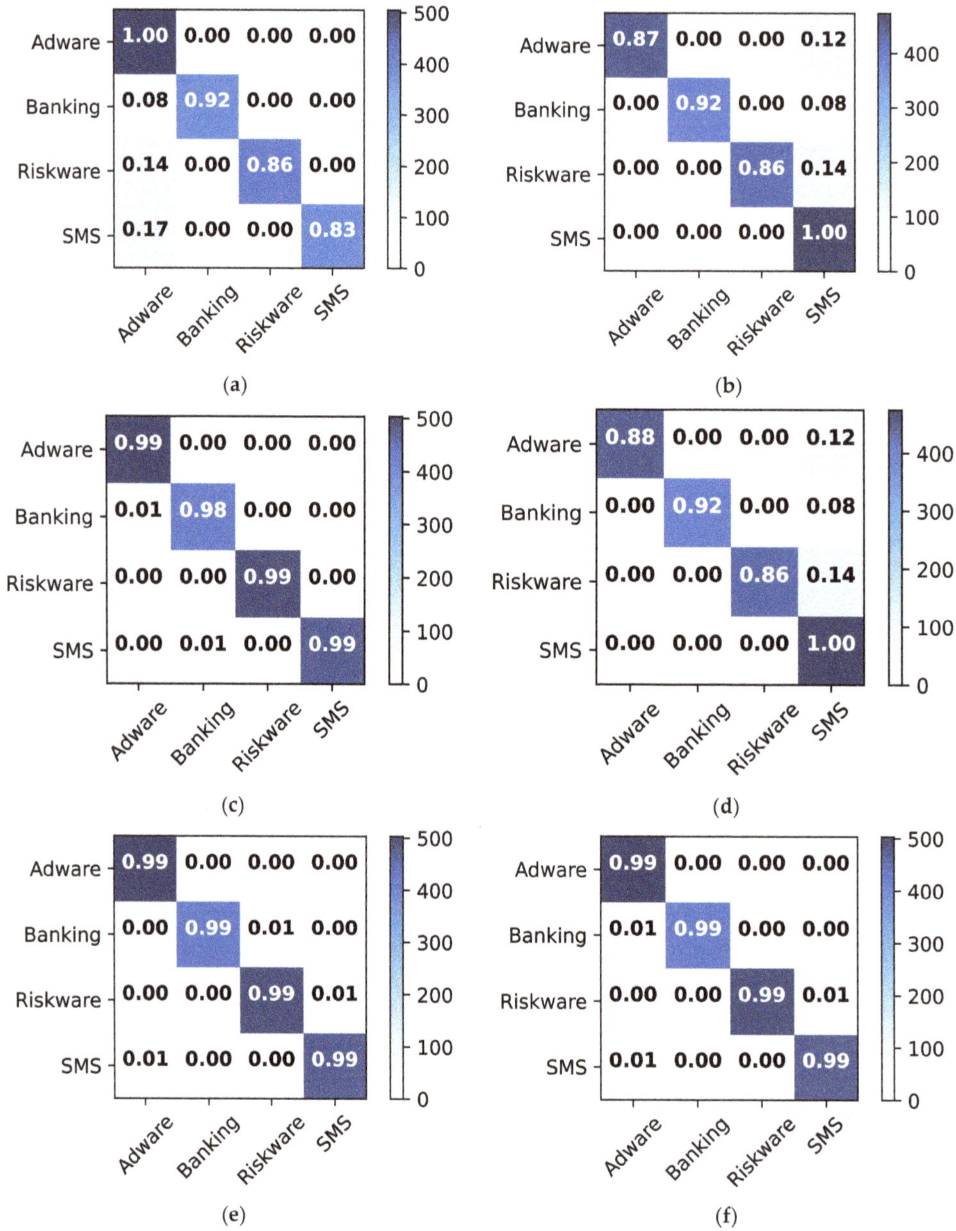

Figure 8. Confusion matrices for malware classification using dataset 2 with 256 × 256. (**a**) GNB; (**b**) SVM; (**c**) DT; (**d**) LR; (**e**) RF; (**f**) ensemble.

Table 10 compares the proposed approach to previously published studies. These studies mostly made use of network traffic to classify Android malware. Aresu et al. [14], showed how analysis of mobile botnets' HTTP traffic can be utilized to classify them into families. To do so, it analyzes HTTP traffic data to create malware clusters. This method also extracts signatures that can be used to detect new clustered malware with an accuracy of 98.66%. Li et al. [20] presented the Droid Classifier, which automatically builds multiple models over a set of annotated malware apps. Each model is built using common identifiers collected from network traffic. Adaptive threshold settings are designed to represent diverse virus traits with an accuracy of 94.66%. Shanshan et al. [38] proposed identifying infected files by their URLs. Multi-view neural networks provide depth and breadth of information when analyzing malware, in addition to creating and distributing soft attention-weighting elements for use with specific data. The accuracy of URL-based malware classification is 95.74%. Shyong et al. [39] combined static authorization with dynamic network monitoring to classify Android apps. During the dynamic evaluation step, malicious network traces are used to obtain various attributes, and Random Forest is then used to identify malware samples. The average Android malware performance is 98.86%. Shanshan et al. [28] presented a method to detect Android malware using URLs. Multi-view neural networks are used to construct malware detection models that focus on feature depth. The weights of the features are dispersed to work on certain inputs. The suggested approach has an accuracy of 98%. Our technique outperforms this method, with a 99% malware detection accuracy.

Table 4. Performance comparisons for malware classification and detection using both datasets with 229 × 229.

	Dataset 1 (229 × 229)				
	Methods	**Precision (%)**	**Recall (%)**	**F1-Score (%)**	**Accuracy (%)**
Classification	GNB	86	90	86	87.21
	SVM	95	89	81	93.11
	DT	98	98	97	98.03
	LR	95	89	91	93.34
	RF	98	97	98	98.03
	Ensemble	**98**	**97**	**98**	**98.18**
Detection	GNB	94	91	92	92.24
	SVM	93	93	92	92.16
	DT	99	99	98	98.94
	LR	93	93	92	92.16
	RF	99	99	99	99.02
	Ensemble	**99**	**99**	**99**	**99.02**
	Dataset 2 (229 × 229)				
Classification	GNB	98	95	96	98.02
	SVM	96	92	94	95.96
	DT	95	95	94	95.04
	LR	98	97	97	97.98
	RF	97	97	97	97.02
	Ensemble	**98**	**98**	**98**	**98.1**
Detection	GNB	94	93	93	93.11
	SVM	93	91	91	91.08
	DT	99	99	99	98.96
	LR	95	93	93	93.1
	RF	**99**	**99**	**99**	**99.04**
	Ensemble	95	93	93	94.16

Table 5. Performance comparisons for malware classification and detection using both datasets with 256 × 256.

	Dataset 1 (256 × 256)				
	Methods	**Precision (%)**	**Recall (%)**	**F1-Score (%)**	**Accuracy (%)**
Classification	GNB	91	84	85	84.98
	SVM	92	91	90	91.02
	DT	96	96	96	96
	LR	92	91	90	91.02
	RF	96	96	96	96
	Ensemble	**96**	**96**	**96**	**96**
Detection	GNB	94	94	94	94.01
	SVM	94	94	94	94.01
	DT	98	98	97	98
	LR	94	95	94	94.08
	RF	**99**	**99**	**99**	**99**
	Ensemble	94	95	94	94.11
	Dataset 2 (256 × 256)				
Classification	GNB	93	90	91	91.14
	SVM	97	98	97	97.21
	DT	96	96	96	96.1
	LR	98	98	97	97.99
	RF	97	97	97	97
	Ensemble	**98**	**98**	**99**	**98.11**
Detection	GNB	93	90	91	90.84
	SVM	93	91	91	91.46
	DT	99	99	99	99
	LR	94	91	92	91.36
	RF	99	99	99	99
	Ensemble	**99**	**99**	**99**	**99**

Table 6. Per-class performance comparisons for malware classification using dataset 1 with 256 × 256.

Class	**Method**	**Precision (%)**	**Recall (%)**	**F1-Score (%)**
Adware	GNB	88	100	94
	SVM	88	100	94
	DT	97	98	98
	LR	88	100	94
	RF	**97**	**100**	**99**
	Ensemble	88	100	94
Gen: Mal	GNB	100	88	93
	SVM	100	88	93
	DT	98	97	98
	LR	100	88	93
	RF	**100**	**98**	**99**
	Ensemble	100	88	93

Table 7. Per-class performance comparisons for malware classification using dataset 2 with 256 × 256.

Class	Method	Precision (%)	Recall (%)	F1-Score (%)
Adware	GNB	73	100	85
	SVM	100	87	93
	DT	98	99	99
	LR	1	88	93
	RF	**99**	**99**	**99**
	Ensemble	98	99	99
Banking	GNB	100	92	96
	SVM	100	92	96
	DT	99	98	98
	LR	1	92	96
	RF	99	99	99
	Ensemble	**100**	**99**	**99**
Riskware	GNB	100	86	93
	SVM	99	86	92
	DT	99	99	99
	LR	1	86	93
	RF	99	99	99
	Ensemble	**100**	**99**	**99**
SMS	GNB	100	83	91
	SVM	74	100	85
	DT	99	99	99
	LR	74	1	85
	RF	**99**	**99**	**99**
	Ensemble	98	99	99

Table 8. Optimum feature analysis.

Method	Dataset 1 (229 × 229)								
	Features								
	100	150	200	250	300	350	400	450	500
NB	90.61	91.53	92.14	**92.24**	91.22	90.52	90.66	89.71	89.62
SVM	91.18	91.74	91.36	**92.16**	91.49	91.92	91.82	90.28	89.12
DT	95.84	96.51	97.36	**98.94**	97.44	96.62	97.88	95.72	95.14
LR	91.56	91.92	91.98	**92.16**	92.14	91.94	92.1	90.82	90.24
RF	97.24	97.54	98.82	**99.02**	98.13	97.76	96.71	96.19	95.96
Ensemble	96.88	97.76	98.96	**99.02**	98.52	97.48	96.98	96.71	96.28

Table 9. Average performance comparison with multiple executions.

Method	Dataset 1 (229 × 229)										
	Execution Times										
	1	2	3	4	5	6	7	8	9	10	Average
NB	92.14	92.24	92.2	91.99	92.22	92.18	92.24	92.22	92.24	91.98	**92.16**
SVM	92.12	92.16	92.16	91.99	92.11	91.98	91.92	92.16	92.12	92.14	**92.09**
DT	98.88	98.94	98.9	98.94	98.44	98.92	98.89	98.72	98.94	98.92	**98.85**
LR	92.16	92.16	92.14	92.04	92.1	91.99	91.96	92.16	92.13	92.16	**92.1**
RF	98.97	**99.02**	98.96	99.02	99.01	98.74	98.91	98.86	99.00	**99.02**	**98.95**
Ensemble	**98.99**	99.00	**98.99**	99.02	**99.02**	**99.00**	**99.00**	**98.98**	**99.02**	99.00	**99.01**

Table 10. Comparison with previously published works.

Work	Year	Method	Dataset	Accuracy (%)
Aresu et al. [14]	2015	Signature-based clustering	Drebin and VirusTotal	96.66
Li et al. [40]	2016	Droid classifier	VirusTotal	94.66
Shanshan et al. [38]	2018	Skip gram with neural network	Malicious URLs (Emulator)	95.74
Shanshan et al. [13]	2019	C4.5 decision tree	Drebin and VirusTotal	97.89
Shyong et al. [39]	2020	Random forest	Drebin	98.86
Shanshan et al. [28]	2020	Multi-view neural network	VirusShare	98.81
Our approach	...	Hybrid features with ensemble learning	CIC-AAGM2017 and CICMalDroid 2020	99

The proposed method is thoroughly compared to existing methods using the same datasets. Table 11 shows a performance comparison with state-of-art methods using the same datasets with different strategies. Texture, text, or a combination of both can be used to classify malware. Furthermore, some researchers used a CNN model to classify malware images without using descriptors to select special features. Alani et al. [21] introduced AdStop, a machine-learning-based method that identifies malware in data traffic. The proposed method classified malware using textual features from the CIC-AAGM2017 dataset and a multi-layer perceptron with an accuracy of 98.02%. Acharya et al. [22] proposed a framework *that* extracts clusters using latent Dirichlet allocation and hierarchical clustering techniques. They used a CNN model, which has a precision of 98.3%, to classify malware without relying on any special features. In [22,24,41,42] CNN and TCN models were used to classify malware with texture features. The proposed deep learning models directly collect the malware images for classification without selecting the special features using descriptors. In [21,23,25] multi-layer perceptron (MLP), gradient boosting, and ensemble methods were used to classify malware with textual features. To classify malware, we propose a method that combines textual and texture features from both datasets. When compared to state-of-the-art methods, the proposed approach outperforms, with a classification accuracy of 99%.

Table 11. Performance comparison with state-of-the-art methods using the same datasets.

Work	Dataset	Strategy	Method	Accuracy (%)
Alani et al. [21]	CIC-AAGM2017	Textual	MLP (DNN)	98.02
Acharya et al. [22]	CIC-AAGM2017	Texture	CNN	98.3
Hadiprakoso et al. [23]	CICMalDroid 2020	Textual	Gradient Boosting	96.35
Mohammad et al. [41]	CICMalDroid 2020	Texture	CNN	96.4
Zhang et al. [24]	CICMalDroid 2020	Texture	TCN	95.44
Mahdavifar et al. [25]	CICMalDroid 2020	Textual	Ensemble	97.84
Peng et al. [42]	CICMalDroid 2020	Texture	CNN	98.6
Our approach	CIC-AAGM2017 & CICMalDroid 2020	Hybrid	Hybrid features with ensemble learning	99

4.3. Model Interpretation and Validation Using Explainable AI and t-SNE

To interpret and validate the proposed approach, we extracted a chunk of the most important features from the embedded matrix. Figure 9 depicts the importance of the features among the 30 features. The feature "F24" is the most effective, indicating that it makes the most contribution to malware classification detection. However, the "F29" feature is the least effective and may perform the worst for the proposed strategy. The "F17" feature is the next most effective feature. Thus, we can readily determine which features are the most and least important. To explain the impact of each feature on the model output, we used the Local Interpretable Model-agnostic Explanation (LIME) and SHapley Additive exPlanations (SHAP) libraries [43]. Figure 10 illustrates the proportionate contribution of features to from the average of samples with a base value of 0 (malware) to the output value of 1 (benign). The values for this sample are indicated by numbers at the bottom of the figure. In our case, the base value is 0.22. The red values are those that are moving underneath the base value, whereas the blue values are those that are moving above the base value. The base value is a threshold, and values less than the base value can contribute to the malware class. Values that are greater than the base value can contribute to the benign class. This allows us to evaluate the contribution of each feature to a specific class. Figure 11 depicts the effect of combined features on model output. The red color represents a higher contribution of each feature, whereas the green color represents smaller contributions. The combined effect of the "F24" feature is significant, whereas that of F15 is the smallest. This allows us to easily describe the impact of each feature on a certain class, such as malware or benign. This experiment evaluates the effectiveness of each feature, providing a clear picture of how each attribute affects the model output.

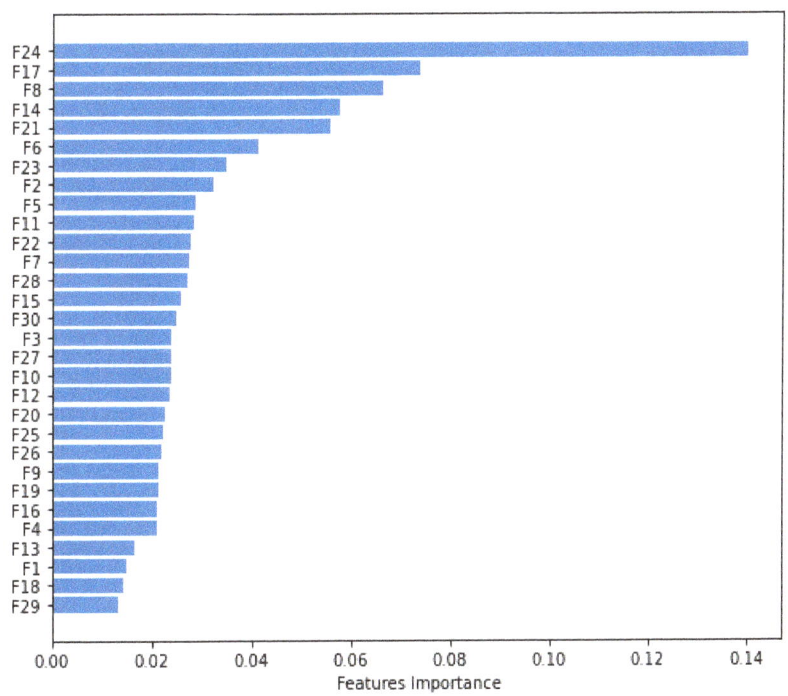

Figure 9. Most significant features.

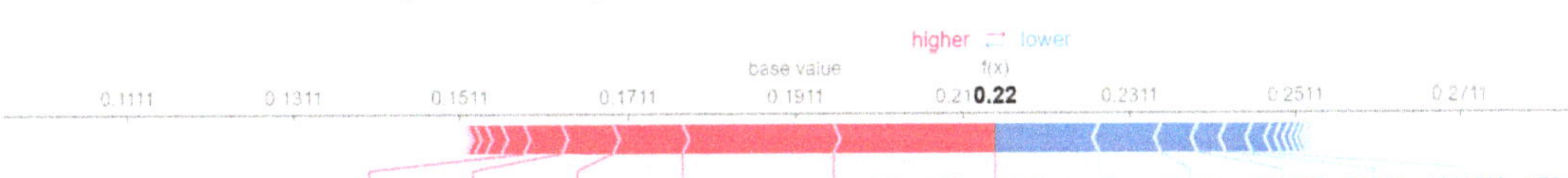

Figure 10. Contribution of features to a certain class based on a threshold value.

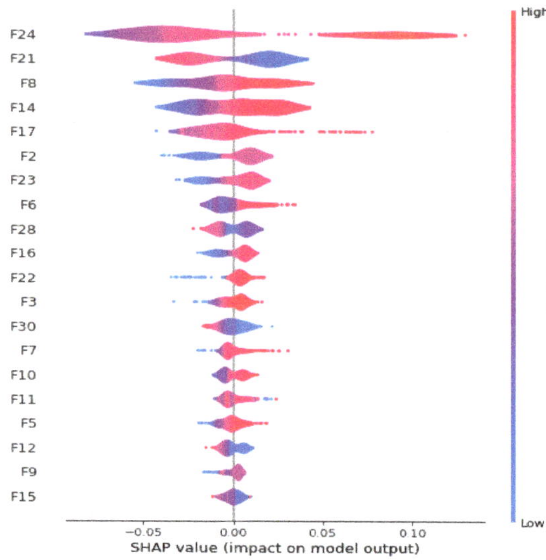

Figure 11. Combined effect of features on model output.

The purpose of the t-distributed stochastic neighbor embedding (t-SNE) visualization method is to identify whether features possess high or sparse knowledge. Furthermore, the t-SNE method is intended to evaluate the efficiency of the suggested approach. Maaten et al. [44] proposed the t-SNE method to visualize high-dimensional data. Figure 12 shows the attentive ratio of semantic and syntactic feature local and global scores for various perplexity values. Using the R programming language, we designed two t-SNE visual studies. In the first experiment, we attempted to determine how much perplexity is required to distinguish between the benign and malicious classes. The best Android malware clusters are distinguished by the highest perplexity scores in the second experiment. For instance, (a,c) have the lowest perplexity values, whereas (b,d) have the highest values. t-SNE makes use of iterations to distinguish between different types of samples. We utilized 400 iterations for each perplexity factor to display the distinct malware and benign groupings. The dataset density has a significant impact on the overall classification results. Because more qualitative data are presented for training, a higher density usually improves accuracy. To improve classification outcomes, the t-SNE visual clusters are better segregated using optimal perplexity settings. A dataset can be divided into sections using an acceptable perplexity value and classified using important hyperparameters. This method is used to demonstrate the efficacy of the presented strategy because semantic aspects can be extracted and classified as malware or benign to improve classification performance.

(a) (b)

Figure 12. *Cont.*

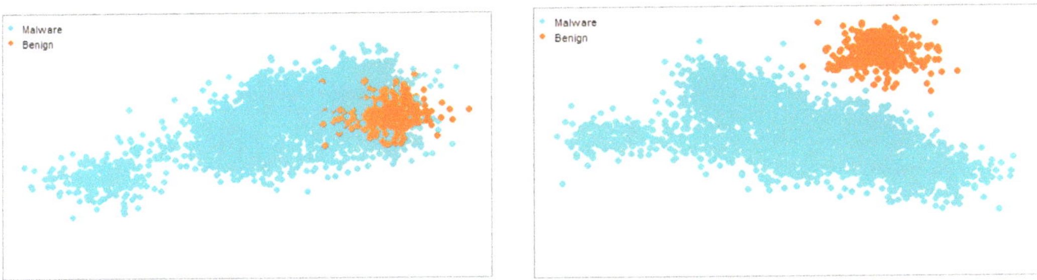

(c) (d)

Figure 12. t-SNE visualization for fused features using minimum (30 and 35) and optimal (50 and 70) perplexity values. (**a**) Perplexity, 30; (**b**) perplexity, 50; (**c**) perplexity, 35; (**d**) perplexity, 70.

5. Conclusions

Mobile apps are susceptible to malicious network activity because of their frequent remote access. Such threats could gather crucial information while adversely affecting commerce, social order, and financial institutions. The malware detection system used in this study takes advantage of the combined influence of textual and textural features, combining the strengths of text and visual elements. We proposed an algorithm for a packet parser that is used to collect HTTP and TCP streams from the encrypted communications generated by malicious traffic. It is possible to recover training vocab features from decoded information using word2vec embeddings. A method for transforming malware images is then developed to examine the byte stream with visual features. We used two standard image sizes (229 × 229) and (256 × 256) to test the proposed approach on features of varying size. The texture features from malware images are combined with trained vocab to classify and detect malware. We designed a voting-based ensemble model for accurate malware classification and detection. The classification and detection rates for dataset 1 with an image size of 229 × 229 are 98.18% and 99.02%, respectively. The classification and detection rates for dataset 2 using a 229 × 229 image size are 98.1% and 99.04%, respectively. Similarly, for a 256 × 256 image size with dataset 1, these values are 96% and 99%, respectively. For dataset 2, these values are 98.11% and 99%, respectively. The first dataset with an image size of 229 × 229 provides better classification results than the second dataset with an image size 256 × 256. The proposed approach outperforms the state-of-the-art methods using the same datasets, as shown in Tables 9 and 11.

In the future, we plan to extract the trained vocab from other pretrained models, such as FastText and BERT. Then, the trained features can be combined with texture features to classify malware. Moreover, the proposed method can be tested with different types of ensembles, such as bagging and stacking.

Author Contributions: F.U. proposed the study, performed simulations, and wrote the manuscript. S.U. wrote the algorithms. M.R.N. made writing suggestions. L.M., S.R. and X.C. reviewed and analyzed the proposed research. All authors have read and agreed to the published version of the manuscript.

Funding: This research received no external funding.

Institutional Review Board Statement: Not applicable.

Informed Consent Statement: Not applicable.

Data Availability Statement: The data that support the findings of this study are openly available in the Canadian Institute for Cybersecurity—Android Adware and General Malware Dataset (CIC-AAGM2017) and CICMalDroid 2020 at https://www.unb.ca/cic/datasets/android-adware.html (accessed on 6 September 2021) and https://www.unb.ca/cic/datasets/maldroid-2020.html (accessed on 6 September 2021), respectively.

Conflicts of Interest: The authors declare no conflict of interest.

References

1. Arshad, S.; Shah, M.A.; Khan, A.; Ahmed, M. Android malware detection & protection: A survey. *Int. J. Adv. Comput. Sci. Appl.* **2016**, *7*, 463–475.
2. Felt, A.P.; Finifter, M.; Chin, E.; Hanna, S.; Wagner, D. A survey of mobile malware in the wild. In Proceedings of the 1st ACM Workshop on Security and Privacy in Smartphones and Mobile Devices, Chicago, IL, USA, 17 October 2011; pp. 3–14.
3. Berman, D.S.; Buczak, A.L.; Chavis, J.S.; Corbett, C.L. A survey of deep learning methods for cyber security. *Information* **2019**, *10*, 122. [CrossRef]
4. Faruki, P.; Bharmal, A.; Laxmi, V.; Ganmoor, V.; Gaur, M.S.; Conti, M.; Rajarajan, M. Android security: A survey of issues, malware penetration, and defenses. *IEEE Commun. Surv. Tutor.* **2014**, *17*, 998–1022. [CrossRef]
5. Egele, M.; Scholte, T.; Kirda, E.; Kruegel, C. A survey on automated dynamic malware-analysis techniques and tools. *ACM Comput. Surv. (CSUR)* **2008**, *44*, 1–42. [CrossRef]
6. Yang, L.; Han, Z.; Huang, Z.; Ma, J. A remotely keyed file encryption scheme under mobile cloud computing. *J. Netw. Comput. Appl.* **2018**, *106*, 90–99. [CrossRef]
7. Ullah, F.; Naeem, M.R.; Mostarda, L.; Shah, S.A. Clone detection in 5G-enabled social IoT system using graph semantics and deep learning model. *Int. J. Mach. Learn. Cybern.* **2021**, *12*, 3115–3127. [CrossRef]
8. Talha, K.A.; Alper, D.I.; Aydin, C. APK Auditor: Permission-based Android malware detection system. *Digit. Investig.* **2015**, *13*, 1–14. [CrossRef]
9. Wang, W.; Wang, X.; Feng, D.; Liu, J.; Han, Z.; Zhang, X. Exploring permission-induced risk in Android applications for malicious application detection. *IEEE Trans. Inf. Forensics Secur.* **2014**, *9*, 1869–1882. [CrossRef]
10. Sanz, B.; Santos, I.; Laorden, C.; Ugarte-Pedrero, X.; Bringas, P.G.; Álvarez, G. PUMA: Permission usage to detect malware in Android. In *International Joint Conference CISIS'12-ICEUTE 12-SOCO 12 Special Sessions, Proceedings of the 5th International Conference (CISIS'12) and EUropean Transnational Education, 3rd International Conference (ICEUTE'12), Ostrava, Czech Republic, 5–7 September 2012*; Springer: Berlin/Heidelberg, Germany, 2013; pp. 289–298.
11. de la Puerta, J.G.; Sanz, B.; Grueiro, I.S.; Bringas, P.G. The evolution of permission as feature for Android malware detection. In Proceedings of the Computational Intelligence in Security for Information Systems Conference, Burgos, Spain, 15–17 June 2015; pp. 389–400.
12. Liu, X.; Liu, J. A two-layered permission-based Android malware detection scheme. In Proceedings of the 2014 2nd IEEE International Conference on Mobile Cloud Computing, Services, and Engineering, Oxford, UK, 8–11 April 2014; pp. 142–148.
13. Wang, S.; Chen, Z.; Yan, Q.; Yang, B.; Peng, L.; Jia, Z. A mobile malware detection method using behavior features in network traffic. *J. Netw. Comput. Appl.* **2019**, *133*, 15–25. [CrossRef]
14. Aresu, M.; Ariu, D.; Ahmadi, M.; Maiorca, D.; Giacinto, G. Clustering Android malware families by HTTP traffic. In Proceedings of the 2015 10th International Conference on Malicious and Unwanted Software (MALWARE), Fajardo, PR, USA, 20–22 October 2015; pp. 128–135.
15. Wang, S.; Yan, Q.; Chen, Z.; Yang, B.; Zhao, C.; Conti, M. TextDroid: Semantics-based detection of mobile malware using network flows. In Proceedings of the 2017 IEEE Conference on Computer Communications Workshops (INFOCOM WKSHPS), Atlanta, GA, USA, 1–4 May 2017; pp. 18–23.
16. Wang, S.; Yan, Q.; Chen, Z.; Yang, B.; Zhao, C.; Conti, M. Detecting Android malware leveraging text semantics of network flows. *IEEE Trans. Inf. Forensics Secur.* **2017**, *13*, 1096–1109. [CrossRef]
17. Wang, S.; Chen, Z.; Zhang, L.; Yan, Q.; Yang, B.; Peng, L.; Jia, Z. Trafficav: An effective and explainable detection of mobile malware behavior using network traffic. In Proceedings of the 2016 IEEE/ACM 24th International Symposium on Quality of Service (IWQoS), Beijing, China, 20–21 June 2016; pp. 1–6.
18. Vierthaler, J.; Kruszelnicki, R.; Schütte, J. Webeye-automated collection of malicious HTTP traffic. *arXiv* **2018**, arXiv:1802.06012.
19. Aniceto, R.C.; Holanda, M.; Castanho, C.; Da Silva, D. Source Code Plagiarism Detection in an Educational Context: A Literature Mapping. In Proceedings of the 2021 IEEE Frontiers in Education Conference (FIE), Lincoln, NE, USA, 13–16 October 2021; pp. 1–9.
20. Ullah, F.; Naeem, M.R.; Bajahzar, A.S.; Al-Turjman, F. IoT-based Cloud Service for Secured Android Markets using PDG-based Deep Learning Classification. *ACM Trans. Internet Technol. (TOIT)* **2021**, *22*, 1–17. [CrossRef]
21. Alani, M.M.; Awad, A.I. AdStop: Efficient flow-based mobile adware detection using machine learning. *Comput. Secur.* **2022**, *117*, 102718. [CrossRef]
22. Acharya, S.; Rawat, U.; Bhatnagar, R. A Low Computational Cost Method for Mobile Malware Detection Using Transfer Learning and Familial Classification Using Topic Modelling. *Appl. Comput. Intell. Soft Comput.* **2022**, *2022*, 4119500. [CrossRef]
23. Hadiprakoso, R.B.; Kabetta, H.; Buana, I.K.S. Hybrid-based malware analysis for effective and efficiency Android malware detection. In Proceedings of the 2020 International Conference on Informatics, Multimedia, Cyber and Information System (ICIMCIS), Jakarta, Indonesia, 19–20 November 2020; pp. 8–12.
24. Zhang, W.; Luktarhan, N.; Ding, C.; Lu, B. Android malware detection using tcn with bytecode image. *Symmetry* **2021**, *13*, 1107. [CrossRef]

25. Mahdavifar, S.; Kadir, A.F.A.; Fatemi, R.; Alhadidi, D.; Ghorbani, A.A. Dynamic Android malware category classification using semi-supervised deep learning. In Proceedings of the 2020 IEEE Intl Conf on Dependable, Autonomic and Secure Computing, Intl Conf on Pervasive Intelligence and Computing, Intl Conf on Cloud and Big Data Computing, Intl Conf on Cyber Science and Technology Congress (DASC/PiCom/CBDCom/CyberSciTech), Calgary, AB, Canada, 17–22 August 2020; pp. 515–522.
26. Chen, Z.; Yu, B.; Zhang, Y.; Zhang, J.; Xu, J. Automatic mobile application traffic identification by convolutional neural networks. In Proceedings of the 2016 IEEE Trustcom/BigDataSE/ISPA, Tianjin, China, 23–26 August 2016; pp. 301–307.
27. David, O.E.; Netanyahu, N.S. Deepsign: Deep learning for automatic malware signature generation and classification. In Proceedings of the 2015 International Joint Conference on Neural Networks (IJCNN), Killarney, Ireland, 12–17 July 2015; pp. 1–8.
28. Wang, S.; Chen, Z.; Yan, Q.; Ji, K.; Peng, L.; Yang, B.; Conti, M. Deep and broad URL feature mining for Android malware detection. *Inf. Sci.* **2020**, *513*, 600–613. [CrossRef]
29. Mikolov, T.; Sutskever, I.; Chen, K.; Corrado, G.S.; Dean, J. Distributed representations of words and phrases and their compositionality. *Adv. Neural Inf. Processing Syst.* **2013**, *26*. [CrossRef]
30. Qiao, Y.; Zhang, W.; Du, X.; Guizani, M. Malware Classification Based on Multilayer Perception and Word2Vec for IoT Security. *ACM Trans. Internet Technol. (TOIT)* **2021**, *22*, 1–22. [CrossRef]
31. Tareen, S.A.K.; Saleem, Z. A comparative analysis of sift, surf, kaze, akaze, orb, and brisk. In Proceedings of the 2018 International Conference on Computing, Mathematics and Engineering Technologies (iCoMET), Sukkur, Pakistan, 3–4 March 2018; pp. 1–10.
32. Lee, W.Y.; Saxe, J.; Harang, R. SeqDroid: Obfuscated Android malware detection using stacked convolutional and recurrent neural networks. In *Deep Learning Applications for Cyber Security*; Springer: Berlin/Heidelberg, Germany, 2019; pp. 197–210.
33. Vasan, D.; Alazab, M.; Wassan, S.; Safaei, B.; Zheng, Q. Image-Based malware classification using ensemble of CNN architectures (IMCEC). *Comput. Secur.* **2020**, *92*, 101748. [CrossRef]
34. Khalilia, M.; Chakraborty, S.; Popescu, M. Predicting disease risks from highly imbalanced data using random forest. *BMC Med. Inform. Decis. Mak.* **2011**, *11*, 51. [CrossRef]
35. Sagi, O.; Rokach, L. Ensemble learning: A survey. *Wiley Interdiscip. Rev. Data Min. Knowl. Discov.* **2018**, *8*, e1249. [CrossRef]
36. Lashkari, A.H.; Kadir, A.F.A.; Gonzalez, H.; Mbah, K.F.; Ghorbani, A.A. Towards a network-based framework for Android malware detection and characterization. In Proceedings of the 2017 15th Annual Conference on Privacy, Security and Trust (PST), Calgary, AB, Canada, 28–30 August 2017; pp. 233–242.
37. Mahdavifar, S.; Alhadidi, D.; Ghorbani, A. Effective and Efficient Hybrid Android Malware Classification Using Pseudo-Label Stacked Auto-Encoder. *J. Netw. Syst. Manag.* **2022**, *30*, 22. [CrossRef]
38. Wang, S.; Chen, Z.; Yan, Q.; Ji, K.; Wang, L.; Yang, B.; Conti, M. Deep and broad learning based detection of Android malware via network traffic. In Proceedings of the 2018 IEEE/ACM 26th International Symposium on Quality of Service (IWQoS), Banff, AB, Canada, 4–6 June 2018; pp. 1–6.
39. Shyong, Y.-C.; Jeng, T.-H.; Chen, Y.-M. Combining static permissions and dynamic packet analysis to improve Android malware detection. In Proceedings of the 2020 2nd International Conference on Computer Communication and the Internet (ICCCI), Nagoya, Japan, 26–29 June 2020; pp. 75–81.
40. Li, Z.; Sun, L.; Yan, Q.; Srisa-an, W.; Chen, Z. Droidclassifier: Efficient adaptive mining of application-layer header for classifying Android malware. In Proceedings of the International Conference on Security and Privacy in Communication Systems, Guangzhou, China, 10–12 October 2016; pp. 597–616.
41. Al-Fawa'reh, M.; Saif, A.; Jafar, M.T.; Elhassan, A. Malware detection by eating a whole APK. In Proceedings of the 2020 15th International Conference for Internet Technology and Secured Transactions (ICITST), London, UK, 8–10 December 2020; pp. 1–7.
42. Peng, T.; Hu, B.; Liu, J.; Huang, J.; Zhang, Z.; He, R.; Hu, X. A Lightweight Multi-Source Fast Android Malware Detection Model. *Appl. Sci.* **2022**, *12*, 5394. [CrossRef]
43. Mathews, S.M. Explainable artificial intelligence applications in NLP, biomedical, and malware classification: A literature review. In Proceedings of the Intelligent Computing—Proceedings of the Computing Conference, London, UK, 16–17 July 2019; pp. 1269–1292.
44. Van der Maaten, L.; Hinton, G. Visualizing data using t-SNE. *J. Mach. Learn. Res.* **2008**, *9*, 2579–2605.

sensors

MDPI

Article

A Proactive Protection of Smart Power Grids against Cyberattacks on Service Data Transfer Protocols by Computational Intelligence Methods

Igor Kotenko [1,*], Igor Saenko [1], Oleg Lauta [2] and Alexander Kribel [1]

[1] Laboratory of Computer Security Problems, St. Petersburg Federal Research Center of the Russian Academy of Sciences (SPC RAS), 39, 14th Liniya, 199178 St. Petersburg, Russia
[2] Department of Integrated Information Security, Admiral Makarov State University of Maritime and Inland Shipping, 5/7 Dvinskaya st., 198035 St. Petersburg, Russia
* Correspondence: ivkote@comsec.spb.ru

Abstract: The article discusses an approach to the construction and operation of a proactive system for protecting smart power grids against cyberattacks on service data transfer protocols. It is based on a combination of computational intelligence methods: identifying anomalies in network traffic by evaluating its self-similarity, detecting and classifying cyberattacks in anomalies, and taking effective protection measures using Long Short-Term Memory (LSTM) and Gated Recurrent Unit (GRU) cells. Fractal analysis, mathematical statistics, and neural networks with long short-term memory are used as tools in the development of this protection system. The issues of software implementation of the proposed system and the formation of a data set containing network packets of a smart grid system are considered. The experimental results obtained using the generated data set demonstrated and confirmed the high efficiency of the proposed proactive smart grid protection system in detecting cyberattacks in real or near real-time, as well as in predicting the impact of cyberattacks and developing efficient measures to counter them.

Keywords: time series; fractal analysis; fractal dimension; Hurst exponent; scaling exponent; cyberattacks

Citation: Kotenko, I.; Saenko, I.; Lauta, O.; Kribel, A. A Proactive Protection of Smart Power Grids against Cyberattacks on Service Data Transfer Protocols by Computational Intelligence Methods. *Sensors* **2022**, *22*, 7506. https://doi.org/10.3390/s22197506

Academic Editor: Arshad Arshad

Received: 18 August 2022
Accepted: 28 September 2022
Published: 3 October 2022

1. Introduction

World trends in information and telecommunication technologies based on digital methods of information transmission, processing, storage, presentation, and protection consist in the mutual penetration and "merging" of information and telecommunication systems not only at the level of technologies for their development and operation, but also their structural and functional association. In this case, the term "data transmission network" (DTN) is widely used [1].

There is an integration and convergence of networks and services. This provides users with access to any service available in multiple networks, due to the flexible possibilities for their processing and management. As a result, on the one hand, the efficiency, reliability, economic benefits, and sustainability of the DTN operation increase. On the other hand, it gives the malefactors the opportunity to act by implementing cyberattacks (CAs) [2].

There are many reasons why it becomes possible to implement CAs. It can be an operating system or other software that has not been updated in time. In addition, outdated security features or vulnerabilities inherent in poorly protected network protocols can lead to attacks. As a result, an attacker can perform various malicious actions, such as blocking network communication, making unauthorized access to DTN devices, controlling traffic, changing network device parameters, and other actions.

The category of dangerous services includes services whose placement on the perimeter carries increased risks: file system access services, Remote Procedure Call (RPC), directory services, printers, virtualization system service interfaces, Virtual Private Network

(VPN), DTN-specific systems, network device services, Telnet, Secure Shell Protocol (SSH), Remote Desktop Protocol (RDP), Virtual Network Computing (VNC), and others [3]. In addition, it should be noted that security flaws in service protocols that lead to traffic redirection and interception of network configuration information, security flaws in the NetBIOS Name Service (NBNS) and Link-Local Multicast Name Resolution (LLMNR) protocols, as well as the use of open (unsecured) data transfer protocols in modern DTNs, have a high level of risk [4]. As practice shows, the vast majority of successful CAs are based on the exploitation of vulnerabilities in some resources that should not be available on the network perimeter [5].

This fully applies to information systems in the energy sector, built according to the Smart Grid (SG) concept. In accordance with this concept, the priority areas for the development of DTN in the energy sector for the coming years include [6]:

- widespread introduction at new and upgraded measurement points of intelligent measuring instruments—"smart" meters with the function of remote control of the load profile of the measured line and measuring transducers with standard communication interfaces and protocols that comply with information security standards;
- installation at each large facility connected to the power grid, advanced automated information-measuring systems operating in real-time;
- creation of a wide network of integrated communications based on various communication lines;
- implementation of automated production management systems in energy companies.

The application of modern information technologies (ITs) makes it possible to significantly increase SGs operation efficiency, making them more reliable and economical, which, in its turn, leads to a reduction in the cost of power reproduced or distributed by them. However, at the same time, there are opportunities to influence SGs by various CAs. A consequence of this impact is the appearance of anomalies in the SG network traffic [6].

Detecting CAs in SGs is quite a complex task. It is necessary to constantly monitor security and control network traffic in order to detect anomalous activity in it. If traffic anomalies are detected, it is necessary to analyze a large number of routes in the network, where sharp fluctuations in traffic, delays in its transmission, or large packet losses appear. At the same time, a high quality of telecommunications service and application service should be ensured. All of this is the motivation for finding and developing new methods and approaches for CAs detection in SGs. Such approaches in this article include an approach that combines several methods of computational intelligence: the use of fractal analysis, statistical methods, and machine learning.

It should be noted that a fairly large number of classification and prediction methods mostly related to anomaly detection [7,8] are currently known and widely used. In particular, regression-based methods have performed well. These include non-parametric regression and classification tree method (CART) [9], multivariate adaptive regression splines (MARS) [10,11], support vector regression (SVR) [12] and others. Regression-based methods demonstrate high classification and prediction performance if their parameters are well-tuned. In some cases (for example, for MARS and SVR), it is proposed to use genetic algorithms to adjust the regression parameters.

However, this does not allow one to speak about the possibility of early detection of CAs. Therefore, it is believed that the most effective method of classification and prediction is the Long Short-Term Memory (LSTM) neural network algorithm. The LSTM property of recurrence allows an Artificial Neural Network (ANN) to "refer" to the results of its work in the past, to analyze predictions. Thus, the content of decisions made to protect SGs from CAs will depend not only on the results of initial training of the LSTM network, but also on the results of further operation of this network in the flow [13,14].

The key parameter of fractal analysis is the Hurst exponent. This measure is used in the analysis of time series. The Hurst exponent shows the amount of delay in the time series between two identical pairs of values. The bigger it is, the smaller this parameter is. To find this parameter, it is first necessary to check the process under study for stationarity.

The presence or absence of stationarity of the process influences the choice of the algorithm by which the scaling index can be calculated.

Fractal properties are more pronounced in non-stationary network traffic, which is predominant in SGs on large data scales. On small amounts of data, or in application layer protocols of the TCP/IP (Transmission Control Protocol/Internet Protocol) model, network traffic can be stationary and show less fractal properties. In this case, machine learning methods are used for further analysis.

Thus, in order to detect and classify the CAs, first, it is necessary to determine whether the traffic is stationary or non-stationary. Next, you should calculate the Hurst exponent (i.e., determine the presence of the self-similarity property in the traffic). In the final stage, anomalies are detected and measures are developed to protect the SG using LSTM [6,15].

The main contribution of this work is as follows: (1) the structures of long-term dependencies in the SG traffic were studied, which made it possible to identify its characteristic features in the interest of the early detection of CAs; (2) a new approach to the detection of CAs based on the study of the fractal properties of traffic has been proposed; (3) the LSTM structure was substantiated, which makes it possible to detect SC with a probability of 0.99; (4) a software prototype was developed that implements the proposed system, and a dataset was generated with SG traffic containing anomalies from the impact of both known and unknown CAs; (5) a comparison was made with other methods of machine learning in identifying the fact of the impact of CAs; (6) an experimental evaluation of the proposed system was carried out, showing its rather high efficiency.

The significance of the new contribution lies in the fact that the detection of CAs is performed using an autoencoder trained on the basis of the reference data of the SG operation and the information exchange in it, taking into account all deviations from the SG regular operation. During operation, the autoencoder is additionally trained by a validated neural network. The result is a generative adversarial network in which neural networks learn from each other. This made it possible to reduce the time for detecting anomalies in network traffic and increase the probability of detecting unknown computer attacks up to 0.8.

The proposed approach has a number of methodological and technical limitations. Methodologically, the approach is limited to the use of the most well-known methods of fractal analysis, which include the Dickey–Fuller test, the rescaled range (R/S) analysis, and the Detrended Fluctuation Analysis (DFA) method, and one of the most promising ANN models, which is the LSTM model. The technical limitations are determined by the computing power of the environment, on which the autoencoder and the neural network are trained, as well as by obtaining a reference sample of the SG operation, taking into account all deviations from the normal operation mode. The training quality of the generative-adversarial network and the detection efficiency of known and unknown CAs depend on the quality of the sample made. Since the autoencoder is additionally trained by the ANN, an incorrect sample can break the ANN operation logic.

The novelty of the results obtained lies in the fact that, based on experimental studies, the best method for determining self-similarity for non-stationary and stationary time series is substantiated, which allows detecting changes in traffic with high accuracy and quickly, and the structure of the LSTM neural network is determined, which provides high accuracy and can sufficiently quickly predict the impact of CAs and allows developing proactive protection measures. This is a significant advantage of the proposed system.

This work is a continuation of the studies published in [6] and is devoted to testing the possibility of using fractal analysis methods for detecting CAs against smart power grids. The difference in this work lies in the addition of neural network analysis methods using LSTM networks to fractal analysis methods. This approach, in contrast to [6], allows one not only to detect anomalies in the SG traffic, but also to identify the types of CAs that are the causes of these anomalies.

For this purpose, we propose the structure of an autoencoder trained on the normal data of the SG network operation, considering possible deviations from the SG normal operation. The complex use of fractal analysis methods, autoencoder, and LSTM network forms the basis of the SG proactive protection system, which is able to detect both known and unknown CAs.

The article has the following further structure. Section 2 is devoted to the analysis of known works in the research field. The theoretical foundations of the proposed proactive CA detection system, which is based on the fractal analysis of the network traffic and their subsequent processing using LSTM networks, are discussed in Section 3. A general description of the proposed system is given in Section 4. Section 5 presents the experimental evaluation results. Section 6 is a discussion of the experimental results and their comparative evaluation. Final conclusions and directions for further research are contained in Section 7.

2. Related Work

Fractal analysis, which studies the properties of self-similarity, is currently in a phase of active development. Fractal analysis is widely used for state monitoring problems, in which time series are investigated. For example, [16] proposes to use the R/S analysis method to analyze the self-similarity of time series. The self-similarity properties of the Voice Over Internet Protocol (VoIP) traffic are modeled and studied in [17]. The fractal dimension, which is an additional measure with respect to the Hurst exponent, is investigated in [18]. The reasons explaining the presence of self-similarity properties in telecommunication traffic are given in [19]. However, the main area of research in all these papers, as a rule, is both VoIP-telephony and economic systems.

At the same time, it should be noted that there are few practical experiments aimed at studying the fractal properties of the network traffic in information and telecommunication systems. Among such works, we can single out works [20–22]. However, [20] considers the mobile communication traffic generated by cellular stations. The authors conclude that the properties of self-similarity are inherent not only in computer and telecommunications networks, but also in the radio waves on which cellular stations operate. Self-similarity of motion is considered in [21,22]. To detect it, it is proposed to use visual cues, which allow one to find similar areas on the motion graph. These areas allow one to identify self-similar processes.

One of the first works, in which the main attention was paid to the self-similarity property of the network traffic, is the work [11]. It significantly changed the existing ideas about the processes taking place in information and telecommunication networks. These issues will be discussed in more detail in the next section. In addition, we should mention some works in which the mathematical models designed to describe self-similarity in network traffic have been proposed and investigated [23,24]. However, these works cannot be considered exhaustive, since they did not consider the issues of CA detection. Consequently, we can assume that our work, on the one hand, further develops the theoretical positions achieved in the study of the fractal properties of the network traffic. On the other hand, it develops the well-known solutions further in the direction of creating a method that makes it possible to detect network traffic anomalies caused by the impact of CAs.

At the same time, it should be noted that when considering threats to SG security, one should be guided by the following two indicators that characterize these threats. The first indicator is the probability of the threat realization. The second indicator is the potential damage that can be incurred by the power company in case of security threat realization [6,14]. Considering and combining these indicators, it is possible to substantiate the choice of the most acceptable threat models for SGs and to create protection systems for them, in which the decisions made would allow one to minimize security risks.

The first group [25–31] summarizes the techniques based on quantitative criteria. Thus, [25] proposes to use the acceptable level of the possible damage from information and technical impact on SG resources and the assessment of the profit factor from investments in protective measures as a measure to rank threat models. Quantitative methods comply with the requirements of ISO 27,001 and 27,002, NIST, and COBIT IV [26,27]. Although these methods take into account the predetermined risk appetite, they do not consider the variability in the construction of the SG protection system [28]. In addition, one of the significant disadvantages of the aforementioned methods is the high cost and complexity of their implementation [29]. At the same time, the complexity of quantitative methods is due to the need to take into account each potential security threat in the formation of options for counteracting CAs and developing solutions to eliminate the consequences of CAs [30]. For these purposes, [31] proposes to perform the ranking of security SG risks. Although this technique is undoubtedly of interest, it contains a number of negative factors associated with the problem of cloud resources.

The second group of methods [32–35] received the generally accepted name of qualitative methods. These methods apply qualitative indicators and criteria for the characterization of SG security threats. The essence of qualitative methods is the search for such a solution, in which the necessary balance is observed between the costs spent on building the protection system and the effect achieved with its help. Such methods form a direction called Cost/Benefit Analysis. In these methods, basically, different positions of the game theory, for example, matrix games are used. Speaking about the disadvantages of qualitative methods, it is necessary to point out their comparatively high computational complexity. It is due to the need to conduct a security risk analysis in order to make an economic justification for the introduction of protection mechanisms and means for various threat models into SG protection systems. Methods using qualitative criteria are similar in essence to the Facilitated Risk Analysis Process (FRAP) method [36,37].

The third approach [38–41] is an integrated one; it rationally combines the first and second groups of methods. Most often, the methods of this group find their application in small and medium-sized energy companies. The disadvantages of these methods include, as a rule, a very small amount of analytical data characterizing the potential damage under the given models of CA realization, as well as insufficiently complete risk assessment.

Besides, the works [42,43] present a structured approach to assessing the threat model for information and telecommunication resources (methods "CRAMM", "MEHARI"). Here an integrated representation of the information security threat parameters is performed, but the specificity of building the SG protection system is practically not considered.

There is a well-known methodology for managing the information security system—Microsoft Security Assessment Tool (MSAT) [44,45]. This tool uses a mechanism for ranking threat models. In addition, the tool provides countermeasures for SG security threats and evaluates their effectiveness. However, the tool is not scalable enough. That is why in SG it is usually implemented in local computing networks or in companies with fewer than 1000 employees. The Risk Management Guide [34] is the basis for this tool's design and operation. Among the main functions performed by the tool, in addition to risk assessment and decision support, one can include performance monitoring and evaluation [13].

Thus, all the considered approaches to CA early detection and prediction are based either on an in-depth analysis of possible risks (probable damage), or on a selective ranking of threats and defenses. In our opinion, these approaches are insufficient to protect SGs from CAs. For this reason, this article discusses the key points of building an improved system for CA early detection, which can be called proactive. The proactivity of the system lies in the fact that it implements anomaly detection in the network traffic, their identification, and classification based on fractal analysis methods, and a neural network with a long short-term memory, which allows one to reduce risks in the implementation of CAs. The consideration of the proposed system is architecture-oriented. On the one hand, it goes beyond an abstract representation, and on the other hand, it does not pay much attention to technical details. We conclude this article with a detailed look at the proposed active security solutions for SGs and their implementation.

3. Theoretical Foundations of the Proposed System

3.1. Stationarity of Temporary Traffic

Consider the autoregressive process in general terms:

$$x_t = C + \sum_{i=1}^{p} \varphi_i X_{t-i} + \in_t + \sum_{i=0}^{q} \theta_i \in_{t-1} \tag{1}$$

where $\varphi_p, \theta_q \neq 0$ are the model parameters, C is a constant, $\in_t$ is a white noise, x_{t-i} is a previous element of the time series.

The model can be interpreted as follows: the current value depends on past values up to lag p and on current and past external shocks up to lag q. To write the autoregressive process, it is convenient to use the lag operator L. The lag operator allows one to obtain the values of the elements of the time series based on several previous values. A lag operator of order i is an operator that shifts the value of the time series x_t by i values back, i.e., $L^i : x_t \to x_{t-i}$. Using lag operators, the autoregressive process can now be written more visually as follows:

$$x_t = C + \sum_{i=1}^{p} \varphi_i L^i X_t + \in_t + \sum_{i=0}^{q} \theta_i L^i \in_t \quad (2)$$

Let us rewrite as follows, moving the autoregressive part to the left side of the equality:

$$\left(1 - \sum_{i=1}^{p} \varphi_i L^i\right) x_t = C + \left(1 + \sum_{i=1}^{q} \theta_i L^i\right) \in_t \quad (3)$$

Now we introduce two polynomials of degree p and q:

$$\varphi(z) = 1 - \sum_{j=1}^{p} \varphi_j z^j = 1 - \varphi_1 z - \varphi_2 z^2 - \ldots - \varphi_p z^p \quad (4)$$

$$\theta(z) = 1 + \sum_{j=1}^{q} \theta_j z^j = 1 + \theta_1 z + \theta_2 z^2 + \ldots + \theta_p z^p \quad (5)$$

where φ_j and θ_j are polynomial coefficients depending on the monomial z, which is a complex number.

Then the autoregressive model can be formally written as $\varphi(L)x_t = C + \theta(L) \in_t$, where $\varphi(L)x$ is the autoregressive part of the polynomial, and $\theta(L) \in_t$ is the moving average part.

The time series is stationary if all roots of the autoregressive polynomial $\varphi(z) = 1 - \varphi_1 z - \ldots - \varphi_p z^p$ lie outside the unit circle of the complex plane $|z_j| > 1$ (that is, they are greater than 1 in absolute value). The inequality $|z_j| > 1$ is satisfied if $|\varphi_j| < 1$. Consequently, the relation $|\varphi_j| < 1$ is a condition of stationarity of the autoregressive process.

In addition, for a stationary process, the average is constant in time $Ex_t \equiv \text{const}$, i.e., the time series does not have a trend, and the covariance between different elements of the time series depends only on how far they are from each other in time. In other words, the covariance depends only on the lag h $\text{cov}(x_t, x_{t+h}) = \gamma(h)$. The value h, which characterizes the difference in time between the elements of the time series, is called a lag variable or delay. Since $\gamma(0) = cov(x_t, x_t) = \text{Var}(x_t)$, the variance of the stationary time series also does not change with time.

Thus, to test the hypothesis of stationarity of the series, the generalized Dickey–Fuller test is used, and to determine anomalous activity in the network, we are guided by the principle of self-similarity for non-stationary traffic, which is violated when anomalous activity occurs. R/S or DFA algorithms are used to calculate the self-similarity property. The first one is faster, and the second one is more accurate. The process is non-stationary if these conditions are violated.

3.2. *Self-Similarity Analysis in Network Traffic*

Many natural processes are characterized by distributions with heavy tails. Such distributions include the Pareto, Cauchy, Levy, and Weibull distributions, as well as the lognormal distribution. An important feature of exponential distributions is the realization of events that deviate strongly from the norm. Such distributions can be applied to model network traffic intensities and rates that have large, theoretically infinite variances.

The lognormal distribution is the earliest model of self-similar traffic. It is used to model network packet arrival intervals and file sizes transmitted [46]. The Weibull distribution is applied to model the arrival processes of FTP protocol blocks. The Pareto distribution is used to model the intervals between requests to web resources and VoIP traffic [17,46].

In our work, network traffic is considered as an aggregation of several flows from different sources. The aggregated flow, jointly transmitted over communication channels with infinite variance, leads to self-similar network traffic, which is described by the model of fractal Brownian motion [47]. If one of the partial flows has self-similarity when aggregating flows, then the resulting aggregate flow will also have self-similarity [24]. In this case, self-similarity is preserved when aggregating flows coming from both homogeneous and heterogeneous traffic sources.

Fractal Brownian motion is easily applicable to modeling self-similar traffic. The process $X(t)$ is called a fractal Brownian motion with the parameter H, $0 \leq H \leq 1$, if the increments of the random process have a Gaussian distribution:

$$P(\Delta X < x) = \frac{1}{\sqrt{2\pi}\delta_0 \tau^H} \int_{-\infty}^{x} \exp\left[-\frac{z^2}{2\delta_0^2 \tau^{2H}}\right] dz \quad (6)$$

where δ_0 is a diffusion coefficient.

Wherein:

(1) $X(0) = 0$;

(2) $\Delta X = X(t_2) - X(t_1)$ has a normal distribution with zero mean and variance—$\delta^2(t_2 - t_1)^{2H}$, $0 \leq H \leq 1$, where $H = 0.5$ indicates a random row. The events are random and uncorrelated. The range of accumulated deviations should increase in proportion to the square root of time.

As noted above, the Hurst exponent H is a measure of the self-similarity of the process. If the process has strongly pronounced fractal properties, then H approaches unity. In the absence of self-similarity $H = 0.5$ [15]. In this case we speak about fractal Brownian motion, which coincides with the classical Brownian motion and imposes a large noise on the time series [16,26,27].

3.3. Detecting Anomalous Bursts Using Machine Learning Techniques

There are many ways to identify anomalies. Figures 1–3 demonstrate the operation of the most popular machine learning algorithms tested on time series generated using an autoregressive integrated moving average model.

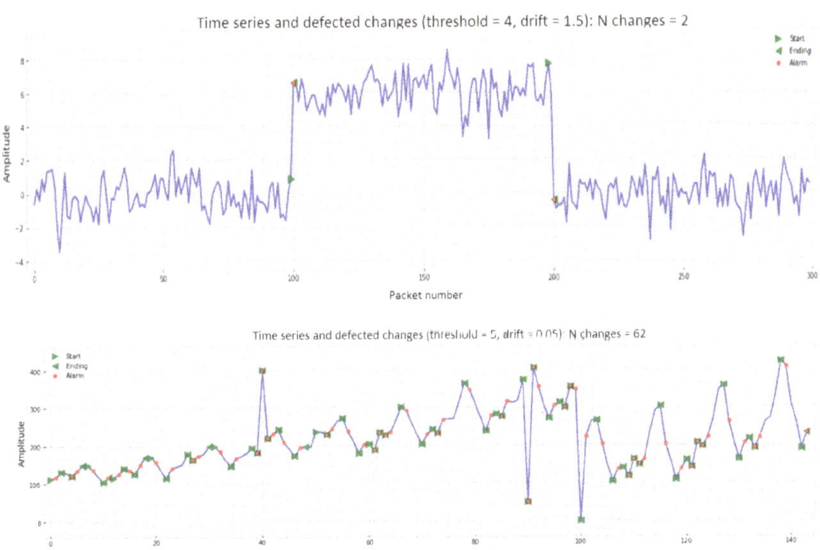

Figure 1. Cumulative sums.

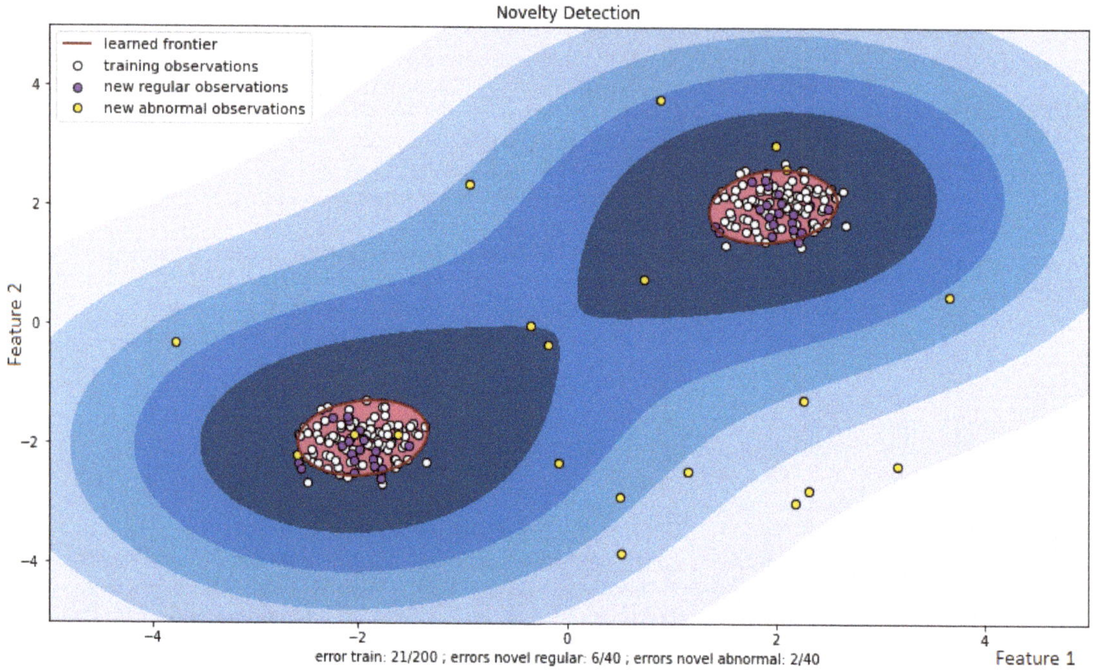

Figure 2. Support Vector Machine.

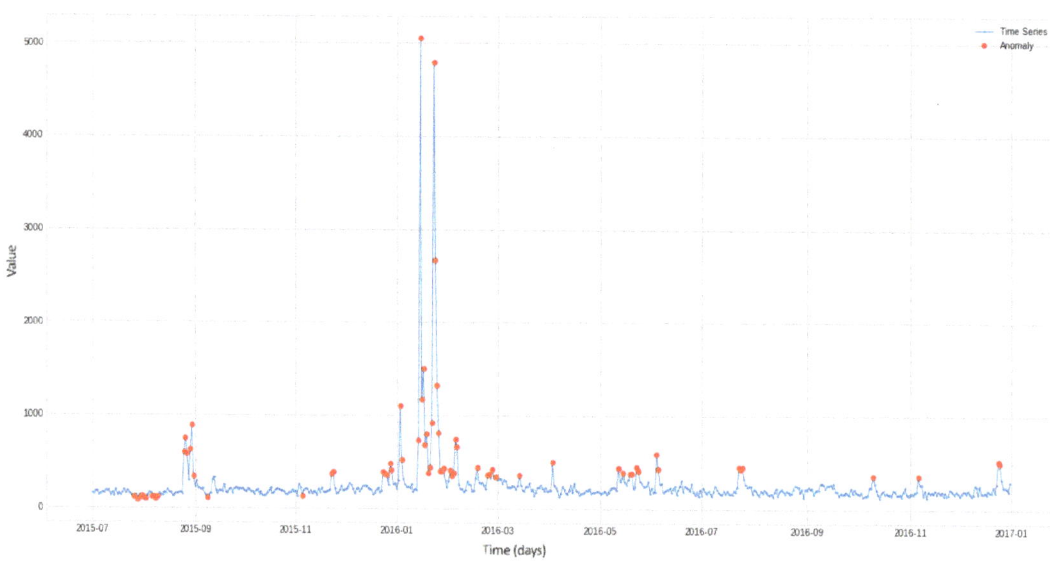

Figure 3. Isolated forest.

Figure 1 shows different variants of time series with different values of the threshold and drift parameters, on which anomalies (changes) are detected using the cumulative sum method. Detected changes are marked with red dots. It can be seen that the number of detected anomalies in the time series can be different (from two to 62).

Figure 2 outlines the results of anomaly detection using the support vector machine. Each observation has two normalized coordinates—Feature 1 and Feature 2. Feature 1 plays the role of the "Packet Delay" characteristic, and Feature 2 displays the "Bandwidth" characteristic. The white dots indicate the observations that were used in the training set. Their boundaries are marked with red lines. Testing a new dataset using a trained Support Vector Machine (SVM) classifier results in a division of observations into normal (purple dots) and anomalous (yellow dots).

Figure 3 depicts the results of anomaly detection in a time series using an isolated forest. At points where there are anomalies (red dots), the time series changes the parameters of its distribution. An isolated forest is good at detecting these changes.

As can be seen from the figures, the algorithms do an excellent job of detecting anomalous outliers. In this case, the anomaly manifests itself in the form of the non-stationarity of some observed time series. These are not only instantaneous jumps in the measurement amplitude, but also slow trends that are practically invisible during the observation period.

However, when testing the above algorithms on real network traffic, it turned out that outliers are not always anomalous. Therefore, to study the features of anomaly detection in SG traffic, a cyber polygon was developed, shown in Figure 4.

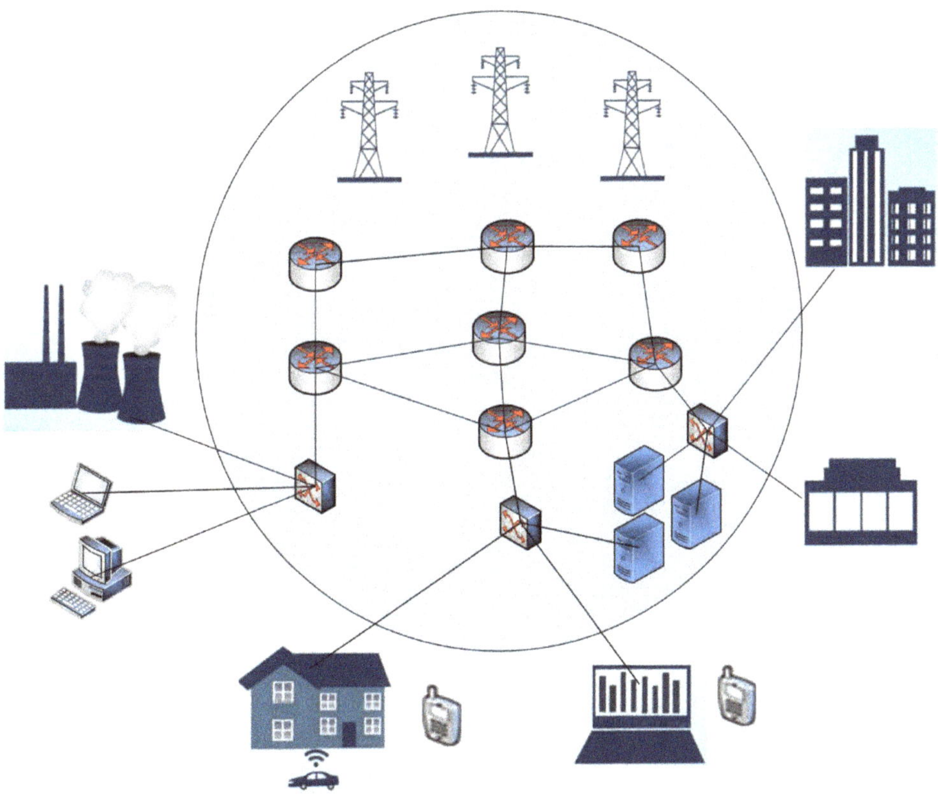

Figure 4. Cyber polygon designed to collect network traffic and analyze its security.

About 30 types of CAs were carried out in the cyber polygon and 40 GB of legitimate traffic was generated. Network traffic was redirected to Security Onion and written to pcap files. From this traffic, a dataset was formed using Netsniff-ng and Bro. The attacks were carried out using the Kali Linux distribution against known vulnerable services deployed in the central part of the scheme. Next, a search for anomalous bursts was carried out using the algorithms of cumulative sums, isolated forest, and SVM. Despite the fact that these algorithms do an excellent job of finding anomalous bursts, it was found that bursts are not always anomalies.

For data packet transmission, modern standards, protocols, and technologies for high-speed electrical networking were considered, such as:

- Fast Ethernet and Gigabit Ethernet suite of standards, which define wired connections and electrical signals at the physical layer, and packet format and medium access control protocols at the data link layer;
- wireless transmission standards based on GSM/EDGE and UMTS/HSPA, which allow data rates of 100 Mbps (with mobile subscribers) and 1 Gbps (with fixed subscribers);
- IEEE 802.11 local wireless networks, which use infrared radiation and radio waves as the physical transmission medium.

The scenario according to which the message packets are transmitted is stationary in this case. Sensors were installed in homes, shops, and offices. The Leningrad Nuclear Power Plant (LNPP) and the South-Western Thermal Power Plant (SWTPP) acted as sources of electricity. The control over the security of the SG network was ensured by the operator (the incident monitoring system).

The intercepted traffic was a data set containing information processed by the operators and dispatching systems of the SG power system. This information included the following parameters:

- equipment state parameters;
- load parameters for transformers;
- parameters of the distributed measurement system;
- power quality parameters;
- information about the locations of damage and denial of service;
- power factor values;
- profiles and forecasts of electricity consumption, as well as some other parameters.

The SG telecommunications network was considered one of the types of computer networks. Therefore, we assumed that the telecommunications SG network has the self-similarity property. Our assumption was later confirmed in the course of experiments.

Based on the fact that the greatest amount of information is stored and transmitted by the operators and dispatchers of the SG power system, the monitoring system, as well as the LNPP and SWTPP data transmission networks with control system were selected as the object for the implementation of the CA.

It was assumed that the ports in the edge network equipment have a bandwidth of 1 Gbit/s [6] and operate over the Ethernet protocol. Traffic generation was performed using the developed simulation model. The GNS3 framework (Galaxy Technologies, LLC., https://www.gns3.com/ (accessed on 20 September 2022)) was used to build this model.

Table 1 shows a list of the main attributes that were included in the dataset generated with GNS3.

The total number of different Flow.ID values in the dataset was 1,522,917. Address 10.200.7.217, corresponding to SWTPP, was used as Source.IP for 7% of all entries. The value 10.200.7.218 corresponding to LNPP was in the Source.IP parameter for 8% of all records. The rest 85% of the entries had other Source.IP values. The generated addresses 10.200.7.7 and 10.200.7.8 were used as Destination.ID field values in 9% of all records. They corresponded to computer networks located in the "Passage" and "Gostiny Dvor" shopping centers. The remaining 82% of the records had other values of the Destination.IP field (their number was 2,939,141).

The self-similarity analysis was performed on the time series formed from the values of the Packet.Length.Mean field. This attribute in the generated dataset had 10,700 unique values. The most frequent values were 267.5 and 243.5 [6].

Table 1. The main attributes included in the dataset.

#	Attribute Name	Comments
1	Bwd.Packet. Length.Max	The maximum packet length (in bytes) in the backward direction
2	Bwd.Packet. Length.Mean	The mean packet length (in bytes) in the backward direction
3	Bwd.Packet. Length.Min	The minimum packet length (in bytes) in the backward direction
4	Bwd.Packet. Length.SD	The standard packet length deviation (in bytes) in the backward direction
5	Destination.IP	The destination IP address
6	Destination.Port	The destination port number
7	Flow.Duration	The total flow duration
8	Flow.ID	A flow identifier. It has the following format: Source.IP-Destination.IP-Source.Port-Destination.Port-Protocol
9	Fwd.Packet. Length.Max	The maximum packet length (in bytes) in the forward direction
10	Fwd.Packet. Length.Mean	The mean packet length (in bytes) in the forward direction
11	Fwd.Packet. Length.Min	The minimum packet length (in bytes) in the forward direction
12	Fwd.Packet. Length.SD	The standard packet length deviation (in bytes) in the forward direction
13	Packet.Length.Mean	The mean length value of the packets registered in the flow (both forward and backward directions)
14	Protocol	The transport layer protocol number identification (value is 6 for the TCP protocol and 17 for the UDP protocol)
15	Source.IP	The source IP address of the flow
16	Source.Port	The source port number
17	Timestamp	Packet capture moment. The value is stored in the following format: Dd/mm/yyyy HH:MM:SS
18	Total.Backward.Packets	The total number of the backward packets
19	Total.Fwd.Packets	The total number of the forward packets
20	Total.Length.of. Backward	The total number of bytes in the backward direction obtained from all the flow (all the packets have been transmitted)
21	Total.Length.of. Fwd	The total number of bytes in the forward direction received from all the flow (all packets have been transmitted)

Two CA types impacted the SG's simulated infrastructure. These attacks were a DDoS attack and a "Network and Vulnerability Scanning" attack. Traffic impacted by the first type of attack was simulated using the IXIA's IP network test equipment. A distributed network and SYN Flood, Ping Flood, and UDP Flood methods were used to implement the first CA type. The second CA type was simulated using IP network scanning tools Nmap and Xspider. The probing method was used to implement this attack. According to this method, Nmap or Xspider network scanner simulates an attack aimed at active exploitation of the analyzed vulnerability.

A SYN Flood attack was simulated as follows. The attacker (client) used the standard way of opening TCP connections. For this purpose, a SYN packet was sent to an open server port. After receiving and processing this packet, the server returned the SYN-ACK packet. The SYN-ACK packet contained client-specific data taken from the Traffic Control Unit (TCU) store. In normal circumstances, the client sends back an ACK packet, which serves as confirmation and allows a TCP connection to be opened. However, in the case of a SYN attack, the attacker generated and sent multiple repeat requests to the server with spoofed IP addresses. The server, being the target of the attack, treated them as legitimate requests. It processed them all and tried to open a TCP connection for them. Ping Flood and UDP Flood attacks were implemented in a similar way.

Thus, under conditions of CAs, the data set included the additional attributes (flags) shown in Table 2.

Table 2. Additional dataset attributes.

#	Attribute Name	Comments
1	ACK.Flag.Count	The number of times the ACK (Acknowledged) flag for packets sent in both directions was 1
2	FIN.Flag.Count	The number of times the FIN flag for sent packets was 1. Normally the operation ends with the transmission of a packet in which the FIN is 1
3	RST.Flag.Count	The number of times the RST (Reset) flag for packets sent in both directions was 1
4	SIN.Flag.Count	The number of times the SIN (Synchronization) flag for packets sent in both directions was 1

The type of attacks being modeled was considered during dataset generation and determined the FIN, SIN, RST, and ACK flags values. For example, the number of single SIN and ACK flag values increased if the "Network and Vulnerability Scanning" attack was simulated. Thus, in the traffic used for the experiments described in this article, single SIN flag values accounted for 20% and single ASK flag values for 60% of all values [6].

Figures 5–8 and Table 3 present the data obtained at the cyber polygon on the protocols and network parameters under study. Table 3 shows the network protocol parameters that were studied. Figure 5 shows statistics on retransmitted or dropped packets. Figure 6 demonstrates the dynamics of changes in the number of connections to the server. Figure 7 depicts how the state of the TCP header parameters changed over time during the lifetime of the IP packet. Figure 7 shows how the packet rate has changed over time.

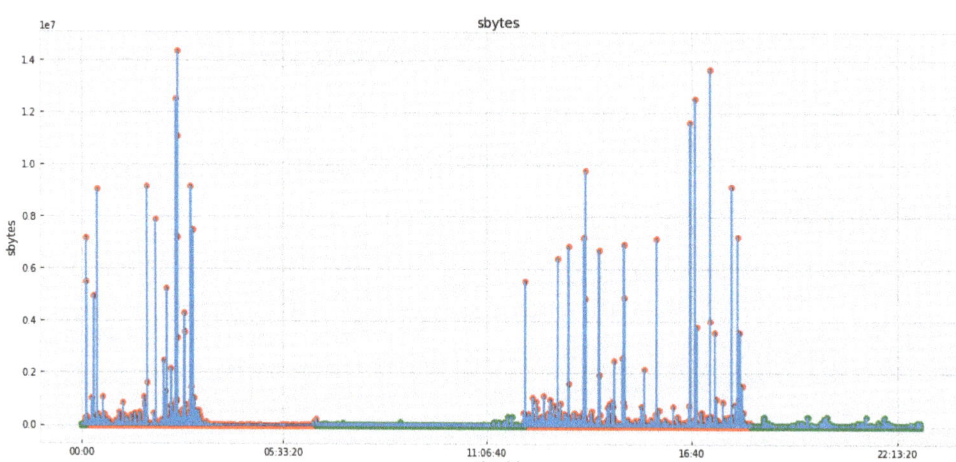

Figure 5. Retransmitted or dropped packets.

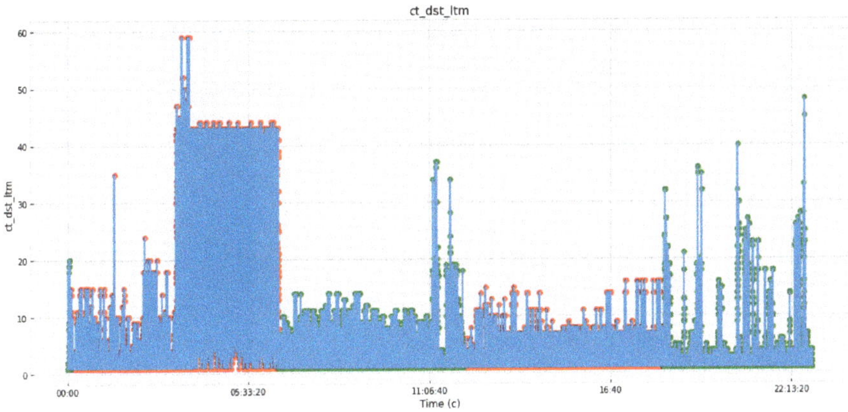

Figure 6. Number of connections to the server.

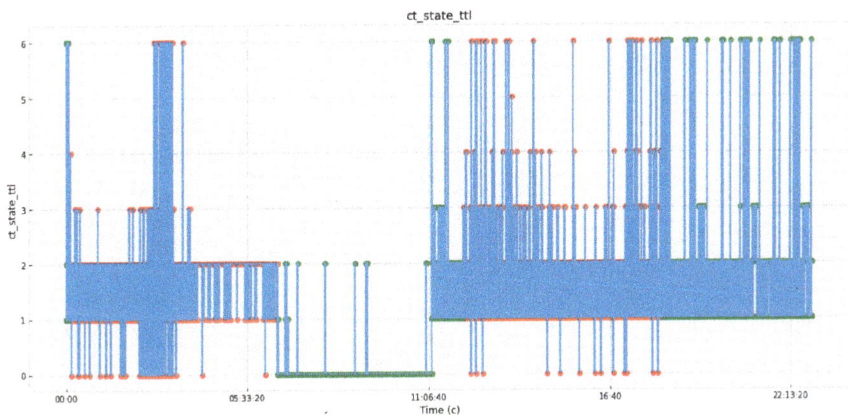

Figure 7. States of TCP header parameters during the lifetime of an IP packet.

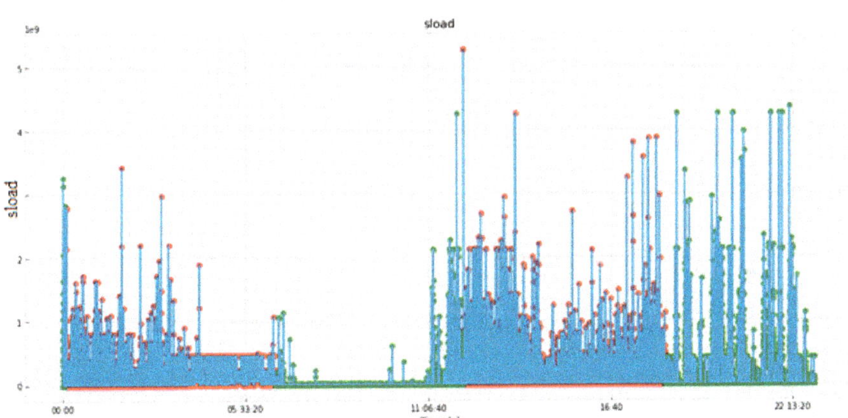

Figure 8. Packet transfer rate, bit/sec.

Table 3. Network protocol parameters.

No.	Name	Type	Description
1	dbytes	integer	Destination to source transaction bytes
2	dintpkt	float	Destination interpacket arrival time
3	djit	float	Destination jitter (mSec)
4	dload	float	Destination bits per second
5	dloss	integer	Destination packets retransmitted or dropped
6	dmeansz	integer	Mean of the packet size sent by destinations
7	dpkts	integer	Destination to source packet count
8	dsport	integer	Destination port number
9	dstip	nominal	Destination IP address
10	dtcpb	integer	Destination TCP base sequence number
11	dttl	integer	Destination to source time to live value
12	dur	float	Record total duration
13	dwin	integer	Destination TCP window advertisement value
14	ltime	timestamp	Record last time
15	proto	nominal	Transaction protocol
16	res_bdy_len	integer	Actual uncompressed content size of data
17	sbytes	integer	Source to destination transaction bytes
18	service	nominal	http, ftp, smtp, ssh, dns, ftp-data, irc and others
19	sintpkt	float	Source interpacket arrival time
20	sjit	float	Source jitter (mSec)
21	sload	float	Source bits per second
22	sloss	integer	Source packets retransmitted or dropped
23	smeansz	integer	Mean of the packet size sent by sources
24	spkts	integer	Source to destination packet count
25	sport	integer	Source port number
26	srcip	nominal	Source IP address
27	state	nominal	Indicates to the state and its dependent protocol
28	stcpb	integer	Source TCP base sequence number
29	stime	timestamp	Record start time
30	sttl	integer	Source to destination time to live value
31	swin	integer	Source TCP window advertisement value
32	synack	float	TCP connection setup time
33	tcprtt	float	TCP connection setup round-trip time
34	trans_depth	integer	Represents the pipeline depth into connection

In Figures 5–8, anomalous packets are marked with red dots and normal (legitimate) packets are marked with green dots. As can be seen from these figures, many bursts are legitimate and, conversely, in many places where there are no bursts, there are anomalies. Therefore, the issue of timely detection of bursts of traffic in SG, identification of anomalous ones from them, as well as classification of detected anomalies in order to predict the fact of the impact of CAs and develop effective countermeasures is an acute issue.

3.4. Anomaly Detection with Classifiers

To evaluate the effectiveness of popular classifiers, a dataset was formed containing correlated parameters with anomalous queries. For this purpose, a correlation matrix was built (Figure 9).

Figure 9. Correlation matrix.

The label parameter (the last parameter in the correlation matrix) is an indicator showing the presence of anomalies. From the parameters presented in the correlation matrix, 20 parameters were selected that are most correlated with anomalies. They are outlined in Figure 10.

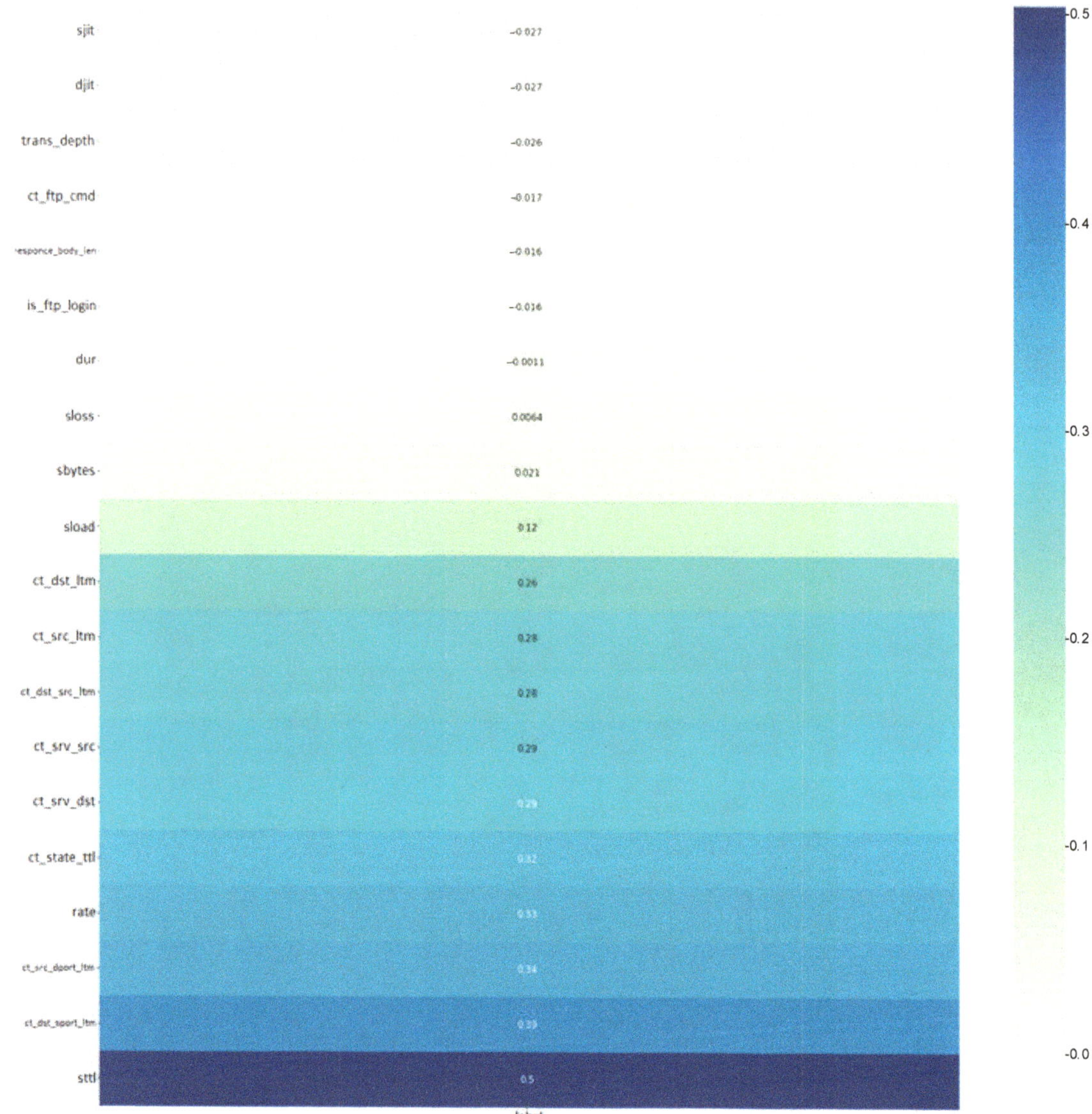

Figure 10. Most correlated parameters.

The sttl parameter, which indicates the lifetime of the packet during its transmission from source to sender, is most affected. The dynamics of this parameter with indications of anomalies (shown in red dots) are demonstrated in Figure 11. The figure shows the dynamics of the value, which tells the local server how long to keep the packet information in the IP protocol.

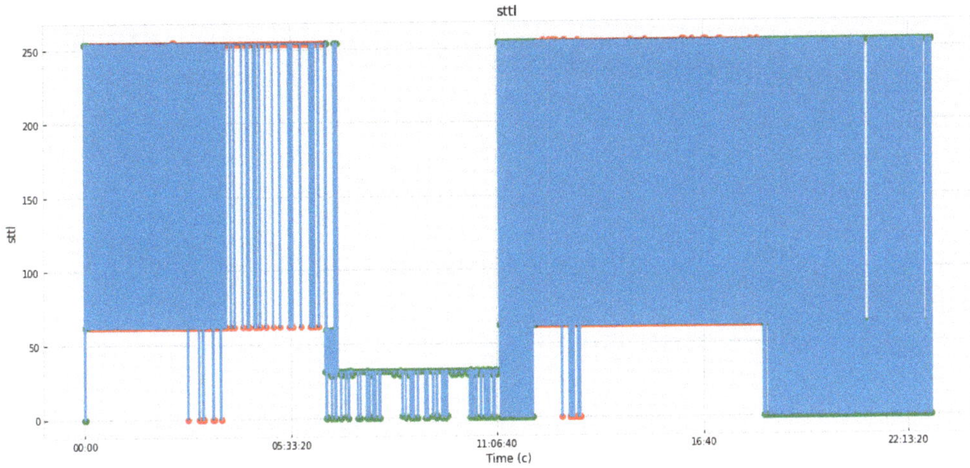

Figure 11. Packet lifetime from source to sender.

Logistic regression, random forest, and decision tree were chosen as classifiers that were used at the cyber polygon. These classifiers are not chosen by chance. They have been widely used in the works of many researchers and in many cases provide a sufficiently high classification efficiency, including in ensembles of classifiers. To evaluate their effectiveness, a confusion matrix was calculated. It was used to determine not only the accuracy, but also the number of false positives. The results of the selected classifiers are shown in Figure 12. It can be seen that despite the high efficiency of the classifiers used at the cyber polygon, they all had a large number of false positives.

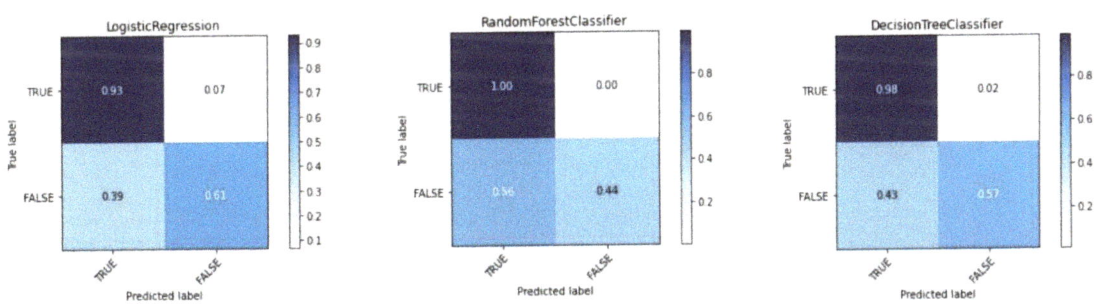

Figure 12. Comparing the efficiency of classifiers.

For Logistic Regression the First Kind Error is 39%. The Random Forest algorithm has 56% of false positives. The Decision Tree algorithm also shows quite aggressive behavior, which is caused by the First Kind Error, equal to 43%.

The obtained results confirm that the main problem of known classifiers is the poor ability to recognize previously unknown anomalies.

4. General Description of the Proposed System

4.1. Stages of System Operation

To detect anomalies in smart power grids from cyber attacks, a proactive protection system is proposed. The scheme of operation of this system contains the following stages (Figure 13):

- collection of network traffic;
- stationarity check;

- preparation of initial data;
- fractal analysis;
- machine learning.

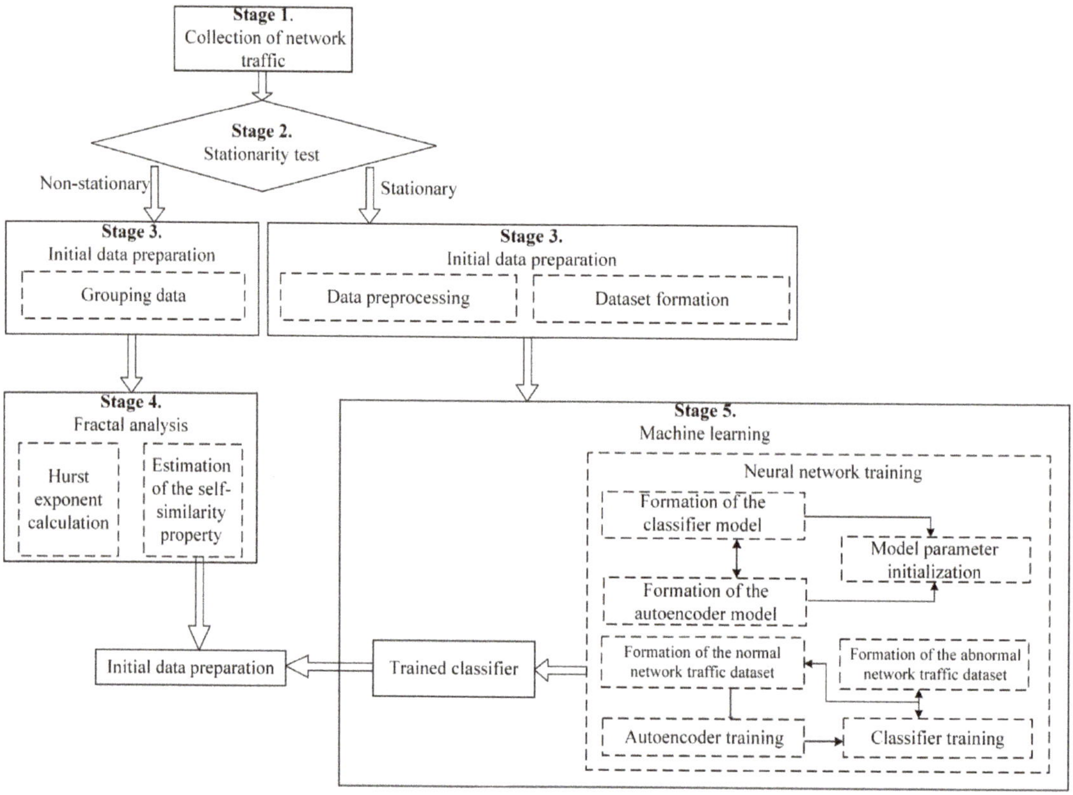

Figure 13. Scheme of the proactive protection system when anomalies are detected.

Initially, after traffic is collected, it is checked for stationarity. To calculate the Hurst exponent in stationary traffic, R/S analysis is used, and in non-stationary noisy traffic with time-varying characteristics, DFA analysis is used. The procedure for estimating the Hurst exponent based on R/S and DFA analysis was considered in detail in [6,48].

Next, the detected anomalies are processed in order to predict the fact of the impact of cyber attacks. To do this, a hybrid neural network consisting of an autoencoder and a classifier is used as a machine learning method.

An autoencoder is a feed-forward neural network that reconstructs the input signal at the output. Inside it has a hidden layer, which is the code that specifies the model. The autoencoder is designed to be able to exactly copy the input to the output.

It is proposed to use cells with LSTM as autoencoder layers in the system. The architecture of the LSTM and the algorithm of its operation were considered in detail in [13,49]. Therefore, further, we will consider in more detail the operation of the developed SG protection system against CAs on service data transfer protocols.

4.2. *Software Implementation*

For the software implementation of the proposed CA detection system (Figure 14), the Python language was chosen. Pandas library, written in the programming languages C, Cython, and Python, was used for data processing and analysis. As a result, Python, despite its high availability, becomes quite a powerful tool for data analysis. It allows one to perform groupings, create pivot tables at a high level, and have easy access to tabular data.

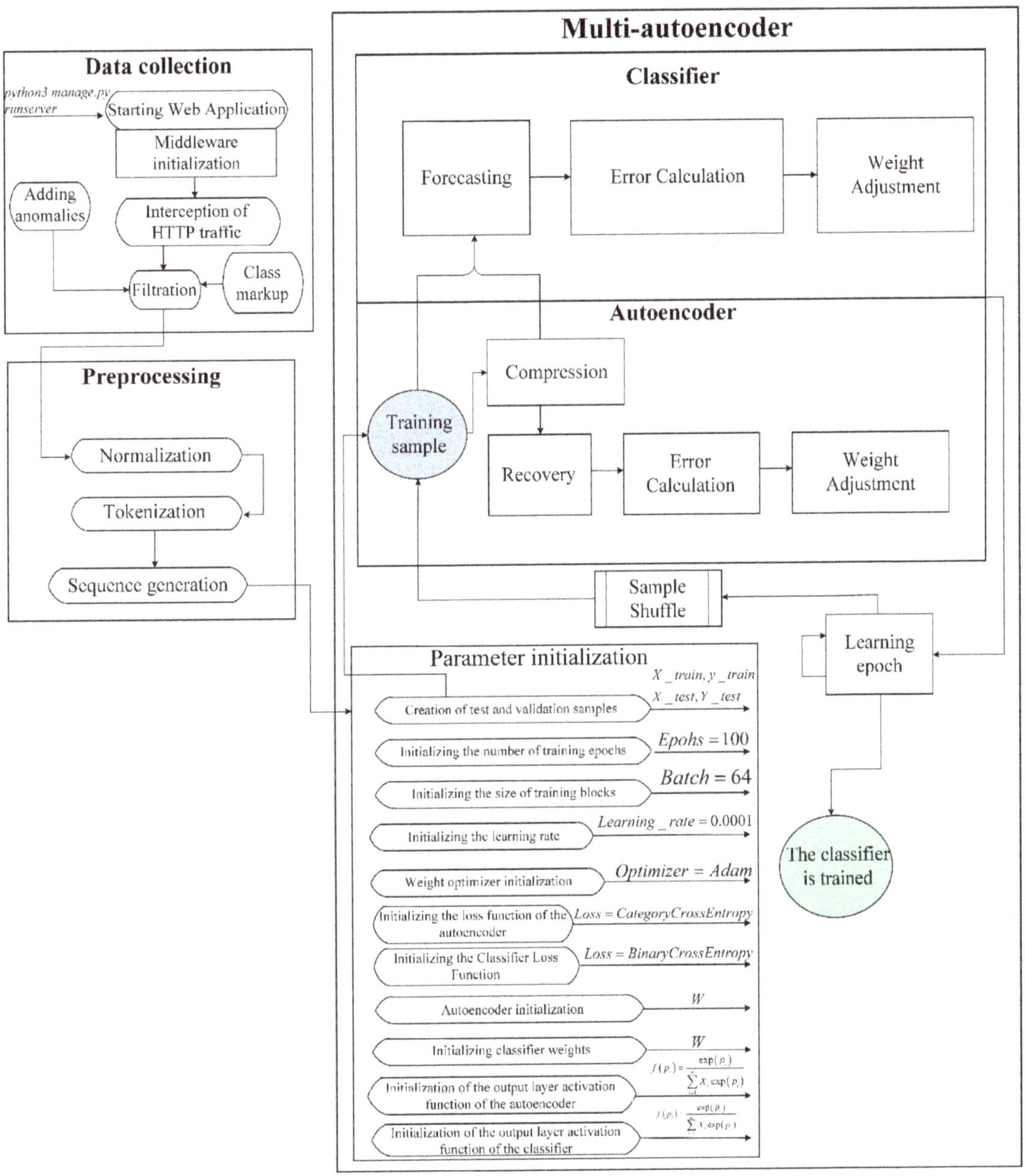

Figure 14. Scheme of functioning of the system for CA detection in stationary traffic.

In addition to the Pandas library, we used the NumPy library, which is a lower-level toolkit that allows one to work with multidimensional arrays (tensors) and high-level mathematical functions. The Matplotlib module was used to build graphs. Necessary calculations were carried out in the integrated development environment Jupiter notebook.

In order to intercept the request, a middleware layer (framework Django) was used. It is a middleware framework that allows one to process requests from the browser before they reach the server, as well as to handle responses before they are returned to the browser.

To test the stationarity of traffic, an experiment was carried out, which consisted in plotting the distribution of lengths between two identical characters and estimating the stationarity of the resulting series using the Dickey–Fuller test.

Next, preprocessing and normalization of the resulting sample were performed. Vector representation of characters was used, since the HTTP protocol is a text-based protocol. To implement this method of representation, all the characters available in the dataset were replaced by numeric equivalents (tokens), which have no independent application. Then the words were translated into a sequence of sequences.

An example of the resulting array of sequences is shown in Figure 15.

```
data_train = pad_sequences(X_train_sequences, maxlen=max_len_str, padding='post')
data_test = pad_sequences(X_test_sequences, maxlen=max_len_str, padding='post')

data_train

array([[24,  4,  2, ...,  0,  0,  0],
       [24,  4,  2, ...,  0,  0,  0],
       [24,  4,  2, ...,  0,  0,  0],
       ...,
       [24,  4,  2, ...,  0,  0,  0],
       [24,  4,  2, ...,  0,  0,  0],
       [24,  4,  2, ...,  0,  0,  0]], dtype=int32)
```

Figure 15. Sequence example.

It was taken into account that all sequences must have the same length. If the request length was less than the sequence length, the missing characters were replaced by zeros.

4.3. Subsystem for Determining the Stationarity of Network Traffic

Using the Dickey–Fuller test, the value of the autoregression coefficient α is checked in the first-order autoregressive equation $AR(1)$:

$$y_t = \alpha \cdot y_{t-1} + \varepsilon_t \,, \tag{7}$$

where y_t is a time series and ε is white noise, $t = 1, \ldots, T$.

1. If $H_1 : \alpha < 1$, then the series y_t will be stationary, $y_t \sim I(0)$ and the Ordinary least squares (OLS) estimator $\hat{\alpha}$ will have a normal distribution with zero mean and variance $\hat{\alpha}$.

To test the unit root hypothesis, an OLS estimator $\hat{\alpha}$ is constructed:

$$\hat{\alpha} = \frac{\sum_{t=1}^{T} y_{t-1} y_t}{\sum_{t=1}^{T} y_{t-1}^2} \tag{8}$$

and the corresponding t-statistic:

$$t_\alpha = \frac{\hat{\alpha} - 1}{S / \sqrt{\sum_{t=1}^{T} y_{t-1}^2}} \tag{9}$$

where $S^2 = T^{-1} \sum_{t=1}^{T} (y_t - \hat{\alpha} y_{t-1})^2$ is the estimated variance of the residuals.

If the value of statistic t_α lies to the left of the critical value at the 5% significance level, i.e., $t_\alpha < t_{critical}^{5\%}$, then the time series is stationary.

2. If $H_0 : \alpha = 1$, then the distribution of this estimate will no longer be normal, and the process y_t will be non-stationary with a time-dependent variance $y_t \sim I(1)$. In this case, to model the dynamics of such a series, it is necessary to use its first difference $\Delta y_t = y_y - y_{t-1}$.

Under the null hypothesis, the normalized bias statistic $T(\hat{\alpha}-1)$ and the t-statistic t_α have non-standard marginal Dickey–Fuller distributions:

$$T(\hat{\alpha}-1) \Rightarrow \frac{\int_0^1 W(r)dW(r)}{\int_0^1 W^2(r)dr} \text{ and } t_\alpha \Rightarrow \frac{\int_0^1 W(r)dW(r)}{\sqrt{\int_0^1 W^2(r)dr}} \tag{10}$$

where $W(r)$ is the standard Wiener process (Brownian motion).

If $t_\alpha > t^{5\%}_{critical}$, then the time series is non-stationary.

4.4. Subsystem of Anomaly Analysis in a Stationary and Non-Stationary Network

To detect anomalies in a stationary network, it is proposed to use a hybrid neural network model (Figure 16) created on the TensorFlow framework using the Python language.

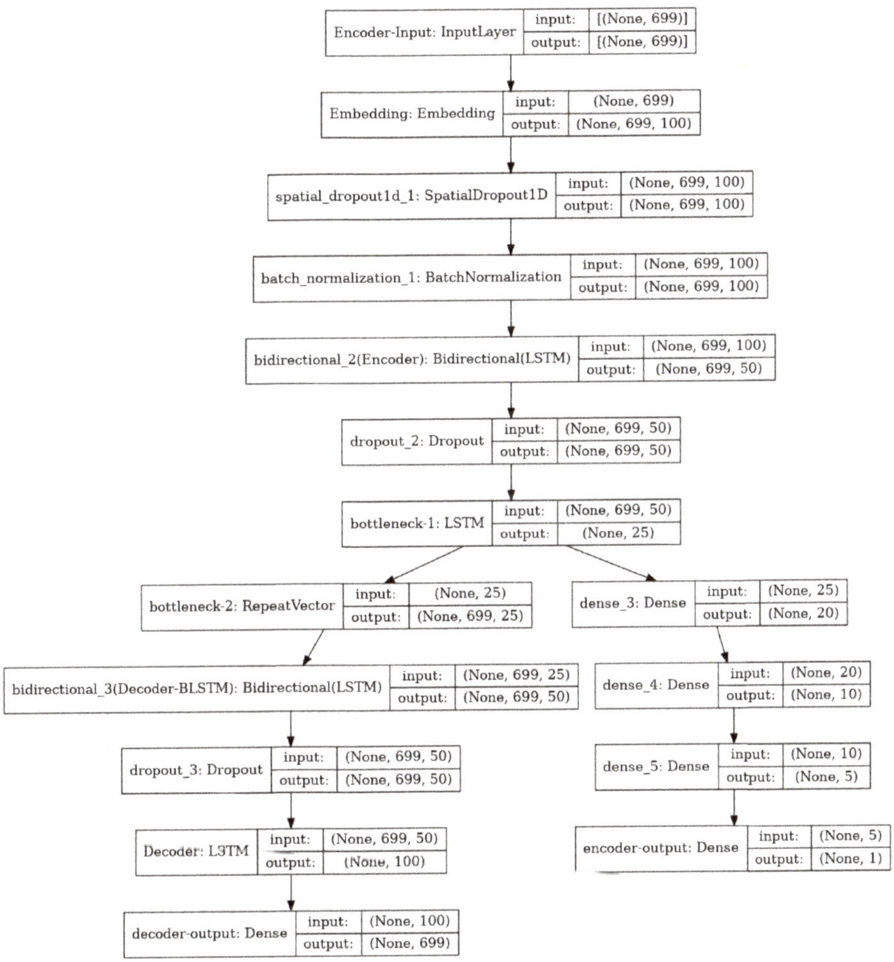

Figure 16. Hybrid neural network model.

The autoencoder model consists of Gated Recurrent Units (GRUs), which are elements of the LSTM neural network. Data up to 699 symbols are fed to the input of the neural network.

The neural network has several output layers. The output layer of an autoencoder has exactly the same dimension as the input layer. The classifier has one output layer. It determines if the request is anomalous or legitimate.

5. Experimental Evaluation of the System

Non-stationary network traffic received using the created cyber polygon was divided into legitimate (Figure 17) and anomalous (Figure 18) samples.

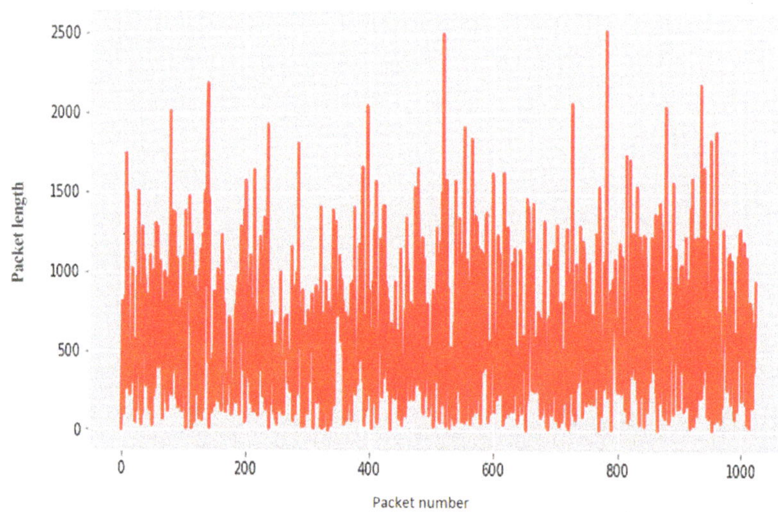

Figure 17. Legitimate traffic.

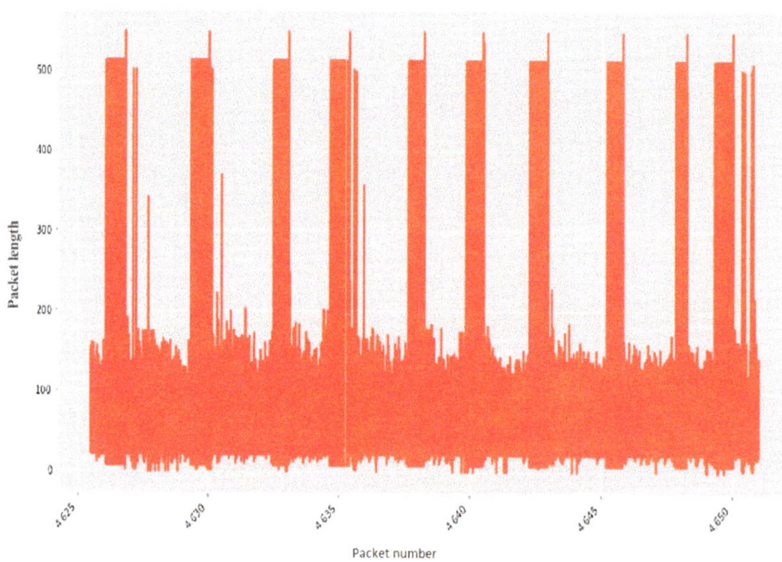

Figure 18. Abnormal traffic.

The analysis showed that in order to detect anomalous behavior in traffic, it is enough to analyze its main parameters. There is no need to study the contents of each packet. Examples of anomalies detected based on traffic telemetry analysis are a sudden increase in traffic from a workstation or a change in its structure compared to normal daily rates for a given network device.

For each sample, the Hurst exponent was calculated using the R/S algorithm.

Figure 19 depicts an example of calculating H for non-stationary traffic, which showed the result $H = 1.378$.

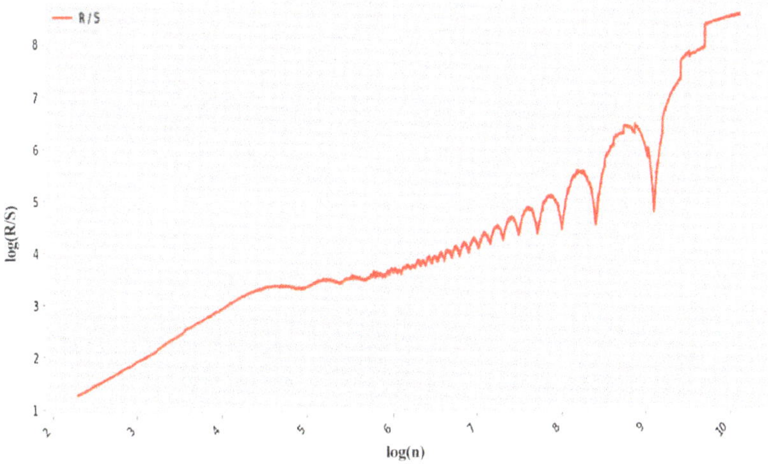

Figure 19. R/S versus time on a logarithmic scale.

In turn, the Hurst exponent exceeding the maximum value of 1 confirms the presence of anomalies in network traffic.

To quickly find anomalies caused by CAs, the network stream is first divided into groups. The Hurst exponent is then calculated for each of the groups. The result of such processing is shown in Figure 20. In this example, 10,000 points were divided into 20 groups.

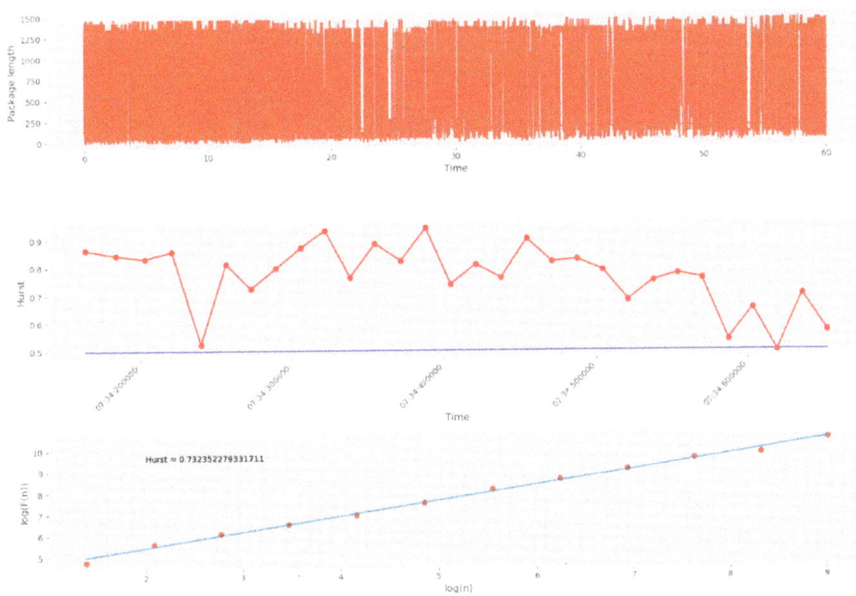

Figure 20. Computing H for legitimate UDP traffic.

The threshold corresponding to the white noise boundary ($H = 0.5$) is indicated by the blue line. The points on the second graph correspond to the number of packet groups (30 points in total). On the third graph, the dots correspond to the number of scales (12 dots in total). The number of scales affects the accuracy and duration of the algorithm. Increasing the number of scales increases accuracy, and decreasing the number of scales decreases accuracy.

Figure 20 shows the Hurst exponent for all groups of packages, which is above the 0.5 mark. This indicates self-similarity properties for each of the network traffic groups. The third graph (logarithmic regression graph) shows the Hurst exponent for all data, which confirms the presence of fractal properties and repetitive processes.

Next, we tested abnormal network traffic received during a DDoS attack and a cyber-attack "Scanning the network and its vulnerabilities". The result of calculating H for this anomalous traffic is shown in Figure 21. It can be seen that in this case the self-similarity property is violated, since the Hurst exponent at each of the intervals has a value less than the threshold of 0.5.

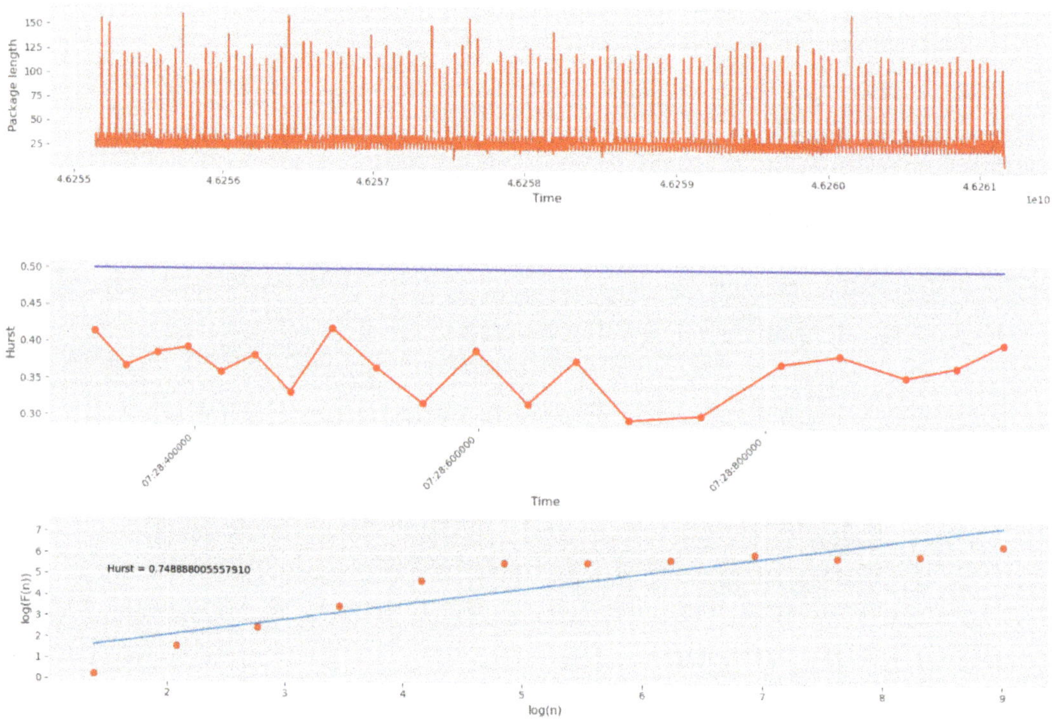

Figure 21. Computing H for abnormal UDP traffic.

The training dataset includes both legitimate and anomalous traffic. Only legitimate traffic was fed to the input of the autoencoder. The classifier input received legitimate and anomalous traffic, as well as hidden latent representations received from the autoencoder after encoding the information. The results of the selection of the neural network parameters are shown in Figure 22. The selection of parameters was carried out in such a way that the loss function during training of the autoencoder decreased, while the accuracy of the classifier grew.

```
Trial 27 Complete [00h 27m 58s]
decoder-output_loss: 19800.93359375

Best decoder-output_loss So Far: 19800.93359375
Total elapsed time: 05h 00m 36s

Search: Running Trial #28

Hyperparameter     |Value              |Best Value So Far
decoder-output     |0.0014             |0.0039
encoder-output     |90                 |45
learning_rate      |1e-06              |1e-05
tuner/epochs       |10                 |10
tuner/initial_e...|0                  |0
tuner/bracket      |0                  |0
tuner/round        |0                  |0

Epoch 1/10
1351/1351 [==============================] - 218s 156ms/step - loss: 92.7079 - encoder-
Epoch 2/10
1351/1351 [==============================] - 208s 154ms/step - loss: 92.4347 - encoder-
Epoch 3/10
1026/1351 [======================>.......] - ETA: 43s - loss: 92.1161 - encoder-output_
```

Figure 22. Selection of neural network hyperparameters.

Figure 23 demonstrates the results of estimating accuracy growth and loss reduction over 30 training epochs.

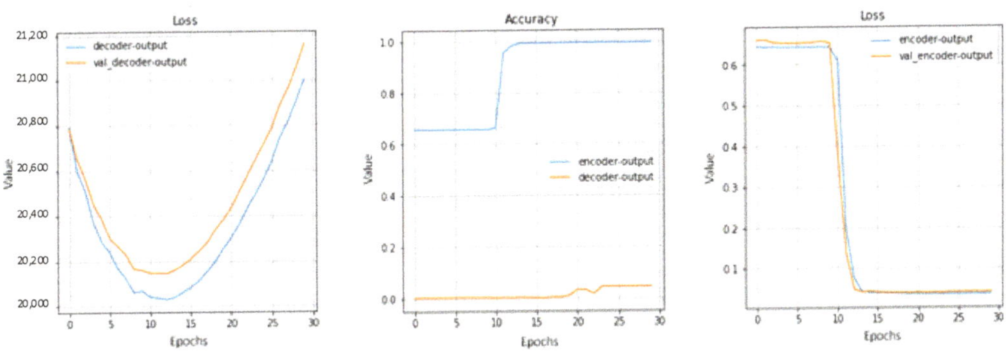

Figure 23. Decoder and classifier training on 30 epochs.

To empirically evaluate the generalizing ability of the neural network, a 10-fold stratified K-Folds cross-validator was used on unique data with the most uniform use of available data (Figure 24).

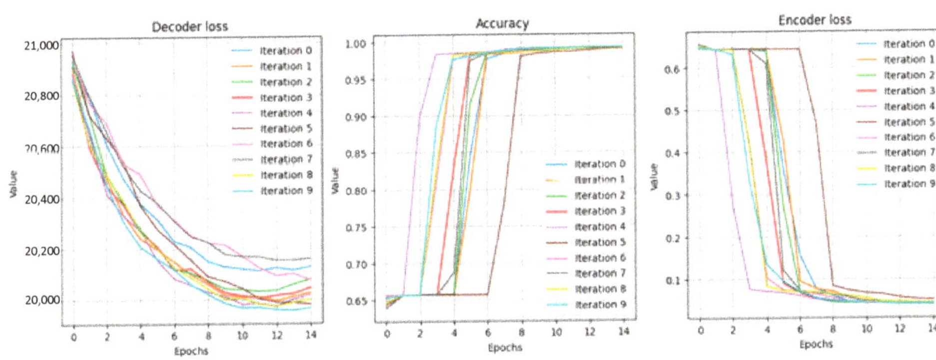

Figure 24. 10-fold cross-validation at 15 epochs.

After training the neural network, an experiment was conducted to assess the accuracy and completeness of the detection of known anomalies. First, a dataset with CAs of the same type was used as in the dataset when training the model. The anomaly detection result showed a value of 96.9%.

Then a new dataset was formed containing CAs previously unknown to the classifier ("0-day" attacks). The algorithm recognized 80% of previously unknown attacks. At the same time, it determined that 99% of legitimate requests are not anomalous.

It has been observed that the system allows false positives. In particular, two requests were dropped by the neural network. Examples of such false positives in the system are shown in Figure 25.

```
In [118]: data_last[(data_last['pred']==True) & (data_last['anomaly']==False)]
Out[118]:
                                                     data  anomaly  pred
 177   <START>POST /post/new/title-Quantum+communicat...         0  True
1118      <START>POST /post/new/title-BlueWillowPlate++t...      0  True

In [121]: data_last['data'][(data_last['pred']==True) & (data_last['anomaly']==False)].to_list()
Out[121]: ['<START>POST /post/new/title=Quantum+communications+at+ITMO+University:+building+100%+secure+data+transfer+systemst
ext=The+wise+Director,+in+whose+nature<STOP>',
 '<START>POST /post/new/title=BlueWillowPlate++text=In+his+self-delusion+and+ignorance<STOP>']
```

Figure 25. Examples of false positives in the system.

Given the fact that the dataset contained 57,000 queries, of which 20,000 were anomalous, the value of 2 is not a significant drawback of the proposed approach.

6. Discussion

Experiments have shown that SG network traffic has fractal properties. In other words, in large volumes this traffic has the property of self-similarity.

In addition, experiments have shown that the proposed proactive SG protection system upon detection of CAs based on the assessment of self-similarity of system functioning parameters using fractal indicators and predicting the fact of the impact of CAs by applying the proposed structure of the LSTM neural network has fairly high efficiency in detecting both known and unknown CAs. The probability of detecting known CAs is 0.96, and "0-day" attacks is 0.8.

A comparative evaluation of the proposed approach was carried out with intrusion detection systems (IDS) and intrusion prevention systems (IPS), which were based on signature [50], statistical [51], and machine learning methods [52,53].

The results of this assessment are shown in Table 4. It demonstrates the detection rate (in seconds) and detection accuracy of known and unknown CAs types. In addition, the table indicates what type of traffic the method is suitable for.

Table 4. Comparative analysis of methods for detecting cyber attacks.

Method Name	Detection Rate (s)	Detection Accuracy		Traffic Type	
		Known Attacks	Unknown Attacks	Stationary	Non-Stationary
Signature methods [50]	5	0.99	0.5	+	-
Statistical methods [51]	30	0.92	0.6	+	-
Machine learning methods [52,53]	28	0.72–0.97	0.8	+	-
Proposed method	**5**	**0.96**	**0.8**	+	+

Table 4 shows that signature methods and the proposed method are the fastest in terms of detection rate. Also, because signature methods use predefined rules, they have the highest accuracy in detecting known attacks. However, their accuracy in detecting unknown attacks is very low. A value of 0.5 indicates that this accuracy corresponds to the law of equiprobability.

Statistical methods lose out to signature methods in terms of detection rate and accuracy, since they use accumulated statistics. However, sometimes they are able to detect unknown attacks.

Machine learning methods are quite diverse and well-developed. Their effectiveness depends on the classification and clustering models they use. In the works [52,53] considered in Table 4, the SVM, Gaussian Naive Bayes, and Decision Tree models were used. In these methods, it is necessary to train models on control samples. Therefore, machine learning methods lose signature methods in detection rate. However, they have higher accuracy in detecting unknown attacks.

The proposed method has a detection rate, similar to the signature methods, and the accuracy corresponds to the values, similar to the machine learning methods. At the same time, it retains its effectiveness when working with non-stationary traffic, which is most typical for SG traffic. The remaining methods work well only in the case of stationary traffic.

It should be noted that, at present, for the continuous controlling of the transfer of technological and other information in SG, the systems built on distributed ledger and blockchain technologies, based on smart contracts, are actively used [54,55]. The use of such solutions makes it possible to protect the information transmitted in SG from CAs aimed at violating its confidentiality and integrity. However, these technologies do not provide early detection of CAs, their classification, and protection of SG network devices from CAs, the implementation of which is aimed at learning the SG structure and the subsequent violation of the performance of the network and its elements.

The proposed approach to proactive protection of SG from CAs can be implemented in many existing IDS and IPS, whose main task is to analyze internal data streams, searching in them for bit sequences that may represent malicious actions or events, as well as monitoring system logs. It increases the probability of detecting unknown CAs by using the autoencoder and LSTM networks, reducing the probability of false positives and the time and the amount of RAM involved in analyzing the network traffic. Thus, the disadvantages of existing IDS and IPS, based on rules, as well as signature and anomaly technologies are leveled.

It should be noted that the conducted studies only demonstrate the effectiveness of the proposed proactive system for predicting and detecting CAs in the SG network. It can be considered as an attack detection system that combines the advantages inherent in signature, statistical, and machine learning methods, and is devoid of their inherent disadvantages. At the same time, it expands the scope of attack detection methods by extending them to a non-stationary type of traffic.

7. Conclusions

The article discusses a new approach to the operation of a system for protecting smart power grids from CAs on service data transfer protocols, based on the detection of anomalies in network traffic by evaluating its self-similarity property, detecting cyber attacks in anomalies in real or near real-time, their classification and acceptance effective protection measures using LSTM and GRU cells. Fractal analysis, mathematical statistics, and neural networks with long short-term memory were used as tools in the development of this protection system.

The proposed system is based on the application of the main provisions of the theory of fractals and the use of self-similarity assessment methods proposed by this theory, such as the Dickey–Fuller test, R/S analysis, and the DFA method. When testing fractal methods that make it possible to study long-term dependencies in network traffic, the DFA method is more efficient than R/S analysis due to its ability to process not only stationary, but also non-stationary series with high accuracy. Its joint application with LSTM networks can significantly increase the probability of detecting CAs.

The experimental evaluation of the proposed approach showed that, compared with many other approaches, one of the main advantages of fractal analysis is its speed, as well as the ability to detect anomalies in traffic of any kind. Only an increase in the number of processed data transfer protocol header parameters (packet length, flags, and others) leads to an increase in the calculation time. At the same time, the proposed system demonstrated a fairly high probability of detecting CAs, reaching a value of 0.96 for known attacks and 0.8 for previously unknown attacks.

The specificity of the proposed system is that the detection of CAs is performed using an autoencoder trained on the basis of the reference data of the SG operation and the information exchange in it, taking into account all deviations from the regular operation of the SG. During operation, the autoencoder is additionally trained by a reasonable neural network, i.e., the result is a generative adversarial network in which neural networks learn from each other.

Further studies are associated with the integration of the proposed system with other known protection systems, as well as with the attack detection methods available in the arsenal of computer security systems.

Author Contributions: I.K. was responsible for conceptualization and methodology; I.S. and O.L. analyzed the data; A.K. conceived and designed the experiment; all authors wrote the article. All authors have read and agreed to the published version of the manuscript.

Funding: This research is being supported by the grant of RSF #21-71-20078 in SPC RAS.

Institutional Review Board Statement: Not applicable.

Informed Consent Statement: Not applicable.

Data Availability Statement: Not applicable.

Conflicts of Interest: The authors declare no conflict of interest.

References

1. Kaur, S.; Goel, R. A Review on Data Transmission Techniques for Energy Efficiency in Wireless Sensor Networks. In Proceedings of the 2016 International Conference on Wireless Communications, Signal Processing and Networking (WiSPNET), Chennai, India, 23–25 March 2016; pp. 699–703. [CrossRef]
2. Vyshnavi, S.B.; Sree, S.R.; Jayapandian, N. Network Security Tools and Applications in Research Perspective. In Proceedings of the 2019 Third International conference on I-SMAC (IoT in Social, Mobile, Analytics and Cloud) (I-SMAC), Palladam, India, 12–14 December 2019; pp. 655–659. [CrossRef]
3. Mellia, M.; Zincir-Heywood, N.; Diao, Y. Overview of Network and Service Management. In *Communication Networks and Service Management in the Era of Artificial Intelligence and Machine Learning*; IEEE: Piscataway, NJ, USA, 2021; pp. 1–18. [CrossRef]
4. Belej, O.; Nestor, N.; Polotai, O.; Sadeckii, J. Features of Application of Data Transmission Protocols in Wireless Networks of Sensors. In Proceedings of the 2019 3rd International Conference on Advanced Information and Communications Technologies (AICT), Lviv, Ukraine, 2–6 July 2019; pp. 317–322. [CrossRef]
5. Uçtu, G.; Alkan, M.; Doğru, İ.A.; Dörterler, M. Perimeter Network Security Solutions: A Survey. In Proceedings of the 2019 3rd International Symposium on Multidisciplinary Studies and Innovative Technologies (ISMSIT), Ankara, Turkey, 11–13 October 2019; pp. 1–6. [CrossRef]
6. Kotenko, I.; Saenko, I.; Lauta, O.; Kribel, A. An Approach to Detecting Cyber Attacks against Smart Power Grids Based on the Analysis of Network Traffic Self-Similarity. *Energies* **2020**, *13*, 5031. [CrossRef]
7. Ageev, S.; Kotenko, I.; Saenko, I.; Kopchak, Y. Abnormal Traffic Detection in Networks of the Internet of Things Based on Fuzzy Logical Inference. In Proceedings of the IEEE International Conference on Soft Computing and Measurements (SCM), St. Petersburg, Russia, 19–21 May 2015; pp. 5–8. [CrossRef]
8. Desnitsky, V.A.; Kotenko, I.V.; Nogin, S.B. Detection of Anomalies in Data for Monitoring of Security Components in the Internet of Things. In Proceedings of the IEEE International Conference on Soft Computing and Measurements (SCM), St. Petersburg, Russia, 19–21 May 2015; pp. 189–192. [CrossRef]
9. Brezigar-Masten, A.; Masten, I. CART-based selection of bankruptcy predictors for the logit model. *Expert Syst. Appl.* **2012**, *39*, 10153–10159. [CrossRef]
10. Ju, X.; Chen, V.C.P.; Rosenberger, J.M.; Liu, F. Fast knot optimization for multivariate adaptive regression splines using hill climbing methods. *Expert Syst. Appl.* **2021**, *171*, 114565. [CrossRef]
11. Ju, X.; Rosenberger, J.M.; Chen, V.C.P.; Liu, F. Global optimization on non-convex two-way interaction truncated linear multivariate adaptive regression splines using mixed integer quadratic programming. *Inf. Sci.* **2022**, *597*, 38–52. [CrossRef]

12. Ju, X.; Liu, F.; Wang, L.; Lee, W.-J. Wind farm layout optimization based on support vector regression guided genetic algorithm with consideration of participation among landowners. *Energy Convers. Manag.* **2019**, *196*, 1267–1281. [CrossRef]
13. Kotenko, I.; Saenko, I.; Lauta, O.; Karpov, M. Methodology for Management of the Protection System of Smart Power Supply Networks in the Context of Cyberattacks. *Energies* **2021**, *14*, 5963. [CrossRef]
14. Kotenko, I.; Saenko, I.; Lauta, O.; Kribel, A. Ensuring the survivability of embedded computer networks based on early detection of cyber attacks by integrating fractal analysis and statistical methods. *Microprocess. Microsyst.* **2022**, *90*, 104459. [CrossRef]
15. Leland, W.E.; Taqqu, M.S.; Willinger, W.; Wilson, D.V. On the self-similar nature of Ethernet traffic. *SIGCOMM Comput. Commun.* **1993**, *23*, 183–193. [CrossRef]
16. Raimundo, M.S.; Okamoto, J., Jr. Application of Hurst Exponent (H) and the R/S Analysis in the Classification of FOREX Securities. *Int. J. Model. Optim.* **2018**, *8*, 116–124. [CrossRef]
17. Dang, T.D.; Sonkoly, B.; Molnar, S. Fractal analysis and modeling of VoIP traffic. In Proceedings of the 11th International Telecommunications Network Strategy and Planning Symposium (NETWORKS 2004), Vienna, Austria, 13–16 June 2004; IEEE: Vienna, Austria, 2004; pp. 123–130. [CrossRef]
18. Sánchez-Granero, M.J.; Fernández-Martínez, M.; Trinidad-Segovia, J.E. Introducing fractal dimension algorithms to calculate the Hurst exponent of financial time series. *Eur. Phys. J. B* **2012**, *85*, 1–13. [CrossRef]
19. Grillo, D.; Lewis, A.; Pandya, R. Personal Communication Services and Teletraffic Standardization in ITU-T. In *The Fundamental Role of Teletraffic in the Evolution of Telecommunications Networks, Proceedings of the 14th International Teletraffic Congress—ITC 14, Antibes Juan-les-Pins, France, 6-10 June 1994*; Labetoulle, J., Roberts, J.W., Eds.; Elsevier: Amsterdam, The Netherlands, 1994; pp. 1–12. [CrossRef]
20. Strelkovskaya, I.; Solovskaya, I.; Makoganiuk, A. Spline-Extrapolation Method in Traffic Forecasting in 5G Networks. *J. Telecommun. Inf. Technol.* **2019**, *3*, 8–16. [CrossRef]
21. Ju, F.; Yang, J.; Liu, H. Analysis of Self-Similar Traffic Based on the On/Off Model. In Proceedings of the 2009 International Workshop on Chaos-Fractals Theories and Applications, Shenyang, China, 6–8 November 2009; pp. 301–304. [CrossRef]
22. Fractal Objects and Self-Similar Processes. Available online: https://archive.physionet.org/tutorials/fmnc/node3.html (accessed on 15 January 2022). [CrossRef]
23. Ruoyu, Y.; Wang, Y. Hurst Parameter for Security Evaluation of LAN Traffic. *Inf. Technol. J.* **2012**, *11*, 269–275. [CrossRef]
24. Ably, P.; Flandrin, P.; Taqqu, M.S.; Veitch, D. Self-Similarity and long-range dependence through the wavelet lens. In *Theory and Applications of Long Range Dependence*; Birkhauser Press: Boston, MA, USA, 2002; pp. 345–379.
25. Canadian Electricity Association. *Canadian Smart Grid Framework*; Canadian Electricity Association: Calgary, AB, Canada, 2010.
26. Federal Office for Information Security. *Protection Profile for the Gateway of a Smart Metering System*; V.1.2; Federal Office for Information Security: Bonn, Germany, 2014.
27. European Network and Information Security Agency (ENISA). *Smart Grid Security: Recommendations for Europe and Member States*; ENISA: Athens, Greece, 2015.
28. *ISO/IEC 27005*; Information Technology—Security Techniques—Information Security Risk Management. ISO: Geneva, Switzerland, 2008.
29. *ISO/IEC TR 27019:2013*; Information Security Management Guidelines based on ISO/IEC 27002 for Process Control Systems Specific to the Energy Utility Industry. ISO: Geneva Switzerland, 2013.
30. Kendrick, D.; Groom, L.; Stewart, J.; Watson, M.; Mulvaney, C.; Casterton, R. "Risk Watch": Cluster randomised controlled trial evaluating an injury prevention program. *Inj. Prev.* **2007**, *13*, 93–99. [CrossRef]
31. Fang, X.; Misra, S.; Xue, G.; Yang, D. Managing smart grid information in the cloud: Opportunities, model, and applications. *IEEE Netw.* **2012**, *26*, 32–38. [CrossRef]
32. Prasad, I. Smart Grid Technology: Application and Control. *Int. J. Adv. Res. Electr. Electron. Instrum. Eng.* **2014**, *3*, 9533–9542.
33. Müller, K.J. Verordnete Sicherheit—Das Schutzprofil für das Smart Metering Gateway. *Datenschutz Datensicherheit* **2014**, *35*, 547–551. [CrossRef]
34. Protection Profile for the Security Module of a Smart Metering System (Security Module PP). Available online: http://www.commoncriteriaportal.org/files/ppfiles/pp0077b_pdf.pdf (accessed on 15 January 2022).
35. Anwar, A.; Mahmood, A. Cyber Security of Smart Grid Infrastructure. In *The State of the Art in Intrusion Prevention and Detection*; CRC Press: Boca Raton, FL, USA, 2014; pp. 139–154. [CrossRef]
36. Bale, J.P.M.; Sediyono, E.; Marwata, M. Risk management in information technology using facilitated risk analysis process (FRAP) (case study: Academic information systems of Satya Wacana Christian University). *J. Theor. Appl. Inf. Technol.* **2014**, *68*, 339–351.
37. Nurul, A.H.; Zaheera, Z.A.; Puvanasvaran, A.P.; Zakaria, N.A.; Ahmad, R. Risk assessment method for insider threats in cyber security: A review. *Int. J. Adv. Comput. Sci. Appl.* **2018**, *9*, 16–19. [CrossRef]
38. Tankard, C. Advanced persistent threats and how to monitor and deter them. *Netw. Secur.* **2011**, *2011*, 16–19. [CrossRef]
39. Lekidis, A. Cyber-Security Measures for Protecting EPES Systems in the 5G Area. In Proceedings of the 17th International Conference on Availability, Reliability and Security (ARES '22), Vienna, Austria, 23–26 August 2022. [CrossRef]
40. Bella, H.K.; Vasundra, S. A study of Security Threats and Attacks in Cloud Computing. In Proceedings of the 2022 4th International Conference on Smart Systems and Inventive Technology (ICSSIT), Tirunelveli, India, 20–22 January 2022; pp. 658–666. [CrossRef]
41. Sterbenz, J.P.G.; Hutchison, D.; Çetinkaya, E.K.; Jabbar, A.; Rohrer, J.P.; Schöller, M.; Smith, P. Resilience and survivability in communication networks: Strategies, principles, and survey of disciplines. *Comput. Netw.* **2010**, *54*, 1245–1265. [CrossRef]

42. El Fray, I. A Comparative Study of Risk Assessment Methods, MEHARI & CRAMM with a New Formal Model of Risk Assessment (FoMRA) in Information Systems. In *Computer Information Systems and Industrial Management. CISIM 2012. Lecture Notes in Computer Science*; Cortesi, A., Chaki, N., Saeed, K., Wierzchoń, S., Eds.; Springer: Berlin/Heidelberg, Germany, 2012; Volume 7564, pp. 428–442. [CrossRef]
43. Syalim, A.; Hori, Y.; Sakurai, K. Comparison of Risk Analysis Methods: Mehari, Magerit, NIST800-30 and Microsoft's Security Management Guide. In Proceedings of the 2009 International Conference on Availability, Reliability and Security, Fukuoka, Japan, 16-19 March 2009; IEEE: New York, USA, 2009; pp. 726–731. [CrossRef]
44. MEHARI. Overview. Available online: http://meharipedia.x10host.com/wp/wp-content/uploads/2019/05/MEHARI-Overview-2019.pdf (accessed on 15 January 2022).
45. Microsoft Security Center of Excellence. Available online: http://www.microsoft.com/rus/technet/security (accessed on 15 January 2022).
46. Downey, A. Lognormal and Pareto distributions in the Internet. *Comput. Commun.* **2005**, *28*, 790–801. [CrossRef]
47. Norros, I. A Storage Model with Self-Similar Input. *Queueing Syst.* **1994**, *16*, 387–396.
48. Kotenko, I.; Saenko, I.; Kribel, A.; Lauta, O. A technique for early detection of cyberattacks using the traffic self-similarity property and a statistical approach. In Proceedings of the 29th Euromicro International Conference on Parallel, Distributed and Network-Based Processing (PDP), Valladolid, Spain, 10–12 March 2021; pp. 281–284. [CrossRef]
49. Kotenko, I.; Lauta, O.; Kribel, K.; Saenko, I. LSTM neural networks for detecting anomalies caused by web application cyber attacks. *Front. Artif. Intell. Appl.* **2021**, *337*, 127–140. [CrossRef]
50. Visoottiviseth, V.; Sakarin, P.; Thongwilai, J.; Choobanjong, T. Signature-based and behavior-based attack detection with machine learning for home IoT devices. In Proceedings of the 2020 IEEE Region 10 Conference (TENCON), Osaka, Japan, 16–19 November 2020; IEEE: New York, USA, 2020; pp. 829–834. [CrossRef]
51. Amma, N.G.B.; Selvakumar, S.; Velusamy, R.L. A Statistical Approach for Detection of Denial of Service Attacks in Computer Networks. *IEEE Trans. Netw. Serv. Manag.* **2020**, *17*, 2511–2522. [CrossRef]
52. Zhe, W.; Wei, C.; Chunlin, L. DoS attack detection model of smart grid based on machine learning method. In Proceedings of the 2020 IEEE International Conference on Power, Intelligent Computing and Systems (ICPICS), Shenyang, China, 28–30 July 2020; IEEE: New York, USA, 2020; pp. 735–738. [CrossRef]
53. Shaukat, S.; Ali, A.; Batool, A.; Alqahtani, F.; Khan, J.S.; Arshad; Ahmad, J. Intrusion Detection and Attack Classification Leveraging Machine Learning Technique. In Proceedings of the 2020 14th International Conference on Innovations in Information Technology (IIT), Al Ain, United Arab Emirated, 17–18 November 2020; IEEE: New York, USA, 2020; pp. 198–202. [CrossRef]
54. Górski, T. Reconfigurable Smart Contracts for Renewable Energy Exchange with Re-Use of Verification Rules. *Appl. Sci.* **2022**, *12*, 5339. [CrossRef]
55. Górski, T. Continuous Delivery of Blockchain Distributed Applications. *Sensors* **2022**, *22*, 128. [CrossRef]

sensors MDPI

Article

Electricity Theft Detection in Smart Grids Using a Hybrid BiGRU–BiLSTM Model with Feature Engineering-Based Preprocessing

Shoaib Munawar [1], Nadeem Javaid [2], Zeshan Aslam Khan [1], Naveed Ishtiaq Chaudhary [3,*], Muhammad Asif Zahoor Raja [3], Ahmad H. Milyani [4] and Abdullah Ahmed Azhari [5]

1 Department of Electrical and Computer Engineering, International Islamic University, Islamabad 44000, Pakistan
2 Department of Computer Science, COMSATS University Islamabad, Islamabad 44000, Pakistan
3 Future Technology Research Center, National Yunlin University of Science and Technology, 123 University Road, Section 3, Douliou, Yunlin 64002, Taiwan
4 Department of Electrical and Computer Engineering, King Abdulaziz University, Jeddah 21589, Saudi Arabia
5 The Applied College, King Abdulaziz University, Jeddah 21589, Saudi Arabia
* Correspondence: chaudni@yuntech.edu.tw

Abstract: In this paper, a defused decision boundary which renders misclassification issues due to the presence of cross-pairs is investigated. Cross-pairs retain cumulative attributes of both classes and misguide the classifier due to the defused data samples' nature. To tackle the problem of the defused data, a Tomek Links technique targets the cross-pair majority class and is removed, which results in an affine-segregated decision boundary. In order to cope with a Theft Case scenario, theft data is ascertained and synthesized randomly by using six theft data variants. Theft data variants are benign class appertaining data samples which are modified and manipulated to synthesize malicious samples. Furthermore, a K-means minority oversampling technique is used to tackle the class imbalance issue. In addition, to enhance the detection of the classifier, abstract features are engineered using a stochastic feature engineering mechanism. Moreover, to carry out affine training of the model, balanced data are inputted in order to mitigate class imbalance issues. An integrated hybrid model consisting of Bi-Directional Gated Recurrent Units and Bi-Directional Long-Term Short-Term Memory classifies the consumers, efficiently. Afterwards, robustness performance of the model is verified using an attack vector which is subjected to intervene in the model's efficiency and integrity. However, the proposed model performs efficiently on such unseen attack vectors.

Keywords: electricity theft detection; smart grids; robustness; smart meters; Tomek links

Citation: Munawar, S.; Javaid, N.; Khan, Z.A.; Chaudhary, N.I.; Raja, M.A.Z.; Milyani, A.H.; Ahmed Azhari, A. Electricity Theft Detection in Smart Grids Using a Hybrid BiGRU–BiLSTM Model with Feature Engineering-Based Preprocessing. *Sensors* **2022**, *22*, 7818. https://doi.org/10.3390/s22207818

Academic Editor: Arshad Arshad

Received: 14 September 2022
Accepted: 9 October 2022
Published: 14 October 2022

1. Introduction

Power generation, transmission and distribution collectively build a power system infrastructure. The power generation phase generates electricity at a high voltage level. The generated electricity is supplied to the end user through transmission lines. The end user is the consumer who consumes the supplied electricity via distribution network [1]. Smart Meters (SMs) are installed on the end users' side by Utility Providers (UPs) in order to monitor the consumed energy [2]. There are two types of losses, Technical Losses (TLs) and Non-Technical Losses (NTLs) [3]. TLs are the network-associated losses, which are confined to the design and material of the infrastructure, while NTLs are the losses which occur due to the interruption of the end consumers to obtain financial benefits by under-reporting the consumed energy. The interruption of the end consumer is basically a malicious activity, which is adopted by the fraudulent consumers. The connected fraudulent consumers tend to tamper the net metering of their consumed energy by adopting various data tampering techniques, such as meter tampering using shunt devices, double tapping of the lines and electronic faults [4]. The effects of such malicious activities over-burden the UPs with huge financial losses, which disrupt the smooth energy flow and demand curve. For instance, the study conducted in [5] reports that the monitored losses have been increased from 11 percent to 16 percent during the last two decades (1980–2000). The increased losses clearly highlight that revenue losses due to NTLs are a conspicuous issue and need special attention. NTLs vary from country to country. The literature in [6] reports that about 20% of the total revenue loss in Indian electricity network is due to the aforementioned malicious activities. Similarly, the United States is also facing a revenue loss of USD 6 billion annually [7,8]. Worldwide, revenue losses of about USD 96 billion are reported due to such malicious activities [9].

In order to investigate the aforementioned problems, the literature suggests various counter measure approaches to reduce such losses. The suggested approaches are advance metering infrastructure (AMI) and Neighborhood Area Network (NAN) [10], which are hardware-based approaches. In AMI, a sequential data is a target parameter, which is analyzed to extract suspicious behavior in order to find out maliciousness. Furthermore, consideration of sequential and non-sequential information enhances the detection of malicious behavior. Sequential data are Time-Series Data of the consumers, whereas non-sequential data are an auxiliary data that contain attributes of geographical, demographical and topographical data. Moreover, NAN and morphological patterning assessment focuses on multiuser network-based detection. A NAN is a multiple consumer network where a master meter is deployed to monitor the total consumed energy. A master meter is connected to a distribution low-voltage side of the transformer, which works as an observer meter to monitor the cluster of the connected SMs. TLs of the distribution lines are numerically adjusted as a beta σ, which is added to the total network's consumption. The data relevancy of the network is observed in order to investigate the maliciousness. Total consumption in addition with the σ factor is related to the observer meter's reading. Furthermore, in morphological patterning analysis, a historic and forecasted data competency is measured, which is correlated based on the error factor. A threshold is set as a monitoring parameter which analyzes the parity check of each of the consumptions and reports malicious activity.

Based on the above analysis, the motivation is to propose a data-oriented approach to detect NTLs. The problem of imbalanced data, defused decision boundary and extraction of abstract features are the main factors to target through data-oriented-based analysis of the Time-Series Data.

2. List of Contributions

The contributions are as follows:

- To tackle the imbalance data issue, theft class data are synthesized using six theft variants. Later on, the synthesized data are oversampled using a K-means synthetic minority oversampling technique (SMOTE).
- A Tomek links technique is used to eliminate cross-pairs across the decision boundary.
- To overcome the data leakage problem, a simple stratified approach is opted for.
- Cumulative and distinct features are engineered using stochastic feature engineering, which enables the model to learn data characterization and uniqueness.
- An integrated hybrid model of Bi-Directional Gated Recurrent Units (Bi-GRU) and bi-directional long-term short-term memory (Bi-LSTM) is used to tackle misclassification and high FPR issues.
- Furthermore, to verify the robustness of the proposed model, an unseen variant of the theft data with temperate randomness is analyzed to acknowledge the stability and integrity.

3. Literature Review

This section overviews Electricity Theft Detection (ETD)-related proposed research activities of various authors in smart metering applications.

3.1. Considering Sequential Data

A major portion of NTLs is due to fraudulent behavior of the consumers' accomplishing an effort to bypass the Utility Provider (UP) surveillance and to under-report the consumed energy. A solution proposed in [11] adopts a data-driven approach which uses a Machine Learning technique, Ensemble Bagged Tree (EBT) algorithm by stacking many Decision Trees to detect NTLs. As time complexity and memory consumption due to large computational complexity have remained formal constrains for Machine Learning (ML) algorithms. To improve both, searching and Weighted Feature Importance (WFI) techniques are deployed to enhance theft detection schematics. A Gradient Boosting Classifier (GBCs)-based detector is used to detect anomalies by considering intentional remedies while non-fraudulent anomaly intervention is ignored. Furthermore, the Gradient Boosting Theft Detector (GBTD) for the classification purposes is pursued by a preprocessing module using WFI. WFI uses stochastic features such as mean, min, max and Standard Deviation in collaboration with the consumption pattern extracted features, which improves performance and reduces time complexity [12]. The author pinpoints the Detection Rate and FPR only, however, a clustering mechanism is required to be considered in order to identify the misclassification due to a sudden drop in the consumption, which is ultimately started before the period of analysis. During training of the model, a problem of data leakage occurs which is not tackled properly. In [13], a maximal overlapped discrete wavelet packet transform is used to extract the abstract features from the dense time-series electricity consumption data, whereas, to tackle the data balancing issue, a random under-sampling boosting (RUSBoost) algorithm is proposed, which eliminates vital information of the data while re-sampling the data samples. Similarly, [14] uses SMOTE for data balancing. The balanced data are then preprocessed using a min–max scalar normalization method to refine the input raw data. A pool of various algorithms is used containing AdaBoost, CatBoost, XGBoost, LGBoost, RF [15] and extra trees to find FPR and Detection Rate, however, SMOTE over-samples the minority class, with confused pairs having trace contents of both classes. The generalization performance of single hidden-layer feed-forward neural networks (SLFN) due to over-training leads to degradation when the back-propagation algorithm performs. To overcome such issues, a hybrid Convolutional Neural Network and Fandom Forest (CNN–RF) is proposed, where the CNN is designed to learn features between different hours of the day [15]. Obtained features are taken as an input by Random Forest (RF) to segregate thieves from honest customers. However, memory elapsing is a serious issue to monitor consumption patterns for long periods of time. The RF module takes a lot of memory, causing over-fitting issues. Significantly, a fast operation is an optimum choice, whereas operating maxpooling is a slower operation and causes greater time of execution. Furthermore, due to the non-availability of real-world theft scenarios, data analyzing classification based only on linear Theft Cases is not a significant investigation scenario. Similarly, a hybrid module integrating Convolutional Neural Network and long-term short-term memory (CNN–LSTM) has been developed [4]. CNNs have the capability of self-learning, whereas LSTM performs better on sequential data, however, memory elapse is still a question for such scenarios. A Semi-Supervised Auto-Encoder (SSEA) is used to learn the advanced features [16]. The input of multiple Time-Series Data is organized as a 1D vector in multiple channels. Moreover, to improve a linear separability of the samples, a distributed stochastic neighbor embedding (t-SNE) is used to localize each data point. Adding a high dimensionality though class separation is a pre-requisite for such a scenario, which is not simply tackled by t-SNE to add dimensionality for the class separation. Data leakage during training of the model and the consideration of non-malicious factors are important aspects, however, [17] pays no attention to these issues. Furthermore, the authors in [18,19] adopt a data-driven approach using a Machine Learning technique, XGBoost, without considering any auxiliary information. The study in [20,21] investigates the impact of imbalanced data. The imbalanced data are balanced through synthesized data. The data reductionality is carried out through Principle Component Analysis (PCA) and hyper parameters are tuned through Bayesian optimizer. An AUC score of 97% is reported using a feed-forward network. The study in [22] uses a hybrid model of graph convolutional network and EU Convolutional Neural Network. CNN is used to capture the latest features. The study in [23] targets the AMI infrastructure to investigate malicious consumers. The benign data are manipulated through cyber attacks. A deep neural network CNNGRU hybrid model is developed to correlate the malicious and benign samples.

3.2. Monitoring Morphological Patterning

An LSTM model is used by [24,25] to investigate pattern morphology. The pattern authentication is investigated by mapping them together. A prediction error is calculated between the real and predicted consumption, which decides the authenticity of the consumed pattern. However, due to excessive computational complexities, LSTM is not a suitable option. The authors in [26] propose a Stacked Sparse Denoising Auto-Encoder (SSDAE), which monitors the reconstruction error of the corresponding consumption pattern based on the extracted features. The extracted key features from the raw samples are provided as an input. A comparative correlation is observed between the samples provided as an input and reconstructed patterns. The similarity index is observed through an Optimized Estimated Threshold (OET). OET decides the sample's class based on the measured value of reconstruction error (RE). However, based on non-sequential attributes, consideration of exogenous variables affects the morphology of consumers' patterns [27]. In addition to short-term vacations, demographical, geographical, SM firmware and EM distort the pattern's morphology, which is beyond the scope of detection, using SSDAE's estimated threshold as a segregating boundary for the classes. Furthermore, the tampering of consumption patterns before installation of SM on customers' premises remains undetected. The tampered pattern reconstruction significantly deceives the SSDAE detector, which causes misclassification. In [28], NTLs are categorically divided based on the time period, including consumers cheating during ON-Peak hours, OFF-Peak hours and malicious customers cheating constantly. The detection model becomes unstable when inconsistent attacks are injected. To monitor such inconsistent variations, categorical variables are incorporated in linear regression to develop a categorical variable linear regression detector. In [29], an Anomaly Pattern Detection Hypothesis Testing (APD-HT) investigates theft activities. A reference and a detection window are used to analyze the data streaming of SMs. The data streaming analysis is based on binomial data distribution. However, variations due to the intervention of non-malicious factors are beyond detection.

3.3. Tampering with Smart Meter Readings

In addition to the data-oriented approaches [30–32], another novel Distributed Generation (DG)-based approach of energy monitoring is proposed. A renewable DG unit consists of Photo-Voltaic (PV) modules, which are installed on consumers' premises. Consumers generate energy according to their needs and sell back the excessive amount of energy to the UPs. A two-metering system is adopted, namely, net metering system and Feed-in Tariffs (FITs) policy. Net metering systems monitor consumed energy provided by the UP, while FITs policy monitors the excessive energy generated by a DG for selling purposes. Manipulating and tampering with injected (sold) readings of DG by malicious customers tends to falsely report over-charging. The work in [33] proposed a solution by deploying Supervisory Control and Data Acquisition (SCADA) metering points to monitor various electrical parameters.

3.4. Investigating Neighborhood Area Networks

Hardware-based infrastructure utilizes network-based topology to enhance detection performance. The authors pinpoint the limitations of misclassification due to manipulation of non-malicious factors and deceiving a detection detector to accept the malicious pattern as a normal one [34]. The authors suggest to deploy an SM on the transformer's side, so that a balancing load flow scenario is overlooked, scrutinizing the discrepancies being caused by the non-malicious factors and smart attackers. A Neighborhood Area Network (NAN) proposes a master meter (MM) approach, which is installed on the distribution transformer side and monitors total supplied energy to the NAN [35]. The total supplied energy is compared with the sum of total individuals' SM readings within the corresponding NAN, where TLs are accommodated by addition of a constant parameter. The inequality within the readings indicates a theft occurrence, while equality in the NAN means a complete benign consumption. A Correlation Analysis for Pinpointing Electricity Theft (CAPET) scheme is introduced, which measures the correlation between total utilized energy in the NAN at the low voltage level side. Inequality and deviation shows malicious activity. However, change in TLs is subjected to environmental conditions; a seasonal change abruptly affects the balanced correlation between MM and SM readings. Inequality in reading of the dispatched side and consumer premises indicates suspicious activity, which is beyond consideration. Similarly, in [36], the author develops an ensemble technique by combining the suspicious ranks obtained from the Maximum Information Coefficient (MIC) and clustering technique. The arithmetic and geometric means of these two ranks are combined using a famous rank product method which decides whether a sample is benign or malicious. The decision is based on the rank's intensity. A high intensity indicates malicious activity. The MIC and clustering technique analyzes the correlation of NTLs and the observer meter, respectively. In order to identify unusual shapes, a degree of abnormality is calculated by clustering technique [37]. However, such correlations are void of consideration for variable TLs and non-sequential auxiliary data aspects.

4. Proposed System Model

Figure 1 shows the proposed system model, while limitations, along with their proposed solutions, are mapped in Table 1.

The system model comprises the data preprocessing module, data augmentation module and classification module. These modules are subdivided into 7 main steps.

- Step (1) is a data preprocessing step, where missing values are filled using a mean-based strategy and outliers are removed. Filling and removing such values is a necessary step of the data preprocessing, as noisy and ambiguous data affect accuracy and degrade the misclassification scenario. A simple imputer is implemented to fill such values.
- In step (2), the preprocessed data are augmented where benign samples are modified and manipulated due to their rare existence. The problems of skewness and bias are observed if the model is trained on such imbalanced data. Therefore, it is a necessary step to balance the data before the training of the model.
- In step (3), benign class data are manipulated and theft class data are generated.
- In step (4), decision boundaries' associated cross-pairs are identified and eliminated. As cross-pair is a combination of the opposite class samples. Henceforth, a Tomek links technique is used. The majority class samples are removed, and minority class samples are retained in order to preserve the data integrity.
- In step (5), the data is stratified in order to inhibit the defusion of the data while splitting.
- In step (6), abstract features are engineered based on stochastic feature engineering.
- In step (7), Time-Series Data are inputted to a developed Bi-GRU [38] and Bi-LSTM [39]. A binary sigmoid function classifies the samples [40]. Bi-LSTM [41] is featured with the handling of high dimensional data, while Bi-GRU is used to avoid the computational complexity due to its fast operating features.

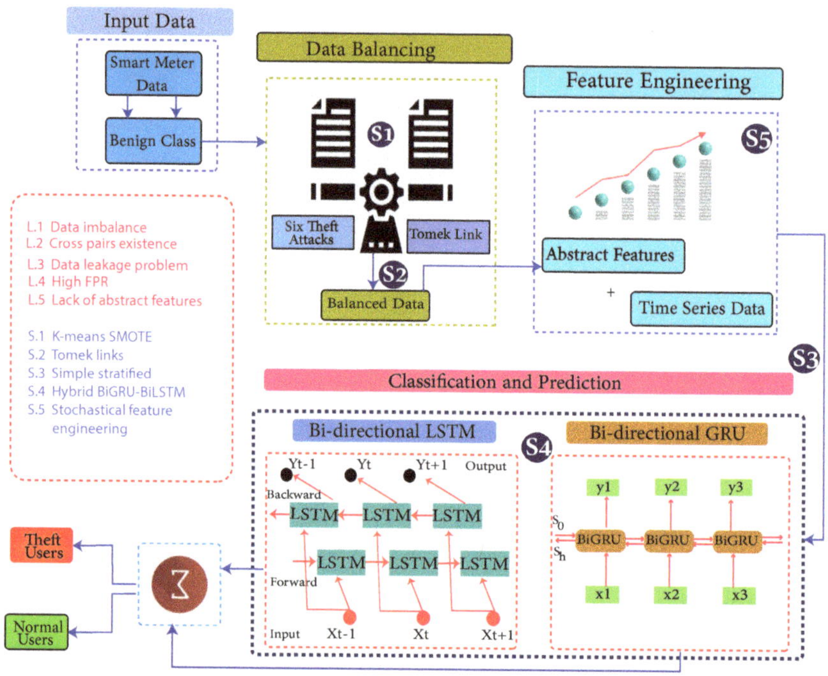

Figure 1. System Model Architecture.

Table 1. Mapping of Limitations and Proposed Solutions.

Limitation Number	Limitation Identified	Solution Number	Solution Proposed	Validations
L1	Data imbalance issue	S1	A K-means SMOTE technique is used to solve the data imbalance issue	V1: Performance comparison of the models
L2	Misclassification due to cross-pairs	S2	A Tomek links technique is used to identify the cross-pairs and remove them accordingly	V2: Table 3 Removal of cross-pairs
L3	Data leakage during training	S3	A simple stratified methodology is used to divide the data based on key attributes into subgroups for training of the model	V3: Equations (1)–(7)
L4	High FPR	S4	A hybrid model of Bi-GRU and Bi-LSTM is used to classify samples precisely and reduce high FPR	V4: Figure 6a,b AUC and PRC curve
L5	Lack of abstract features	S5	A stochastic feature engineering approach is opted to generate abstract features	V5: Table 5

This paper is an extension of [9]. Algorithm 1 presents the BiGRU–BiLSTM-based scheme for the detection of the anomalies in smart grids. It consists of seven steps. Initially, data are segregated based on distinct characterizations. Later on, six data manipulating techniques are appertained on the honest consumers' data, which are pursued by concatenation and data balancing techniques. Moreover, data are preprocessed and cross-pairs are removed. Furthermore, stratified sampling and feature engineering are accomplished.

Algorithm 1: Bi-GRU- and Bi-LSTM-based Detection Scheme.

Step 1:
Input: Benign Consumers B_C, Output: Fraudulent Consumers F_C
Step 2: Generating Theft Samples
$T1 = B_C * random(0.1, 0.9)$;
$T2 = B_C * x_t$ where $(x_t = random(0.1, 0.9))$;
$T3 = B_C * random[0, 1]$;
$T4 = mean\ (B_C) * random(0.1, 1.0)$;
$T5 = Mean(S)$ for each column;
$T6 = S(T) - t$ revesing a time sequence;
Step 3: concatenation
Concat (B_C + F_C);
Step 4: Balancing Data
$B_C = F_C$;
Step 5:
S_{ith} of majority class having smaller EU Distance with decision boundary is removed;
Step 6: Data Leakage
$p(s) = C_i + C_j$;
$C_i \subseteq p(s)$;
$C_j \subseteq p(s)$;
$S_{j1}, S_{j2}, S_{j3}, \ldots, S_{jn}\ \varepsilon\ C_j$;
$S_{i1}, S_{i2}, S_{i3}, \ldots, S_{in}\ \varepsilon\ C_i$;
$S_i \notin S_j$;
$C_i(S_{i1,\ldots,n}) \notin C_j(S_{j1,\ldots,n})$;
Step 7: Feature Engineering
$F1 = Mean$ of P_s against each row;
$F2 = Std$ of P_s against each row;
$F3 = Min \in C_i$ against each row;
$F4 = Max \in C_j$ against each row;
Output: Honest Consumers ε B_C , Fraudulent Consumers ε F_C.

4.1. Dataset

A realistic electricity consumption dataset, namely, the State Grid Corporation of China (SGCC), is used in this paper. It is administered during the 2014–2016 period and is supposed to be one of the most extensive datasets of SMs. It is structured as Time-Series Data, which are collected after every 24 h. Each consumer has a unique household ID. The consumption volume of each consumer is recorded against their household ID along with the date and time. It is a dataset of 1035 days and 42,372 consumers. We are using 1500 benign consumers' data of six months due to the limited resources of our machine. Machine specifications are Intel(R) core (TM) M-5y10c, CPU@ 0.80 GHz 1.00 GHz, RAM 4 GB. Moreover, The simulator is Google CoLab. The meta information of the SGCC dataset is shown in Table 2.

Generally, in a power system, the electricity consumption data of end users are collected through SMs. The collected data are acquired using various sensors of the SMs. A data communication network aggregates the data at a specific central location. However, certain complications such as the malfunctioning of the sensors, failure of the SMs, errors in data transmission and storage servers generate inherent erroneous and ambiguous data. Discarding such data shrinks the size of the dataset considerably, and thus authentic analysis of the data becomes onerous.

Table 2. Metadata Information of SGCC Dataset.

Description	Value
Administering years of the dataset	2014–2016
Total number of benign consumers	38,756
Total number of fraudulent consumers	3616

4.2. Data Leakage

The population is divided into mutually exclusive subgroups using stratified sampling. It is a homogeneous division and known as strata. The purpose of using stratified sampling is to clearly classify each strata of the samples' population. The SGCC dataset is divided into training and testing data. The training and testing samples are segregated into subgroups by opting stratified sampling in order to avoid misclassification due to extensive diversity in the data. Training and testing samples are confined to their specific operations only. Training samples are used to train the model, whereas testing samples are exploited to validate classification and prediction. In this way, data leakage of training into testing and vice versa is reduced, which results in a good generalization. The mathematical representation of the data leakage is as follows:

$$p(s) = C_i + C_j \tag{1}$$

$$C_i \subseteq p(s) \tag{2}$$

$$C_j \subseteq p(s) \tag{3}$$

$$S_{j1}, S_{j2}, S_{j3}, \ldots, S_{jn} \; \varepsilon \; C_j \tag{4}$$

$$S_{i1}, S_{i2}, S_{i3}, \ldots, S_{in} \; \varepsilon \; C_i \tag{5}$$

$$S_i \notin S_j \tag{6}$$

$$C_i(S_{i1,\ldots,n}) \notin C_j(S_{j1,\ldots,n}) \tag{7}$$

where p, s and C represent Population of the Samples, Number of Samples and samples' unique class, respectively, whereas i and j are the mutual binary classes.

4.3. Data Preprocessing

Data is preprocessed where raw data are transformed into affine usable data. As the consumption data are highly complex in nature and dimensionality, tackling such large data manually is an impractical task, which takes much time to execute. Such complex data results in high FPR and low accuracy. Missing values in raw data are filled by applying a simple imputer, where a mean-based strategy is applied for such ambiguous values.

4.4. Data Augmentation and Balancing

Due to the rare existence of the malicious samples, the benign class samples' are modified and manipulated to synthesize malicious class data, which are inputted to ML and Deep Learning (DL) models. Such random data distribution causes skewness and bias problems. To tackle such issues, over-sampling techniques are used. Under-sampling techniques discard the majority class, which disrupts the important information, while oversampling techniques synthesize the duplicate samples of the minority class, which are prone to over-fitting. In our scenario, the balanced data are synthesized by six theft variants to cope with the realistic theft data. Manipulating techniques used for the synthesis of the data are as follows [42–46]:

$$T1(s_t) = s_t * rand(0.1, 0.9) \tag{8}$$

$$T2(s_t) = s_t * x_t(x_t = random(0.1, 0.9)) \tag{9}$$

$$T3(s_t) = s_t * (random[0, 1]) \tag{10}$$

$$T4(s_t) = mean(s_t) * random(0.1, 1.0) \tag{11}$$

$$T5(s_t) = mean(s_t) \tag{12}$$

$$T6(s_t) = S_{T-t}\ (Where\ T\ is\ consumption\ time) \tag{13}$$

- In data manipulation technique 1, as shown in Figure 2a, a random number is multiplied with benign class Time-Series Data in order to manipulate fair consumption.
- The data manipulating technique 2 is shown in Figure 2b. To capture the consumption's discontinuity, a random number is multiplied to manipulate the honest consumption's data. Random number multiplication is a series-based discontinuity in the consumption pattern.
- The data manipulating technique 3 is shown in Figure 3a. A random multiplication of 1 and 0 with Time-Series Data shows either the original consumption or a complete zero consumption. There is no ramping function in between 1 and 0. It is a straightforward switching ON, OFF operation with a complete connected load or the cut off. The multiplication is a mode to copy the historic consumption project, and it is not confined to a continuous Time-Series Data.
- In Theft Case 4, total consumption is aggregated into a mean which is multiplied by a random number in between (0.1, 1.0), as shown in Figure 3b.
- The data manipulating technique 5 is shown in Figure 4a. The aggregated mean is multiplied with a random number. It is a two-part manipulation. The average value is a centered value of continuous Time-Series Data, where maximum consumption is under-reported. In the second part, the same aggregated value is multiplied with a random number in between (0.1–0.9), where the average value is under-reported as well in an extra exploitation.
- The data manipulating technique 6 is shown in Figure 4b. A continuous swapping of the low consumption and peak consumption hours is practiced, where a couple slabs of consumed energy are shifted from ON-Peak hours to OFF-Peak hours and vice versa. In such manipulating techniques, the consumer pays the charges for the consumed energy, however, the vigilant swapping does not affect the UPs extensively.

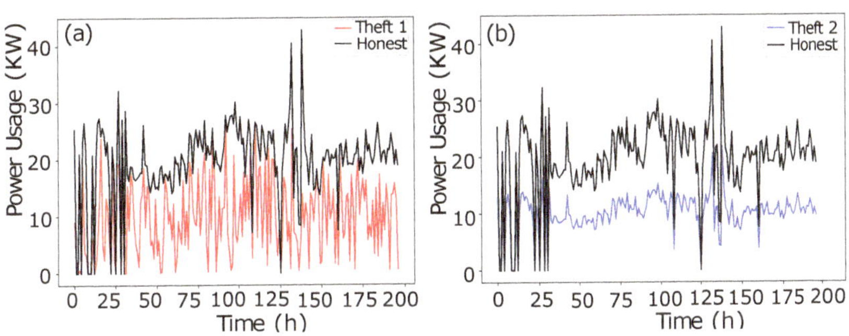

Figure 2. (**a**) Theft Case 1. (**b**) Theft Case 2.

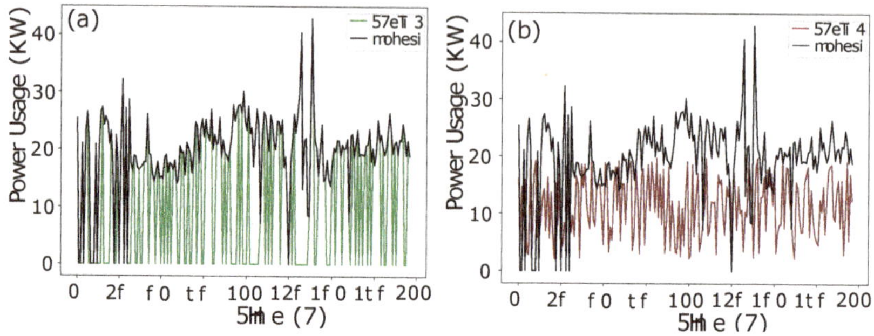

Figure 3. (**a**) Theft Case 3. (**b**) Theft Case 4.

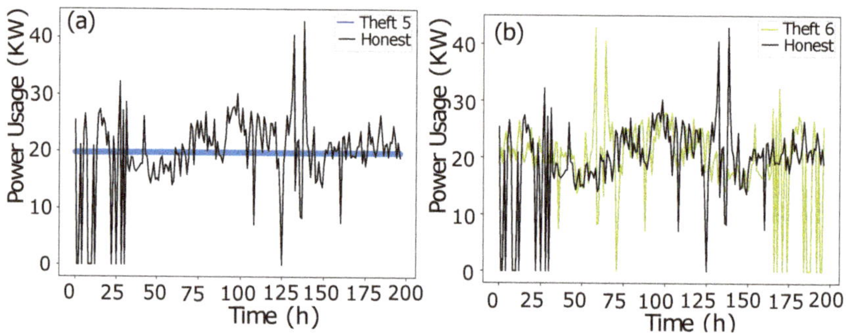

Figure 4. (**a**) Theft Case 5. (**b**) Theft Case 6.

4.5. Bi-Directional LSTM

To resolve the problem of vanishing gradients in RNNs [47], Bi-LSTM is developed to preserve information for a long time period. Bi-LSTM infrastructure consists of two LSTMs, which operate parallel in the forward and backward direction. Past and future Time-Series Data are processed through forward and backward direction gates, respectively. The input data are fed in the forward direction, and the reverse copy of the same inputted data are fed in the backward direction as well. Such nature of the inputted data with a reverse copy increases the data compatibility. The compatibility limits the gates to function accordingly as needed. The architecture contains two hidden layers, and the output layer is concatenated afterwards.

4.6. Feature Engineering

Synthetic features are helpful to improve the performance of the model. Four various types of synthetic stochastic features are generated, namely, mean, min, max and standard deviation. Time-Series Data of SGCC are analyzed on a monthly usage basis. The generation of the stochastic features creates a subset of available features, which reduces noise and improves DR slightly. However, FPR is reduced to a larger extent. The stochastic features are numeric features. Weighted Feature Importance (WFI) of these features is classifier-dependent. Certain features may not be of default importance to obtain a suitable DR and low FPR. The stochastic features are the principal important features, which contribute in our scenario. To confirm the validation, we iteratively tested and trained the classifiers on the SGCC dataset. Mathematical representation of the generated features is as follows:

$$y(t) = \{y_t; t = 0, 1, 2, 3, 4, \ldots, n\} \tag{14}$$

$$\mu = \sum_{i}^{n} \frac{O_n}{T_O} \quad (15)$$

$$\sigma = \sqrt{\frac{\sum_{i=0}^{n} (O_i - \mu)^2}{P_y}} \quad (16)$$

$$Minimum = O_{sv}[y\{t_i\}] \quad (17)$$

$$Maximum = O_{hv}[y\{t_i\}] \quad (18)$$

where, $y(t), t, O, T, n, u, sv, hv$ and P show Time-Series Data containing various numbers of features, time spans, observations, total number of observations of a specific time sequence, number of observations, mean, smallest value, highest value and total population of the dataset, respectively. Figure 5 shows the complete flow diagram of the overall classification scenario.

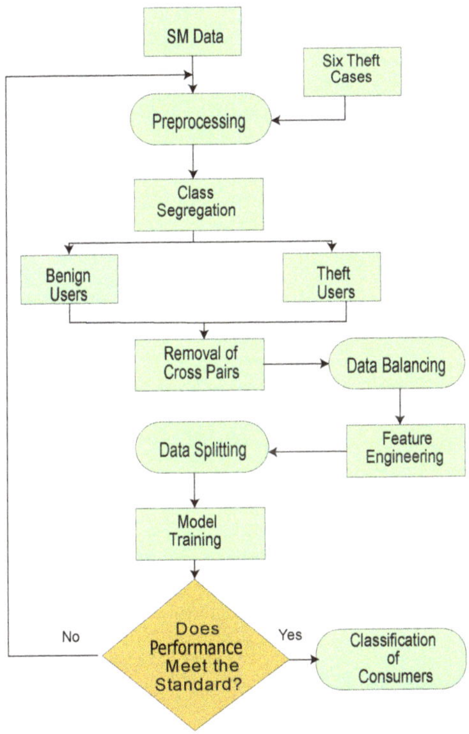

Figure 5. Methodology outline for detection of NTLs.

5. Performance Evaluation

To evaluate the performance of our developed hybrid model, we use DR, FPR and AUC scores and accuracy [48]. The origin of all of the aforementioned parameters is a confusion matrix. Parametric division of the dataset is observed based on the confusion matrix in shapes of True Positive (TP), FP, True Negative (TN) and False Negative (FN). TP and TN correctly analyze the honest user as honest and malicious as malicious, respectively. FP and FN wrongly classify the samples. Similarly, a model's detection and sensitivity are monitored by DR, which is referred to as TPR in the literature as well. Basically, DR is the representation of the model's sensitivity and detection, which is mathematically shown in Equation (19).

$$DetectionRate = \frac{TruePositive}{(TruePositive + FalseNegative)} \tag{19}$$

FPR is a vital evaluation factor in a detection and classification scenario to monitor the competency of a model which shows false alarms. A false alarm is an incorrect classification of positive samples as negative ones and vice versa. Such alarming parameters are quite expensive, which requires on-site inspection to verify, and it results in a huge monitory loss. To mitigate huge revenue losses, high FPR needs to be reduced. Mathematically, it is shown in Equation (20) [49].

$$FPR = \frac{FalsePositive}{(FalsePositive + TrueNegative)} \tag{20}$$

Moreover, the accuracy is the measure of the correctly predicted instances. Mathematically, it is represented as in Equation (21).

$$Accuracy = \frac{(TP + TN)}{(TP + TN + FP + FN)} \tag{21}$$

A suitable and good classifier is one having low FPR, high DR and high accuracy as well.

6. Simulation Results

The exploited data (SGCC) are a real-time residential consumer's data. Similar indexing pattern-based morphology classifies the consumers into two classes, in perspective of their consumption, which are properly labeled. A staging numeric binary is placed for each individual consumer's consumption pattern. Label 0 indicates a fair consumer, whereas 1 indicates a fraudulent consumer. The monitored and reordered patterns are recorded after every 24 h for each consumer. Benign class data are manipulated in order to synthesize malicious data for each of the theft variants. Later on, both classes' data are concatenated. However, a data balancing technique is required to reduce the class bias issue due to the skewness of the model towards the majority class. K-means SMOTE is deployed to balance the data. Before provision of the data to a model for training, both classes are segregated through an affine decision boundary, where cross-pairs are removed, which degrades model detection and classification accuracy. The Tomek links technique identifies and removes the in-rushed cross-pairs across the decision boundary. The number of identified and removed samples is shown in Table 3.

Table 3. Cross-Pairs Identification and Removal.

Total Samples (Before)	Removal of Cross-Pairs	Remaining Samples
10,500	105	10,395

In Figure 6a, the performance of the proposed BiGRU–BiLSTM is compared with an existing CNN–LSTM model [32]. The curves in Figure 6a indicate the AUC of the CNN–LSTM, proposed and ML-based models. Initially, at an AUC score of 0.50, both of the classifying models comparatively perform quite well, where high TPR and the lowest FPR are achieved, as shown in Figure 7a. The initial assessment based on the AUC curve shows that the CNN–LSTM model [32] classifies the samples efficiently with the recorded lowest FPR when the inputted samples passed are fewer in number. However, a small spike in the AUC curve at 0.60 shows that the data complexity moderately confuses the CNN–LSTM classification and results in an increasing FPR. The increasing FPR behavior is fluctuated in a range of AUC scores from 0.60–0.82, while during the defined ranged our proposed hybrid model Bi-GRU–Bi-LSTM performs much better to learn the data complexity and reduce FPR. The maximum AUC score of 0.93 is achieved by our proposed model with a high sensitivity rate (TPR) as compared with the opponent model. Moreover, performance of the proposed model is analyzed using a PRC curve. Figure 7b shows the performance curve of PRC, which ensures that a low PRC rate is not an optimal factor due to the high misclassification rate. Misclassification of the consumers spikes FPR and burdens the UPs due to the on-site inspection for the conformation of the consumers' nature, which is expensive in practice due to the revenue loss.

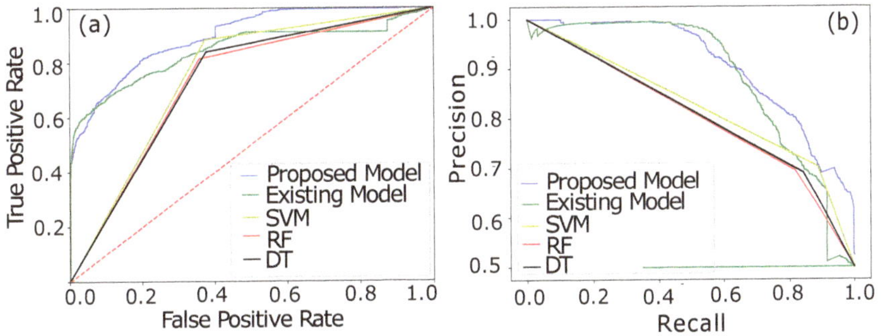

Figure 6. (**a**) AUC Analysis of the proposed and CNN–LSTM models. (**b**) PRC analysis of both models.

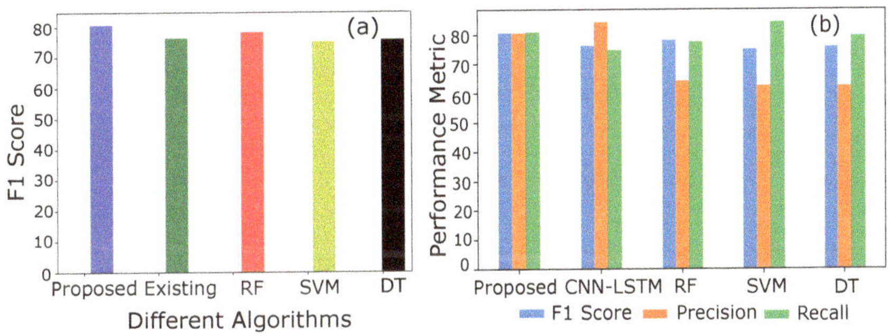

Figure 7. (**a**) F1 Score of different models. (**b**) Comparison of F1 Score, precision and recall.

Similarly, accuracy is not a good metric to evaluate the results of the whole classification scenario. Accuracy-based performance analysis of different models is shown in Figure 7a,b. Accuracy is the number of correct predictions over the total number of predictions. However, the prediction sometimes goes wrong and misclassifies the samples mistakenly. Figure 7b shows that CNN is a dumb classifier, and it takes advantage of the skewness of available data. To overcome the issue and to evaluate the performance of the classifier, F1 and precision scores are plotted.

The leading diagonal of the confusion matrix contains FP and FN, which are referred to as mistakes of the classifiers. A perfect classifier has the zero leading diagonal. Fluctuations in precision and recall are formally due to these two aforementioned factors.

Precision- and recall-based performance of a model is integrated into a single matrix called an F1 score. It is the harmonic mean of the precision and recall. Only a significant increase in both, i.e., precision and recall, can cause an increase in F1 score. Figure 7b shows an equilibrium in precision and recall, which results in a high F1 score, while the existing model has a low F1 score due to imbalance increase in precision and recall. Moreover, the bench mark models such as SVM, RF and DT depict the same scenario of the existing model with high fluctuations in F1 scores.

A comparative analysis in Table 4 shows a subsequent improvement in classification between the honest and fraudulent consumers. In addition, feature engineering improves the accuracy of the proposed detection model as shown in Table 5. It is observed that the accuracy is increased from 88.7% to 95%.

Table 4. Performance mapping of the executed models.

Models	F1 Score	Precision	Recall	Accuracy
Proposed	80.7	80.6%	80.9%	88.7%
Existing [33]	76.3	84.3%	74.7%	83.1%
SVM	75.0	62.5%	84.3%	72.5%
DT	75.7	62.3%	79.5%	76.3%
RF	78.2	64.2%	77.6 %	73.6%

Table 5. Performance improvement of the proposed model against stochastic feature engineering.

Models	Without Feature Engineering	With Stochastic Features
Proposed Model	88.7%	95%

7. Robustness Analysis

Robustness shows the effectiveness of a classifier against unseen and independent samples of a similar dataset whenever it is tested on such type of data. The unseen and independent data are referred to as the worst case of noisy data due to their distinctive characterization. In our case, Theft Case 3's data are taken to verify the robustness of the model. Theft Case 3 presents the most irregular consumption patterns as compared with the other Theft Cases due to a temperate randomness in consumption patterns, which is caused by the multiplication of the patterns with 1 and 0. The irregular and distinct patterns mimic changes as directives of inevitable factors, which proscribe the changes as suspected ones. A high-degree patterns' variation disrupts models' decision making. However, the proposed model survives to generalize completely on unseen data, as shown in Table 6.

Table 6. Robustness Performance of Proposed Model against Unseen Theft Attacks.

Models	Accuracy	AUC Score	F1 Score
Proposed Model	88.3%	57.6	54.9
Existing Model	86.9%	54.9	53.6.7

Table 6 depicts the observed accuracy, AUC and F1 scores. The statistics in Table 6 show that a higher DR is achieved with a high FPR. However, the high FPR is within an acceptable range as compared with the existing model.

8. Computational Complexity

To analyze the computational complexity of the proposed model, execution time is considered. Table 7 shows the execution time of the proposed and existing models. It is observed that the execution time of the proposed model is slightly greater as compared with the existing model. However, our major concern is high FPR. The proposed model beats the existing model in high the FPR perspective, which is an expensive parameter. High FPR burdens the UP and results in excessive monitory costs, whereas the computational complexity is a time-oriented parameter, which can be compromised.

Table 7. Computational Complexity Analysis.

Input Batch Size	Execution Time Proposed Model (s)	Execution Time Existing Model (s)
50	218	62
100	165	88
150	159	48
200	159	87
250	166	87
300	152	88

9. Performance Validation

In order to validate the effectiveness of our proposed model, a random testing on unseen theft class data is tested. The unseen theft class data are manipulated data of Theft Case 3, as shown in Equation (10). The observed AUC score of 57% validates the performance of the proposed model. Moreover, variation in the testing data due to the addition of the stochastic features challenges the performance, where an AUC score of 95% is observed. An AUC score of 95% is a good achievement and validates the performance of the proposed model.

10. Conclusions

This research proposes a hybrid model of BiLSTM and BiGRU in order to detect NTLs. Initially, benign and fraudulent consumers are segregated by defining an affine decision boundary through the Tomek Links techniques. Cross-pairs are identified and transformed into majority samples, where the majority class samples are removed and reduce the misclassification of the defused data across a decision boundary, which results in a low FPR. Furthermore, to synthesize theft variants, honest consumption is modified and manipulated by using six different data manipulating techniques. Six numbers of manipulated readings are synthesized for a single benign sample, which requires data balancing. For provision of the balanced benign class data, K-means SMOTE is used. K-means SMOTE over-samples the benign class using a clustering mechanism. The balanced data are inputted to the hybrid architecture of Bi-GRU–Bi-LSTM. The classification analysis is carried out on unseen data samples and achieves an AUC score of 0.93. Similarly, a competitive model of CNN–LSTM is trained and tested on the same data, which fails in the provision of a precise and accurate classification as compared with our proposed model.

Author Contributions: Conceptualization, S.M. and N.J.; methodology, S.M.; software, S.M.; validation, N.J.; writing—original draft preparation, S.M.; writing—review and editing, N.J., Z.A.K., N.I.C. and M.A.Z.R.; supervision, Z.A.K. and N.I.C.; project administration, M.A.Z.R., A.H.M. and A.A.A.; funding acquisition, A.H.M. and A.A.A. All authors have read and agreed to the published version of the manuscript.

Funding: This research received no external funding.

Institutional Review Board Statement: Not applicable.

Data Availability Statement: Not applicable.

Acknowledgments: M.A.Z. Raja like to acknowledge the support of the National Science and Technology Council (NSTC), Taiwan under grant NSTC 111-2221-E-224-043-.

Conflicts of Interest: The authors declare no conflict of interest.

Abbreviations

The following abbreviations are used in this manuscript:

AMI	Advanced Metering Infrastructure
APD-HT	Anomaly Pattern Detection Hypothesis Testing
Bi-GRU	Bi-directional Gated Recurrent Unit
AUC	Area Under the Curve
Bi-LSTM	Bi-directional Long Short-Term Memory
CatBoost	Categorical Boosting
CNN	Convolutional Neural Network
DTKSVM	Decision Tree Combined K-Nearest Neighbor and Support Vector Machine
EBT	Ensemble Bagged Tree
ETD	Electricity Theft Detection
DT	Decision Tree
DR	Detection Rate
DG	Distributed Generation
XGBoost	Extreme Gradient Boosting
Fits	Feed-in Tariffs
FN	False Negative
FP	False Positive
FPR	FP Rate
GBCs	Gradient Boosting Classifiers
LGBoost	Light Gradient Boosting
MIC	Maximum Information Coefficient
ML	Machine Learning
NaN	Not a Number
NAN	Neighborhood Area Network
NTLs	Non-Technical Losses
PV	Photo Voltaic
PRC	Precision Recall Curve
RUSBOOST	Random Under Sampling Boosting
RF	Random Forest
SSEA	Semi-Supervised Auto-Encoder
SGCC	State Grid Corporation of China
SMs	Smart Meters
SSDAE	Stacked Sparse Denoising Auto-Encoder
SCADA	Supervisory Control and Data Acquisition
SVM	Support Vector Machine
TLs	Technical Losses
TN	True Negative
TP	True Positive
UP	Utility Provider
WFI	Weighted Feature Importance
C	Sample's Unique Class
O	Observations
p	Population of the Samples
S	Number of Samples
S_t	Time-Series Data
T	Theft Case
σ	Standard Deviation
μ	Mean

References

1. Grigsby, L.L. *Electric Power Generation, Transmission, and Distribution*; CRC Press: Boca Raton, FL, USA, 2007.
2. Yu, X.; Cecati, C.; Dillon, T.; Simoes, M.G. The new frontier of smart grids. *IEEE Ind. Electron. Mag.* **2011**, *5*, 49–63. [CrossRef]
3. Depuru, S.S.S.R.; Wang, L.; Devabhaktuni, V. Electricity theft: Overview, issues, prevention and a smart meter based approach to control theft. *Energy Policy* **2011**, *39*, 1007–1015. [CrossRef]

4. Buzau, M.M.; Tejedor-Aguilera, J.; Cruz-Romero, P.; Gó mez-Expó sito, A. Hybrid deep neural networks for detection of Non-Technical Losses in electricity Smart Meters. *IEEE Trans. Power Syst.* **2019**, *35*, 1254–1263. [CrossRef]
5. World Bank. *World Development Report 2004: Making Services Work for Poor People*; The World Bank: Washington, DC, USA, 2003.
6. Gaur, V.; Gupta, E. The determinants of electricity theft: An empirical analysis of Indian states. *Energy Policy* **2016**, *93*, 127–136. [CrossRef]
7. Agüero, J.R. Improving the efficiency of power distribution systems through technical and Non-Technical Losses reduction. In Proceedings of the PES T&D 2012, Orlando, FL, USA, 7–10 May 2012; pp. 1–8.
8. Viegas, J.L.; Esteves, P.R.; Melicio, R.; Mendes, V.M.F.; Vieira, S.M. Solutions for detection of Non-Technical Losses in the electricity grid: A review. *Renew. Sustain. Energy Rev.* **2017**, *80*, 1256–1268. [CrossRef]
9. Munawar, S.; Asif, M.; Kabir, B.; Ullah, A.; Javaid, N. Electricity Theft Detection in Smart Meters Using a Hybrid Bi-directional GRU Bi-directional LSTM Model. In *Conference on Complex, Intelligent, and Software Intensive Systems*; Springer: Cham, Switzerland, 2021; pp. 297–308.
10. Salinas, S.; Li, M.; Li, P. Privacy-preserving energy theft detection in smart grids. In Proceedings of the 2012 9th Annual IEEE Communications Society Conference on Sensor, Mesh and Ad Hoc Communications and Networks (SECON), Seoul, Korea, 18–21 June 2012; pp. 605–613.
11. Saeed, M.S.; Mustafa, M.W.; Sheikh, U.U.; Jumani, T.A.; Mirjat, N.H. Ensemble bagged tree based classification for reducing Non-Technical Losses in multan electric power company of Pakistan. *Electronics* **2019**, *8*, 860. [CrossRef]
12. Punmiya, R.; Choe, S. Energy theft detection using gradient boosting theft detector with feature engineering-based preprocessing. *IEEE Trans. Smart Grid* **2019**, *10*, 2326–2329. [CrossRef]
13. Avila, N.F.; Figueroa, G.; Chu, C.C. NTL detection in electric distribution systems using the maximal overlap discrete wavelet-packet transform and random undersampling boosting. *IEEE Trans. Power Syst.* **2018**, *33*, 7171–7180. [CrossRef]
14. Adil, M.; Javaid, N.; Qasim, U.; Ullah, I.; Shafiq, M.; Choi, J.G. LSTM and bat-based RUSBoost approach for Electricity Theft Detection. *Appl. Sci.* **2020**, *10*, 4378. [CrossRef]
15. Li, S.; Han, Y.; Yao, X.; Yingchen, S.; Wang, J.; Zhao, Q. Electricity Theft Detection in power grids with deep learning and Random Forests. *J. Electr. Comput. Eng.* **2019**, *2019*, 4136874. [CrossRef]
16. Liu, Y.; Liu, T.; Sun, H.; Zhang, K.; Liu, P. Hidden electricity theft by exploiting multiple-pricing scheme in smart grids. *IEEE Trans. Inf. Forensics Secur.* **2020**, *15*, 2453–2468. [CrossRef]
17. Kong, X.; Zhao, X.; Liu, C.; Li, Q.; Dong, D.; Li, Y. Electricity Theft Detection in low-voltage stations based on similarity measure and DT-KSVM. *Int. J. Electr. Power Energy Syst.* **2021**, *125*, 106544. [CrossRef]
18. Yan, Z.; Wen, H. Electricity Theft Detection base on extreme gradient boosting in AMI. *IEEE Trans. Instrum. Meas.* **2021**, *70*, 1–9. [CrossRef]
19. Gunturi, S.K.; Sarkar, D. Ensemble Machine Learning models for the detection of energy theft. *Electr. Power Syst. Res.* **2021**, *192*, 106904. [CrossRef]
20. Lepolesa, L.J.; Achari, S.; Cheng, L. Electricity Theft Detection in Smart Grids Based on Deep Neural Network. *IEEE Access* **2022**, *10*, 39638–39655. [CrossRef]
21. Yao, R.; Wang, N.; Ke, W.; Chen, P.; Sheng, X. Electricity Theft Detection in unbalanced sample distribution: A novel approach including a mechanism of sample augmentation. *Appl. Intell.* **2022**, 1–20. [CrossRef]
22. Liao, W.; Yang, Z.; Liu, K.; Zhang, B.; Chen, X.; Song, R. Electricity Theft Detection Using Euclidean and Graph Convolutional Neural Networks. *IEEE Trans. Power Syst.* **2022**, 1–13. [CrossRef]
23. Gu, D.; Gao, Y.; Chen, K.; Junhao, S.; Li, Y.; Cao, Y. Electricity Theft Detection in AMI with Low False Positive Rate Based on Deep Learning and Evolutionary Algorithm. *IEEE Trans. Power Syst.* **2022**, *37*, 4568–4578. [CrossRef]
24. Zheng, K.; Chen, Q.; Wang, Y.; Kang, C.; Xia, Q. A novel combined data-driven approach for Electricity Theft Detection. *IEEE Trans. Ind. Inform.* **2018**, *15*, 1809–1819. [CrossRef]
25. Aslam, Z.; Javaid, N.; Ahmad, A.; Ahmed, A.; Gulfam, S.M. A Combined Deep Learning and Ensemble Learning Methodology to Avoid Electricity Theft in Smart Grids. *Energies* **2020**, *13*, 5599. [CrossRef]
26. Huang, Y.; Xu, Q. Electricity Theft Detection based on stacked sparse denoising autoencoder. *Int. J. Electr. Power Energy Syst.* **2021**, *125*, 106448. [CrossRef]
27. Fenza, G.; Gallo, M.; Loia, V. Drift-aware methodology for anomaly detection in smart grid. *IEEE Access* **2019**, *7*, 9645–9657. [CrossRef]
28. Yip, S.C.; Wong, K.; Hew, W.P.; Gan, M.T.; Phan, R.C.W.; Tan, S.W. Detection of energy theft and defective Smart Meters in smart grids using linear regression. *Int. J. Electr. Power Energy Syst.* **2017**, *91*, 230–240. [CrossRef]
29. Park, C.H.; Kim, T. Energy Theft Detection in Advanced Metering Infrastructure Based on Anomaly Pattern Detection. *Energies* **2020**, *13*, 3832. [CrossRef]
30. Hu, J.; Li, S.; Hu, J.; Yang, G. A Hierarchical Feature Extraction Model for Multi-Label Mechanical Patent Classification. *Sustainability* **2018**, *10*, 219. [CrossRef]
31. Hasan, M.; Toma, R.N.; Nahid, A.A.; Islam, M.M.; Kim, J.M. Electricity Theft Detection in smart grid systems: A CNN-LSTM based approach. *Energies* **2019**, *12*, 3310. [CrossRef]
32. Khalid, R.; Javaid, N.; Al-Zahrani, F.A.; Aurangzeb, K.; Qazi, E.U.H.; Ashfaq, T. Electricity load and price forecasting using Jaya-Long Short Term Memory (JLSTM) in smart grids. *Entropy* **2020**, *22*, 10. [CrossRef]

33. Rostampour, V.; Keviczky, T. Probabilistic energy management for building climate comfort in smart thermal grids with seasonal storage systems. *IEEE Trans. Smart Grid* **2018**, *10*, 3687–3697. [CrossRef]
34. Jokar, P.; Arianpoo, N.; Leung, V.C. Electricity Theft Detection in AMI using customers' consumption patterns. *IEEE Trans. Smart Grid* **2015**, *7*, 216–226. [CrossRef]
35. Buzau, M.M.; Tejedor-Aguilera, J.; Cruz-Romero, P.; Gómez-Expósito, A. Detection of Non-Technical Losses using smart meter data and supervised learning. *IEEE Trans. Smart Grid* **2018**, *10*, 2661–2670. [CrossRef]
36. Biswas, P.P.; Cai, H.; Zhou, B.; Chen, B.; Mashima, D.; Zheng, V.W. Electricity theft pinpointing through correlation analysis of master and individual meter readings. *IEEE Trans. Smart Grid* **2019**, *11*, 3031–3042. [CrossRef]
37. Ismail, M.; Shaaban, M.F.; Naidu, M.; Serpedin, E. Deep learning detection of electricity theft cyber-attacks in renewable Distributed Generation. *IEEE Trans. Smart Grid* **2020**, *11*, 3428–3437. [CrossRef]
38. Zhu, Q.; Zhang, F.; Liu, S.; Wu, Y.; Wang, L. A hybrid VMD–BiGRU model for rubber futures time series forecasting. *Appl. Soft Comput.* **2019**, *84*, 105739. [CrossRef]
39. Bhagat, R.C.; Patil, S.S. Enhanced SMOTE algorithm for classification of imbalanced big-data using Random Forest. In Proceedings of the 2015 IEEE International Advance Computing Conference (IACC), Bangalore, India, 12–13 June 2015; pp. 403–408.
40. Sun, J.; Shi, W.; Yang, Z.; Yang, J.; Gui, G. Behavioral modeling and linearization of wideband RF power amplifiers using BiLSTM networks for 5G wireless systems. *IEEE Trans. Veh. Technol.* **2019**, *68*, 10348–10356. [CrossRef]
41. Hussain, S.; Mustafa, M.W.; Jumani, T.A.; Baloch, S.K.; Alotaibi, H.; Khan, I.; Khan, A. A novel feature engineered-CatBoost-based supervised Machine Learning framework for Electricity Theft Detection. *Energy Rep.* **2021**, *7*, 4425–4436. [CrossRef]
42. Ullah, A.; Munawar, S.; Asif, M.; Kabir, B.; Javaid, N. Synthetic theft attacks implementation for data balancing and a gated recurrent unit based Electricity Theft Detection in smart grids. In *Conference on Complex, Intelligent, and Software Intensive Systems*; Springer: Cham, Switzerland, 2021; pp. 395–405.
43. Asif, M.; Kabir, B.; Ullah, A.; Munawar, S.; Javaid, N. Towards Energy Efficient Smart Grids: Data Augmentation Through BiWGAN, Feature Extraction and Classification Using Hybrid 2DCNN and BiLSTM. In *International Conference on Innovative Mobile and Internet Services in Ubiquitous Computing*; Springer: Cham, Switzerland, 2021; pp. 108–119.
44. Asif, M.; Ullah, A.; Munawar, S.; Kabir, B.; Khan, A.; Javaid, N. Alexnet-AdaBoost-ABC based hybrid neural network for Electricity Theft Detection in smart grids. In *Conference on Complex, Intelligent, and Software Intensive Systems*; Springer: Cham, Switzerland, 2021; pp. 249–258.
45. Kabir, B.; Ullah, A.; Munawar, S.; Asif, M.; Javaid, N. Detection of Non-Technical Losses Using MLP-GRU Based Neural Network to Secure Smart Grids. In *Conference on Complex, Intelligent, and Software Intensive Systems*; Springer: Cham, Switzerland, 2021; pp. 383–394.
46. Dash, S.K.; Roccotelli, M.; Khansama, R.R.; Fanti, M.P.; Mangini, A.M. Long Term Household Electricity Demand Forecasting Based on RNN-GBRT Model and a Novel Energy Theft Detection Method. *Appl. Sci.* **2021**, *11*, 8612. [CrossRef]
47. Khan, Z.A.; Adil, M.; Javaid, N.; Saqib, M.N.; Shafiq, M.; Choi, J.G. Electricity Theft Detection using supervised learning techniques on smart meter data. *Sustainability* **2020**, *12*, 8023. [CrossRef]
48. Javaid, N.; Javaid, S.; Asif, M.; Javed, M.U.; Yahaya, A.S.; Aslam, S. Synthetic Theft Attacks and Long Short Term Memory-Based Preprocessing for Electricity Theft Detection Using Gated Recurrent Unit. *Energies* **2022**, *15*, 2778.
49. Gul, H.; Javaid, N.; Ullah, I.; Qamar, A.M.; Afzal, M.K.; Joshi, G.P. Detection of Non-Technical Losses using SOSTLink and bidirectional gated recurrent unit to secure Smart Meters. *Appl. Sci.* **2020**, *10*, 3151. [CrossRef]

sensors

MDPI

Article

Intelligent Sensors for dc Fault Location Scheme Based on Optimized Intelligent Architecture for HVdc Systems

Muhammad Zain Yousaf [1,*], Muhammad Faizan Tahir [2], Ali Raza [3], Muhammad Ahmad Khan [4] and Fazal Badshah [1]

1 School of Electrical and Information Engineering, Hubei University of Automotive Technology, Shiyan 442002, China
2 School of Electric Power, South China University of Technology, Guangzhou 510630, China
3 School of Electrical Engineering, University of Engineering and Technology, Lahore 39161, Pakistan
4 School of Electrical and Information Engineering, Tianjin University, Tianjin 300072, China
* Correspondence: zain.yousaf@huat.edu.cn

Abstract: We develop a probabilistic model for determining the location of dc-link faults in MT-HVdc networks using discrete wavelet transforms (DWTs), Bayesian optimization, and multilayer artificial neural networks (ANNs) based on local information. Likewise, feedforward neural networks (FFNNs) are trained using the Levenberg–Marquardt backpropagation (LMBP) method, which multi-stage BO optimizes for efficiency. During training, the feature vectors at the sending terminal of the dc link are selected based on the norm values of the observed waveforms at various frequency bands. The multilayer ANN is trained using a comprehensive set of offline data that takes the denoising scheme into account. This choice not only helps to reduce the computational load but also provides better accuracy. An overall percentage error of 0.5144% is observed for the proposed algorithm when tested against fault resistances ranging from 10 to 485 Ω. The simulation results show that the proposed method can accurately estimate the fault site to a precision of 485 Ω and is more robust.

Keywords: Levenberg–Marquardt backpropagation; protection sensor; Bayesian optimization; modular multilevel converter

Citation: Yousaf, M.Z.; Tahir, M.F.; Raza, A.; Khan, M.A.; Badshah, F. Intelligent Sensors for dc Fault Location Scheme Based on Optimized Intelligent Architecture for HVdc Systems. *Sensors* **2022**, *22*, 9936. https://doi.org/10.3390/s22249936

Academic Editor: Arshad Arshad

Received: 7 November 2022
Accepted: 13 December 2022
Published: 16 December 2022

1. Introduction

To date, China celebrates the completion of 30,000 km of ultra-high-voltage lines connecting six regional grids with a total transmission capacity of close to 150 gigawatts [1]. However, power engineers struggle to manage and regulate the impact of dc-link faults in hybrid ac/dc systems [2]. Let us say the 8 GW dc-link from Gansu reports a fault unexpectedly, and the protection algorithm cannot locate it. The power outage might start a chain reaction, resulting in widespread blackouts throughout Hunan and beyond. As a result, ensuring accurate fault location is beneficial to minimize the threat of possible failure and is a prerequisite for the successful and safe operation of dc transmission systems [3]. Furthermore, accurate fault location estimation is important for maintaining the voltage stability of the power system [4] and operating the electricity market efficiently [5].

The prediction of correct fault sites in dc transmission systems has been shown to be reliable by frequency extraction, fault signal analysis, and travelling-wave (TW) approaches in previous studies [2,3,6]. Currently, TW methods based on the concept of travelling-wave reflections are preferred in dc transmission projects since they are highly accurate, reliable, and have high fault resistance [7]. The advancement of TW theory has led to the development of several signal processing techniques, such as wavelet transformation (WT) [8], S and Hilbert–Huang transform [6,9], empirical mode decomposition (EMD) [10], etc. A waveform's characteristics are analyzed using approximate or detailed coefficients in WT to predict fault locations. However, conventional TW methods require a very high sampling frequency to accurately predict fault location, which leads to expensive computation in the power grid [11].

Researchers have begun exploring intelligent algorithms to rectify computation burden and noise handling capabilities [12]. With the alienation coefficient and the Wigner distribution function [13], an effective transmission line protection mechanism for underground cables is proposed in [14] and implemented for a renewable-energy-based grid. In addition, machine-learning tools such as the radial basis function neural network (RBFNN) [15], support vector machine (SVM) [16,17], extreme machine learning [18], k-means cluster [19], etc., are also utilized. These algorithms can self-learn and modify weights and thresholds while training with historical data. Therefore, they are suitable for complex networks such as multiterminal high-voltage direct-current (MMC-MT-HVdc) systems, where constructing a feasible intelligent algorithm can be challenging.

Because of the enlargement of the structure and the extraordinary growth in the number of learning parameters in MMC-MT-HVdc networks, these intelligent algorithms experience slow convergence and computational burden [20]. Downsampling learning parameters may resolve the above issue but may eliminate valuable features. It is, therefore, imperative to find an algorithm for fault location with high accuracy, minimal computational burden, and low sensitivity to noise [21,22]. With this in mind, this work combines the discrete wavelet transform (DWT) [23], Bayesian optimization (BO) [24], and multilayer feedforward neural network (FFNN) to locate the dc-link faults.

A multilayer FFNN model based on BO is trained and evaluated using the selected features. BO is well-known as a powerful technique for optimizing black-box functions when closed-form expressions or surrogate models are unavailable [24]. A study in the literature found that BO provided a higher convergence rate than standard tuning methods after the neural network was adjusted [25,26]. The multi-stage BO introduced in this work reduces the computational burden by reducing the number of simulations needed to find the optimal design for any given neural network. As a result, it provides a well-tuned multilayer FFNN that can achieve improved response with better accuracy.

To simulate numerous fault scenarios, a four-terminal MT-HVdc system is developed in PSCAD/EMTDC, and the proposed model is investigated in a Matlab® environment. Meanwhile, the proposed algorithm is compared to intelligent adversaries such as backpropagation neural networks (BP-NNs) and conventional FFNNs. The test results show that the suggested model performs with the highest fault localization accuracy.

Under the circumstances above, the main contribution of this study is:

- Our initial goal is to create a learning-based algorithm that relies on only one end of the communication link for fault location. Hence, eliminating reliance on the communication link.
- In general, a signal detected by a sensor is invariably interfered with by the surrounding environment or modified by the detecting equipment during the detection process, increasing failure chances. The DWT-based signal analysis model is used to eliminate interference from the observed signal to improve signal analysis and recognition.
- The energy or norm of the current and voltage signals at each frequency band gives a unique signature for different fault locations and has been found to be robust against noise. Therefore, it is used as an extracted feature for pattern recognition.
- The proposed algorithm must be able to locate internal faults with high fault impedances at further distances.

The remainder of the paper is organized in the following way.

Section 2 discusses the mathematical model derivation from a simple backpropagation algorithm to the improvised backpropagation algorithm. It also discusses the implementation of the proposed framework as well. Meanwhile, Section 3 introduces the conditions and properties of the chosen system model, which has been developed to capture fault data under dynamic fault scenarios. Section 4 covers the methodologies utilized to analyze input features extracted under dynamic fault scenarios. It also covers a denoising scheme that is used to denoise features before training and data preprocessing. Section 5 presents comparisons and analyses against adversaries. Finally, Section 6 concludes with a summary of the proposed algorithm. An overview of the proposed method is presented in the next section.

2. Proposed Framework

Figure 1 illustrates the architecture with the proposed methodology. During fault localization, the proposed method has two stages. In the first one, the captured fault window is filtered using discrete wavelets to rectify the noise issue, set to 10 ms. After denoising, the measured time-domain segment is transformed into a time-frequency domain by splitting it into low- and high-frequency components with a DWT-based multi-resolution analysis (MRA) technique.

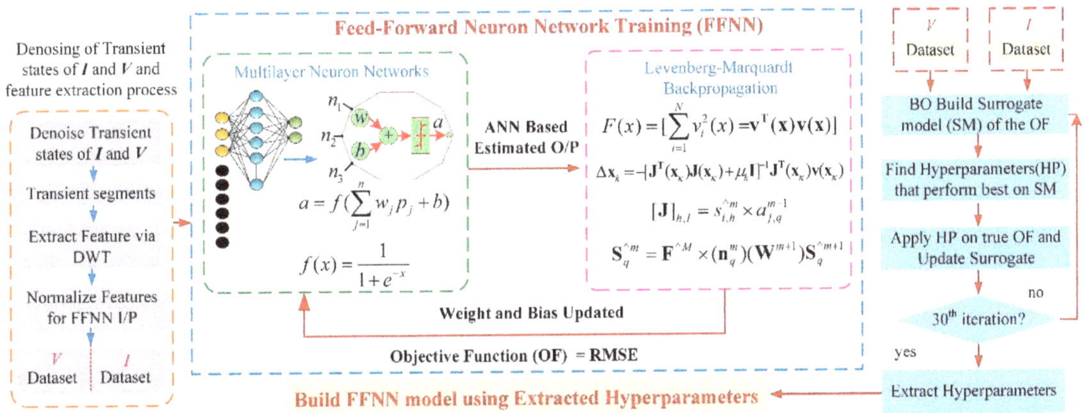

Figure 1. Proposed architecture.

The multilayer neural network receives decluttered information from voltage and current signals as inputs. It then estimates the fault site using the activation function. Levenberg–Marquardt backpropagation (LMBP) is implemented instead of a standard backpropagation algorithm with a performance function. It is a function of the ANN regression model and ground truth of fault sites. The Levenberg–Marquardt method is used to update the weight and bias. The Jacobian matrix of the performance function with respect to the weight and bias variables is calculated via the proposed backpropagation algorithm. After updating the weight and bias, the multilayer ANN is applied to determine the fault site. The multi-stage BO procedure is conducted prior to the update, aiming to increase accuracy during training and to provide an optimal multilayer FFNN by optimizing hyperparameters. The hyperparameters, unlike internal parameters (weights, bias, etc.), are set before the neural network is trained, and they influence the neural network's performance. Regulating them via the trial-and-error method lengthens the training set-up time and may reduce accuracy. Hence, optimizing these hyperparameters enhances the accuracy and convergence speed [24]. In the following sub-section, the detailed architecture of the proposed algorithm is described.

2.1. *Feedforward Neural Network (FFNN)*

This study uses a feedforward neural network with a single hidden layer to model because a neural network with a single hidden layer can handle the most complex functions (i.e., one input layer, one output layer, and one hidden layer) [27]. In a multilayer FFNN, the basic building block is a neuron that mimics a biological neuron's functions and behavior [27]. The schematic structure based on the neuron is shown in Figure 2.

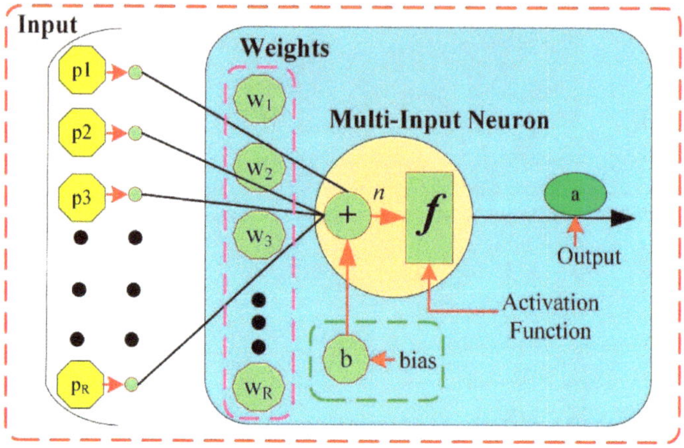

Figure 2. Structure of the multi-input neuron.

Usually, a neuron has multiple inputs. Each element of the input vector $p = [p_1, p_2, K, p_R]$ is weighted by elements w_1, w_2, K, w_j of the weight matrix W. Next, the bias of each neuron is summed with the weighted inputs to form the net-input n, expressed as:

$$n = \sum_{j=1}^{R} w_j p_j + b = W_p + b. \quad (1)$$

Following that, net-input n is sent via an activation function f, which results in the neuron's output a. Mathematically expressed as:

$$a = f(n) \quad (2)$$

In this work, the activation function is based on the hyperbolic tangent sigmoid transfer function. The following equation presents it.

$$f(x) = \frac{2}{1+e^{-2x}} - 1 \quad (3)$$

With reference to Figure 2, the multi-input FFNN executes the following equation:

$$a^2 = f^2\left(\sum_{i=1}^{s} w_{1,i}^2 f^1\left(\sum_{j=1}^{R} w_{i,j}^1 p_j + b_i^1\right) + b^2\right) \quad (4)$$

The output of the neural network is represented by a^2. R stands for the number of inputs, the number of neurons in the hidden layer is denoted by S, and the jth input is represented by p_j. The activation functions of the output and hidden layers are represented by f^2 and f^1, respectively. The bias of the ith neuron is defined by b_i^1, whereas the bias of the neuron in the output layer is represented by b^2. The weight $w_{i,j}^1 p_j$ represents the connection between the jth input and the ith neuron of the hidden layer. Meanwhile, the weight connecting the ith hidden layer source to the output layer neuron is denoted by $w_{1,i}^2$.

2.2. Backpropagation Algorithm

Following the definition of the FFNN, the next step is to create an algorithm for training such networks. To train the established multilayer FFNN, an error backpropagation algorithm based on the steepest descent technique is typically utilized [28]. For the proposed three-layer FFNN, we now express the function that represents the output of unit i in layer $m + 1$ as:

$$a^{m+1} = f^{m+1}\left(n^{m+1}(i)\right) \quad (5)$$

Then to propagate the function and generate net-input ($n^{m+1}(i)$) to unit i, the neuron in the first layer receives extracted features from the MT-HVdc system to provide an initial condition for Equation (5):

$$a^0 = p \tag{6}$$

Equation (5) is further translated in matrix form for an M number of layers in a neural network as:

$$a^{m+1} = f^{m+1}\left(W^{m+1}a^m + b^{m+1}\right), \ldots, m = 0, 1. \tag{7}$$

where a^{m+1} and a^m are the outputs of the network's (m+1)th and mth layers. b^{m+1} reflect the bias vector of the network's (m+1)th layer. Here, external inputs passing to the network via Equation (7), the overall network's outputs are equal to the outputs of the neurons in the last layer:

$$a = a^{\ M} \tag{8}$$

The objective of this study is to locate the dc-link faults. Therefore, the proposed multilayer FFNN requires a set of input–output pairs that characterize the behavior of an MT-HVdc system under faulty settings. Mathematically expressed as:

$$\left[(p_1, t_1), (p_2, t_2), (p_3, t_3), \ldots, \ldots, \ldots, (p_Q, t_Q)\right],\ p_q \text{ is input and } t_q \text{ is the relevant target of the network that uses for training.} \tag{9}$$

After each input propagates through the multilayer FFNN during training, the network output is compared to the target. While doing so, the performance index for the backpropagation algorithm is the mean-square error (MSE), which is to be reduced by modifying the network parameters, given as:

$$F(\boldsymbol{x}) = E[(e^2)] = E[(t-a)^2] \tag{10}$$

In the FFNN, $\boldsymbol{x}$ is the vector matrix containing the network weights and biases. However, in our case, the proposed network has multiple outputs. Therefore, Equation (10) generalized to:

$$F(\boldsymbol{x}) = E\left[\left(e^T e\right)\right] = E\left[(t-a)^T(t-a)\right] \tag{11}$$

Since the steepest descent rule is utilized for the standard backpropagation algorithm, the performance index $F(\boldsymbol{x})$ can be approximated as follows:

$$F^{\wedge}(\boldsymbol{x}) = E\left[\left(t(k)-a(k)\right)^T\left(t(k)-a(k)\right))\right] = e^T(k)e(k) \tag{12}$$

The squared error replaces the expectation of the squared error in Equation (11) at iteration step k. The steepest (gradient) descent algorithm for the estimated MSE is then:

$$w^m_{i,j}(k+1) = w^m_{i,j}(k) - \propto \frac{dF^{\wedge}}{dw^m_{i,j}} \tag{13}$$

$$b^m_i(k+1) = b^m_i(k) - \propto \frac{dF^{\wedge}}{db^m_i} \tag{14}$$

$\propto$ is the learning rate, similar to the number of neurons (S); it is also a hyperparameter. Defined:

$$s^m_i = \frac{dF^{\wedge}}{dn^m_i} \tag{15}$$

as the performance index ($F^\wedge$) sensitivity (s_i^m) that measures the changes in the net input of the ith element in layer m. Next, based on the chain rule, the derivate of Equations (13) and (14 using Equations (5), (12) and (15) can be simplified as:

$$\frac{dF^\wedge}{dw_{i,j}^m} = \frac{dF^\wedge}{dn_i^m} * \frac{dn_i^m}{dw_{i,j}^m} = s_i^m * a_j^{m-1} \tag{16}$$

$$\frac{dF^\wedge}{db_i^m} = \frac{dF^\wedge}{dn_i^m} * \frac{dn_i^m}{db_i^m} = s_i^m \tag{17}$$

Now with the definition of gradient, the steepest descent algorithm is approximated as:

$$w_{i,j}^m(k+1) = w_{i,j}^m(k) - \propto *s_i^m * a_j^{m-1} \tag{18}$$

$$b_i^m(k+1) = b_i^m(k) - \propto *s_i^m \tag{19}$$

The following recurrence relation in matrix form can be satisfied by the sensitivity [29,30]:

$$\mathbf{s}^m = \overline{\mathbf{F}}^m(\mathbf{n}^m)\left(\mathbf{W}^{m+1}\right)^T \mathbf{s}^{m+1}, \text{ for. } m = M-1, \ldots, 2, 1. \tag{20}$$

Equation (20) expresses the step used to propagate the sensitivities backward through a neural network. Mathematically, the sensitivities propagate backward across the network as:

$$\mathbf{s}^M \rightarrow \mathbf{s}^{M-1} \rightarrow \cdots \rightarrow \mathbf{s}^2 \rightarrow \mathbf{s}^1 \tag{21}$$

where

$$\overline{\mathbf{F}}^m(\mathbf{n}^m) = \begin{bmatrix} \overline{f}^m(n_{1^m}^m) & 0 & \mathrm{K} & 0 \\ 0 & \overline{f}^m(n_{2^m}^m) & & 0 \\ \mathrm{M} & \mathrm{M} & & \mathrm{M} \\ 0 & 0 & \mathrm{K} & \overline{f}^m(n_{s^m}^m) \end{bmatrix} \tag{22}$$

And

$$\overline{f}^m\left(n_{j^m}^m\right) = \frac{df^m\left(n_j^m\right)}{dn_j^m} \tag{23}$$

Whereas a recurrence relation is initialized at the final layer as:

$$\mathbf{s}^M = -2\overline{\mathbf{F}}^m\left(\mathbf{n}^M\right)(t-a) \tag{24}$$

Now, we can summarize the overall backpropagation (BP) based on the steepest descent algorithm as (1): First, use Equations (6)–(8) to propagate the input through the network. (2): Next, using Equations (20) and (24), backpropagate the sensitivity. (3): Finally, using Equations (18) and (19), update the weights and biases.

2.3. *Levenberg–Marquardt Backpropagation*

The backpropagation algorithm exhibits asymptotic convergence properties while training the multilayer FFNN, which causes a slow convergence rate due to minor weight changes around the solution. Meanwhile, Levenberg–Marquardt (LM) backpropagation [29] is a variant of Newton's method, which inherits the stability of the steepest descent algorithm and the speed of the Gauss–Newton algorithm [27,29,30]. Now, suppose we want to optimize performance index $F(x)$; then, Newton's method is:

$$x_{k+1} = x_k - A_k^{-1} g_k. \tag{25}$$

where $A_k \cong \nabla^2 F(x)|_{X=X_k}$, plus $g_k \cong \nabla F(x)|_{X=X_k}$. Note that $\nabla^2 F(x)$ represents the Hessian matrix, and $\nabla F(x)$ denotes the gradient. Let us assume that $F(x)$ is a sum-of-squares function, then:

$$F(x) = \sum_{i=1}^{N} v_i^2(x) = \mathbf{v}^T(x)\mathbf{v}(x). \tag{26}$$

Then the gradient and Hessian matrix are expressed in matrix form as:

$$\nabla F(x) = 2\,\mathbf{J}^T(x)\mathbf{v}(x). \tag{27}$$

$$\nabla^2 F(x) = 2\,\mathbf{J}^T(x)\mathbf{J}(x) + 2\,\mathbf{S}(x). \tag{28}$$

$\mathbf{J}(x)$ denotes the Jacobian matrix as:

$$\mathbf{J}(x) = \begin{bmatrix} \frac{dv_1(x)}{dx_1} & \frac{dv_1(x)}{dx_2} & \cdots & \frac{dv_1(x)}{dx_n} \\ \frac{dv_2(x)}{dx_1} & \frac{dv_2(x)}{dx_2} & \cdots & \frac{dv_2(x)}{dx_n} \\ \mathrm{M} & \mathrm{M} & \cdots & \mathrm{M} \\ \frac{dv_N(x)}{dx_1} & \frac{dv_N(x)}{dx_1} & \cdots & \frac{dv_N(x)}{dx_n} \end{bmatrix} \tag{29}$$

$$\mathbf{S}(x) = \sum_{i=1}^{N} v_i(x)\nabla^2 v_i(x) \tag{30}$$

Assume that $\mathbf{S}(x) \approx 0$, then Equation (30) (Hessian matrix) approximate as $\nabla^2 F(x) \cong 2\,\mathbf{J}^T(x)\mathbf{J}(x)$. Next, Equation (25) updates after substituting Equation (27) and the approximation of Equation (28) as:

$$\Delta x_k = x_{k+1} - x_k = -\left[\mathbf{J}^T(x_k)\mathbf{J}(x_k)\right]^{-1} * \mathbf{J}^T(x_k)\mathbf{v}(x_k). \tag{31}$$

The matrix ($\mathbf{H} = \mathbf{J}^T\mathbf{J}$) may not be invertible using the Gauss–Newton method. This issue can be fixed by making the following changes to the approximation Hessian matrix:

$$\mathbf{G} = \mathbf{H} + \mu\mathbf{I} \tag{32}$$

This modification to the Gauss–Newton method eventually leads to the LM algorithm [29]:

$$\Delta x_k = -\left[\mathbf{J}^T(x_k)\mathbf{J}(x_k) + \mu_k\mathbf{I}\right]^{-1}\mathbf{J}^T(x_k)\mathbf{v}(x_k). \tag{33}$$

Now, using the Δx_k direction, recalculate the approximated $F(x)$. If a smaller number is obtained, then the computation procedure is repeated, but the parameter μ_k is divided by a factor ($\alpha > 1$). If the value of $F(x)$ does not decrease, then the value of μ_k for the next iteration in the step is multiplied by α.

The calculation of the Jacobian matrix is an essential step in the LM method. The elements of the Jacobian matrix are calculated using a slight modification to the BP algorithm to address the NN mapping difficulty [29]. For better understanding, similar to Equation (12) for the BP algorithm, Equation (26) is a performance index for the mapping problem in the LM algorithm, where the error vector is $\mathbf{v}^T = [v_1\ v_2\ \mathrm{K}\ v_N] = \left[e_{1,1}\ e_{2,1}\ \mathrm{K}\ e_{s^M,1}\ e_{2,1}\mathrm{K}\ e_{s^M,Q}\right]$, and the vector x parametric values are $x^T = [x_1\ x_2\ \mathrm{K}\ x_N] = \left[w_{1,1}^1, w_{1,2}^1\ \mathrm{K}, w_{S,1R}^1 \cdot b_1^1,\ \mathrm{K},\ b_{S1}^1 . w_{1,1}^2,\ \mathrm{K}, b_{SM}^M\right]$, subscript N defined as $N = Q * S^M$.

Similarly, the n subscript is defined as $n = S^1(R+1) + S^2(S^1+1) + \ldots + S^M(S^{M-1}+1)$ in the Jacobian matrix. Now making all these substitutions in Equation (29) of the Jacobian matrix as:

$$\mathbf{J}(\boldsymbol{x}) = \begin{bmatrix} \frac{de_{1,1}}{dw^1_{1,1}} & \frac{de_{1,1}}{dw^1_{1,2}} & \cdots & \frac{de_{1,1}}{dw^1_{S^1,R}} & \frac{de_{1,1}}{db^1_1} & \cdots \\ \frac{de_{2,1}}{dw^1_{1,1}} & \frac{de_{2,1}}{dw^1_{1,2}} & \cdots & \frac{de_{2,1}}{dw^1_{S^1,R}} & \frac{de_{2,1}}{db^1_1} & \cdots \\ \mathrm{M} & \mathrm{M} & & \mathrm{M} & \mathrm{M} & \\ \frac{de_{S^M,R}}{dw^1_{1,1}} & \frac{de_{S^M,R}}{dw^1_{1,2}} & \cdots & \frac{de_{S^M,R}}{dw^1_{S^1,R}} & \frac{de_{S^M,R}}{db^1_1} & \cdots \\ \frac{de_{1,2}}{dw^1_{1,1}} & \frac{de_{1,2}}{dw^1_{1,2}} & \cdots & \frac{de_{1,2}}{dw^1_{S^1,R}} & \frac{de_{1,2}}{db^1_1} & \cdots \\ \mathrm{M} & \mathrm{M} & \cdots & \mathrm{M} & \mathrm{M} & \cdots \end{bmatrix} \quad (34)$$

Until now, the standard BP algorithm has been used to calculate the Jacobian matrix terms as follows:

$$\frac{dF^{\wedge}(\boldsymbol{x})}{d\boldsymbol{x}_I} = \frac{d\boldsymbol{e}_q^T \boldsymbol{e}_q}{d\boldsymbol{x}_I} \quad (35)$$

Meanwhile, in the LM algorithm, the terms for the elements of the Jacobian matrix can be calculated using the following:

$$[\mathbf{J}]_{h,I} = \frac{dv_h}{d\boldsymbol{x}_I} = \frac{de_{k,q}}{dw_{i,j}} \quad (36)$$

Thus, rather than computing the derivatives of the squared errors as in standard backpropagation, we are calculating the derivatives of the errors in this modified Levenberg–Marquardt algorithm. Similar to the concept for standard backpropagation sensitivities, a new Marquardt sensitivity is defined as follows:

$$[\mathbf{J}]_{h,I} = \frac{de_{k,q}}{dw^m_{i,j}} = \frac{de_{k,q}}{dn^m_{i,j}} * \frac{dn^m_{i,j}}{dw^m_{i,j}} = \hat{s}{}^m_{i,h} * a^{m-1}_{j,q} \quad (37)$$

if $\boldsymbol{x}_I$ is a bias,

$$[\mathbf{J}]_{h,I} = \frac{de_{k,q}}{db^m_i} = \frac{de_{k,q}}{dn^m_{i,q}} * \frac{dn^m_{i,q}}{db^m_i} = \hat{s}{}^m_{i,h} \quad (38)$$

As previously stated, the Marquardt sensitivity can be determined using the same recurrence relation as the standard sensitivities. However, toward the conclusion of the final layer, there is only one modification for calculating the new Marquardt sensitivity:

$$\hat{s}{}^M_{i,h} = \frac{de_{k,q}}{dn^M_{i,q}} = \frac{d\left(t_{k,q} - a^M_{k,q}\right)}{dn^M_{i,q}} = \frac{da^M_{k,q}}{dn^M_{i,q}} \quad (39)$$

for $i = k$, it is

$$\hat{s}{}^M_{i,h} = -f^{\wedge M}\left(n^M_{i,q}\right) \quad (40)$$

for $i \neq k$, it is equal to zero. Note that $f^{\wedge M}$ and its matrix can be defined with the help of Equations (22) and (23). In the proposed model, when extracted features from the MT-HVdc network are applied to the multilayer FFNN as an input (p_q) and the corresponding output (a^M_q) is processed, the LMBP algorithm is initialized with the following:

$$\boldsymbol{S}^{\wedge M}_q = -\boldsymbol{F}^{\wedge M}\left(\boldsymbol{n}^M_q\right) \quad (41)$$

Each column of the matrix in Equation (41) is a sensitivity vector that must propagate back through the network to generate one row of the Jacobian matrix. The columns are propagated backward as follows:

$$\hat{S}_q^m = \hat{F}^m\left(n_q^m\right)\left(W^{m+1}\right)\hat{S}_q^{m+1} \tag{42}$$

The augmentation that follows then obtains all of the Marquardt sensitivity matrices for the overall layers.

$$\hat{S}^m = [\hat{S}_1^m \vdots \hat{S}_2^m \vdots \hat{S}_3^m \vdots \hat{S}_4^m \vdots K \vdots \hat{S}_Q^m] \tag{43}$$

The proposed algorithm based on Levenberg–Marquardt's backpropagation algorithm for fault allocation is given for clarity in Table 1.

Table 1. LMBP algorithm.

LMBP Algorithm for the Fault Location Process
a. With initial weights and bias (randomly generated), all extracted features should be fed into the FFNN as inputs. The outputs of the corresponding features are computed in the network using Equations (6) and (7), followed by error prediction using $e_q = t_q - a_q^M$.
b. Using $F(x) = \sum_{q=1}^{Q} (t_q - a_q)^T (t_q - a_q)$, calculate the sum of squared errors for all inputs with the Q targets in the training set.
c. After initializing with Equation (41), calculate the sensitivity using Equation (42) and augment the individual matrices into the Marquardt sensitivities using Equation (43). Meanwhile, Equations (37) and (38) are used to determine elements of the Jacobian matrix.
d. Then, to obtain Δx_k, update Equation (33) to adjust weights and biases.
e. Using $x_k + \Delta x_k$ recalculate the total of the squared errors. If the newly generated error value is less than the previous one, then divide μ_k by α and return to step a with $x_{k+1} = x_k + \Delta x_k$. If the recalculated value does not decrease, then multiply μ_k by α and return to step c with the new weights.

2.4. Parameter Optimization

Hyperparameters should be distinguished from internal parameters such as weights and biases that are taken into account by the Levenberg–Marquardt backpropagation algorithm in the FFNN model. However, finding values for hyperparameters is a non-convex optimization process for optimal fitting. This is because, like the MT-HVdc system, most existing systems do not have linear responses to their control parameters. From the standpoint of optimization, the problem can be presented as follows:

$$\min_{x \epsilon X^d} f(x), \quad where\ x \epsilon X \subset \Re \tag{44}$$

x is the input vector (control parameters) of dimension d. $f(x)$ is an objective function that depicts a multiscale system with high dimensional control parameters functioning under high-speed channels, such as an FFNN-based relaying model under dynamic conditions to protect the MT-HVdc grid. It is not a simple task to create a precise and accurate model of such systems in this situation. As a result, it is necessary to approach the problem in Equation (44) using the black-box settings shown in Figure 3.

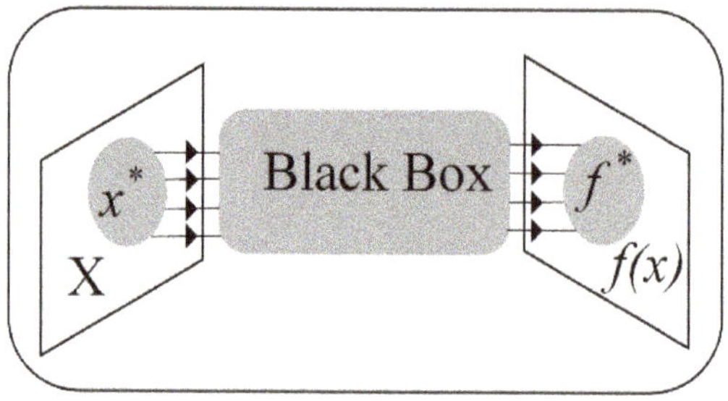

Figure 3. Optimization of black-box systems.

2.4.1. Black-Box Settings

In most black-box systems, including MT-HVdc grids relaying models, it is not easy to acquire $f(x)$ gradient information at an arbitrary value of x. However, gradient information is not required when employing BO based on Gaussian processes (GPs) [31]. As a result, it is a promising and appropriate candidate for black-box optimization. While optimizing, BO is an active learning method that chooses the next observation to maximize the reward for solving Equation (45). Its foundation is Bayes' Theorem.

$$P(f|D_{1:t}) \propto P(D_{1:t}|f)P(f) \tag{45}$$

$P(f)$, $P(f|D_{1:t})$ and $P(D_{1:t}|f)$ are probabilities of prior, posterior, and likelihood based on the current observations, i.e., $D_{1:t} = [(x_1, y_1), (x_2, y_2), .., (x_t, y_t)]$. Various predictive and distributional models can be used as priors in BO, but the GP is preferred due to its practical and theoretical advantages [31].

2.4.2. Gaussian Process (GP)

In the GP, the surrogate model replicates the behaviors of the expensive underlying function. While doing this, the underlying function $f(x)$ that requires optimization is represented in BO as a collaborative and multidimensional Gaussian process. The mean (μ) and covariance $(\mathcal{K})$ functions are calculated using:

$$f_{1:t} = \mathcal{N}(\mu(x_{1:t}), \mathcal{K}(x_{1:t})) \tag{46}$$

In BO, Equation (46) illustrates the process in which the predictive GP is trained. It is worth noting that, unlike other machine-learning algorithms, the goal of BO is to properly forecast where global extrema are situated in the sample space based on previous observations rather than to develop predictors that cover the entire sample space. Furthermore, the problem in Equation (44) is solved using black-box settings, implying that we do not have any prior information about the underlying function. Therefore, to improve the regression quality of the GP, we use a popular kernel/covariance function called the automatic relevance determination Matern 5/2 function in conjunction with a zero-mean GP for $P(f)$, given as:

$$K(x) = \begin{bmatrix} k(x_1, x_1) & \cdots & k(x_1, x_t) \\ \vdots & \ddots & \vdots \\ k(x_t, x_1) & \cdots & k(x_t, x_t) \end{bmatrix} \tag{47}$$

$$k(x_i, x_j) = \sigma_f^2(1 + \sqrt{5r} + \frac{5}{3}r^2)e^{-\sqrt{5r}} \tag{48}$$

where $r = (\sum_{d=1}^{D} \left(x_{i,d} - x_{j,d}\right)^2 / \sigma_d^2))^{1/2}$; σ_f and σ_d are hyperparameters of $K(x)$. These hyperparameters are modified throughout the training phase to reduce the GP's negative-log marginal likelihood using the global or local method. Each parameter in an ARD-type kernel has a scaling parameter that must be set. If the σ_d of one parameter is larger than the others after the GP-based predictive model has been trained, then it can be assumed that a change in this parameter has less sensitivity on the prediction. Furthermore, if a certain parameter has a greater effect, then the proposed solution in BO will alter the training process to reduce σ_d of that parameter in comparison to others. These advantages make the underlying function more interpretable and serve as an implicit sensitivity analysis.

2.4.3. Acquisition Function

Since the original function $f(x)$ is hard to estimate, based on a predefined strategy and auxiliary optimization, an acquisition function $u(x)$ is obtained to find the next point x_{t+1} of the solution. It is worth noting that $u(x)$ does not require any additional points; instead, it relies on past sample knowledge to make predictions at candidate points.

$$\mu(x_{t+1}) = k^T \mathcal{K}^{-1} f_{1:t} \tag{49}$$

$$\sigma^2(x_{t+1}) = k(x_{t+1}, x_{t+1}) - k^T \mathcal{K}^{-1} k \tag{50}$$

Then predictive distribution at the next point is given as:

$$P(f_{t+1}|D_{1:t}, x_{t+1}) \sim \mathcal{N}\left(\mu(x_{t+1}), \sigma^2(x_{t+1})\right) \tag{51}$$

The most prominent acquisition functions in BO are the probability of improvement, upper confidence bound and expected improvement per second. However, we propose an expected improvement per second-plus in this paper. In comparison, it allows for faster model building and optimization, and the term 'plus' prevents a region from overexploiting (more search for a global minimum). Expected improvement (EI) is given as:

$$\mu(EI) = \left(\mu(x) - f^{\wedge *} - \zeta\right)\Phi(Z) + \sigma(x)\phi(Z) \tag{52}$$

where $f^{\wedge *}$ is the best point observed so far. ζ is a hyperparameter for $\mu(EI)$, $Z = \left(\mu(x) - f^{\wedge *} - \zeta\right)/\sigma(x)$, $\phi(.)$ and $\Phi(.)$ are the probability density function and cumulative distribution function of normal distribution. Further interpreted in EI per second ($EIpS$) as:

$$EIpS(x) = \mu(EI)/\mu_S(x) \tag{53}$$

where $\mu_S(x)$ is the posterior mean of the timing Gaussian process model, respectively. The next sampling point x_{t+1} is found by minimizing the expected improvement per second-plus $EIpSp(x)$ acquisition function.

$$x_{t+1} = argmin\ EIpSp\left(\mu(x_{t+1}), \sigma^2(x_{t+1})\right) \tag{54}$$

In doing so, the proposed acquisition function escapes the local objective function minimum and searches for a global minimum by setting $\sigma_f(x)$ to be the posterior objective function ($P(f|D_{1:t})$) standard deviation at point x. Let σ_{NP} be the additive noise posterior standard deviation so that $\sigma_Q^2(x) = \sigma_F^2(x) + \sigma_{NP}^2$. The positive exploration ratio is denoted by $t_{\sigma NP}$. After each iteration, the acquisition function evaluates if the next point x satisfies $\sigma_f(x) < t_{\sigma NP}\sigma_{NP}$. If this is the case, then the acquisition function will announce that x is overexploiting and adjust its kernel function by multiplying θ by the number of iterations [32]. When compared to $EIpS(x)$, this adjustment increases the variation σ_Q for points between observations. It then creates a new point using the newly fitted kernel function. However, if the new point x is still being overexploited, then the function multiplies θ by a factor of ten and tries again. This process is repeated five times, with the goal of generating a point x that is not overexploited. The new x is accepted as the next exploration ratio by the proposed acquisition function. As a result, it manages the tradeoff between examining new points, searching for a better global solution, and focusing on nearby already investigated points. The whole process optimizes the FFNN structure in a much faster and more efficient manner with a reduced computation burden.

2.4.4. Implementation of Proposed Framework

The steps to train the FFNN model with the LMBP algorithm and optimize network hyperparameters with the Bayesian algorithm are demonstrated in Figure 4.

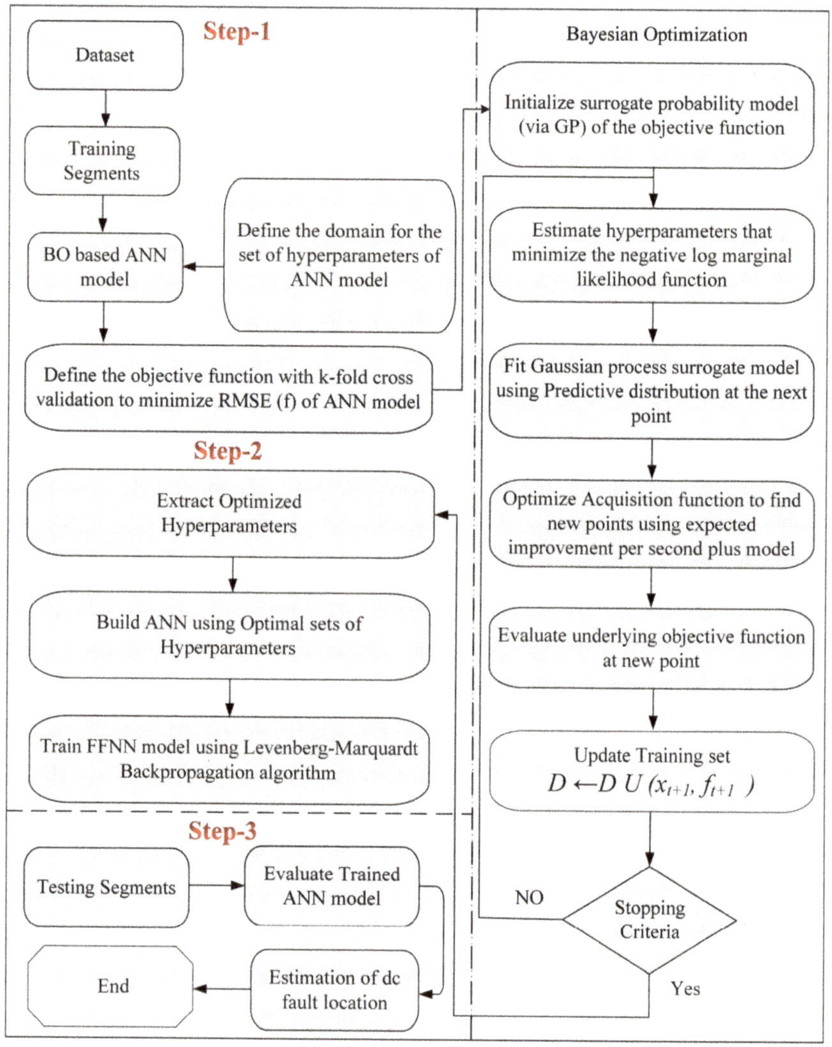

Figure 4. Proposed Framework.

In step 1, fault location and impedance are modified to create the training and testing datasets for several simulations. Additionally, data events are labeled and normalized according to criteria to improve the training process in this mode. The ANN hyperparameters are determined by feeding the training dataset into BO's AI model until the maximum number of iterations is reached. The AI model is updated each time the maximum number of iterations is reached. In step 2, the optimal hyperparameters of the ANN, which gives the minimum root-mean-square error (RMSE), are selected by BO, and the FFNN is trained for the given training data with the help of the LMBP algorithm. In step 3, the trained ANN model is evaluated on a different testing dataset from the training dataset.

To prevent overfitting, K-fold cross-validation was used during the assessment with K = 5. RMSE $=\sqrt{\frac{1}{R_z}\sum_{n=1}^{R_z}(y_n - y_n^{\circ})}$, R_z stands for the data size, y_n for the actual output, and y_n° represents the predicted output. The proposed framework can now be implemented; a system model will be presented in the next section, which enables the collection and analysis of input features for fault types, matching the theoretical foundation to real-world fault scenarios, and using intelligent computation to train and evaluate the framework's effectiveness.

3. System Model

The electrical power from two offshore wind farms is transferred to two onshore converters through dc transmission, as shown in Figure 5 [33]. A boundary is defined by installing current limiter inductors at the end of a dc line. Other test grid settings and MMC parameters are provided in Tables 2 and 3. The cable specifications are provided in Table 4. It is a single-end scheme, which means that information will be gathered near circuit breakers and inductor lines.

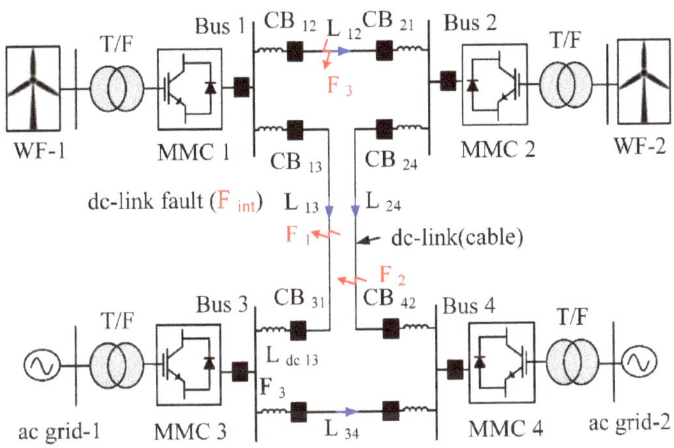

Figure 5. Configuration of MMC-based dc grid.

Table 2. Converter parameters.

Station	Rated dc Voltage [kV]	Rated Capacity [MVA]	Arm Capacitance C_{arm} (μF)	Arm Inductance L_{arm} [mH]	Arm Resistance R_{arm} [Ω]	Bus Filter Reactor [mH]
MMC1	±320	900	29.3	84.8	0.885	10
MMC2	±320	900	29.3	84.8	0.885	10
MMC3	±320	900	29.3	84.8	0.885	10
MMC4	±320	1200	39.0	63.6	0.67	10

Table 3. AC/dc System parameters.

dc System	Link12	Link13	Link34	Link24
Length [km]	100	200	100	150
Inductance [mH]	100	100	100	100
ac system	AC 1	AC 2	AC 3	AC 4
Rated voltage [kV]	400	400	400	400
Reactance X_{ac} [Ω]	17.7	17.7	17.7	13.4
Resistance R_{ac} [Ω]	1.77	1.77	1.77	1.34
Transformer μ_k [pu]	0.15	0.15	0.15	0.15

Table 4. Cable parameters.

Cable	Outer Radius [mm]	[Ωm]	ϵ_{re1} [-]	μ_{re1} [-]	Link34
Core	19.5	1.7×10^{-8}	–		1
Insulation	48.7	–	2.3	150	1
Sheath	51.7	2.2×10^{-7}	–	100	1
Insulation	54.7	–	2.3	AC4	1
Armor	58.7	1.8×10^{-7}	–	400	10
Insulation	63.7	–	2.3	13.4	1

Model Output

As shown in Table 5, the examined system model has several outputs that can be used to determine fault distance from a relay contact point. Additionally, it shows fault resistances and fault types along with a total of 714 dc-link fault scenarios (k) for training. By doing so, dc-link faults are categorized into pole-to-pole (PTP) and pole-to-ground (PTG) faults. It is important to note that a dc-link problem is an internal fault, so the criteria (dV_{dc}/dT) should be applied when an internal failure occurs. By activating this criterion, the trained algorithm begins sampling relevant values for the 10 ms time window and estimating the fault distance. Note that the fault detection strategy is selective in nature.

Table 5. Internal fault scenarios for training data.

Transient Period	Training Samples	Fault Resistance (Ω)	Fault Distance (km)	Noise (dB)
10 ms	357	0.01, 25, 50, … , 375, 400	1, 10, 20, … , 180, 190, 198	20, 25, 30

Total faulty sample = 357/each fault type; dc-link faults are first classified into two parts: pole to pole and pole to ground fault. Therefore, total training samples = k = ($\mathbf{F_{int}}$ = 357 ∗ 2) = 714. Fault distance is noted from MMC1 to MMC 3 and MMC1 to MMC2, respectively.

4. Data Processing

The fact that the initial travelling waves of the voltage and current induced by the dc-link faults from the system above contain helpful information about fault distance is exploited in this study [7]. However, noise interference is expected, considering the dynamic disturbances associated with the MT-HVdc system. Therefore, the following sub-section discusses the noise suppression mechanism before processing data for the regression model.

4.1. Signal Processing

The implementation of the DWT to suppress noises from a measured signal is shown in Figure 6 [34].

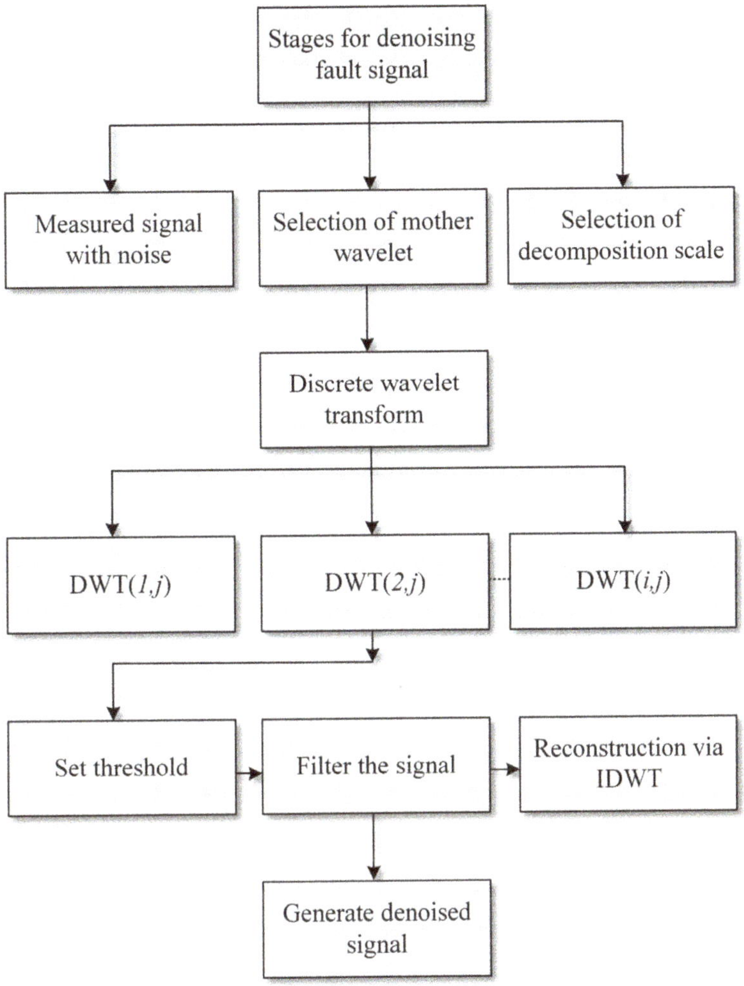

Figure 6. Signal denoising.

4.1.1. Setting Numbers of Decomposition Layers

Transforming discrete wavelets into more decomposition layers helps separate noise from the original signal, resulting in better signal filtering. We have chosen eight levels to keep the balance between signal processing burden and robustness against noise, corresponding to the frequency band of 195.3–390.6 Hz at a sampling frequency of 50 kHz.

4.1.2. Selection of Mother Wavelet Function

The next critical step in the denoising scheme is choosing a mother wavelet. A literature review and practical results presented in the previous studies show that Daubechies (dB) is an appropriate mother wavelet for analyzing fault signals [35]. It is suggested that, in this study, the Pearson correlation coefficient be used to determine the correlation between the Daubechies wavelet function and the cable fault signals in order to determine the best mother wavelet function. The mother wavelet function is written as follows:

$$\varnothing = \sum (X-\overline{X})(Y-\overline{Y}) / \sqrt{\sum (X-\overline{X})^2 (Y-\overline{Y})^2} \quad (55)$$

where X is the original fault signal, $\overline{X}$ denotes the original fault signal's average, Y denotes the noise-eliminated fault signal, and $\overline{Y}$ denotes the noise-eliminated fault signal's average.

4.1.3. Set the Threshold and Filter the Signal

After selecting the mother wavelet, the noise from the fault signal can be filtered out. The Universal threshold is multiplied by the median of each decomposition layer after wavelet decomposition to automatically set the threshold, as expressed:

$$\lambda_j = \frac{\sigma_j}{0.6745} * \sqrt{2\log n_j} \quad (56)$$

λ_j is the threshold of the jth decomposition layer, σ_j is the median of the jth decomposition layer, and n_j is the signal length of the jth decomposition layer. After setting the threshold, the noise is filtered out through the thresholding process. This thresholding process usually includes soft and hard thresholds [35]. However, in this study, a hard threshold is set to filter out the noise.

$$\delta_\lambda{}^{Hard} = \begin{bmatrix} x(t), & if|x(t)| > \lambda \\ 0, & otherwise \end{bmatrix} \quad (57)$$

This equation demonstrates that the hard threshold retains a larger wavelet coefficient while the coefficient below the threshold is set to zero. Finally, using inverse DWT (IDWT), the signal processed by the hard threshold can be configured layer by layer into a noise-free signal. The implementation of the proposed denoising approach with a 20 dB signal-to-noise ratio (SNR) is shown in Figure 7.

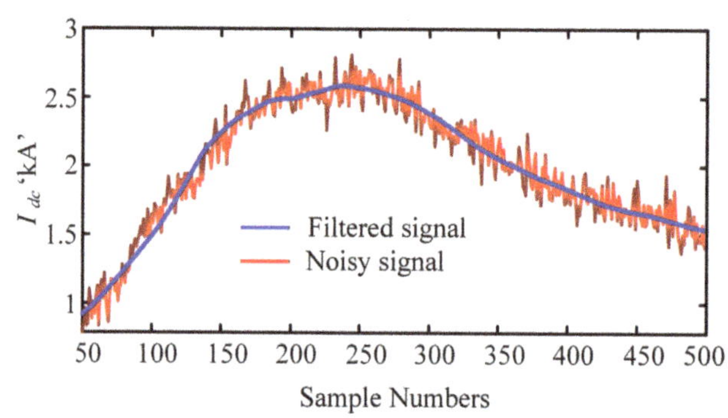

Figure 7. Effect of the denoised solution on the contaminated signal of 20 dB signal-to-noise ratio (SNR).

4.2. Feature Extraction Set-Up

After selecting and denoising the signal, the feature extraction stage is critical for data-driven-based fault detection and location estimation problems. Extracted features are measurable data taken from the transient of the current- and voltage-filtered signals to create a feature vector. This feature vector should be dimensionally compact to successfully implement the learning and generalization processes in the estimation algorithms for fault location. The feature extraction stage is divided into two sub-stages. The first stage involves decomposing all generated samples for each fault location up to eight levels using DWT-MRA to obtain wavelet coefficients. The wavelet coefficients are A_j approximation and D_j detail levels. For each type of fault location, vectors of D1–D8 and A8 coefficients are obtained. The second stage of feature extraction involves providing effective and appropriate statistical parameters for feature vector creation to reduce the collected data and improve estimation performance.

4.2.1. Feature Extraction Results

When a large number of high-frequency components of voltage and current signals are fed for training, several learning tools face problems due to a limitation on the input space dimension. These learning tools lack the capability to provide suitable learning patterns with a large number of features. This is due to the enlargement of the structure and an extreme increase in the number of learning parameters [11]. The regression model used in this study is designed to train with the second norm (referred to as the norm) of the wavelet coefficients. In general, the decomposed signal's norm for wavelet coefficients is determined as follows:

$$norm_{D_j} = \sqrt{\sum_{i=1}^{n} | D_{i,j} |^2} \tag{58}$$

$$norm_{A_j} = \sqrt{\sum_{i=1}^{n} | A_{i,j} |^2} \tag{59}$$

j denotes the decomposition level, and the maximum level of decomposition is N. The detail and approximate coefficients have n values at level j. Overall, the proposed energy vector obtained from the MRA-based DWT for any current or voltage signal from a given time window is represented as

$$x = \left[norm_{D_1}, norm_{D_2}, \ldots, norm_{D_8}, norm_{A_8} \right] \tag{60}$$

Using the MRA-based DWT, norm values of current for ground faults at various sites are calculated and presented in Figure 8, respectively. There is a distinct difference in the approximate norms between the given fault locations at levels D6 through D8. These differences in norms indicate that the obtained features contain distinct fingerprints for estimating ground faults at various places. Figure 9 shows the obtained features of the voltage signal for ground faults between locations 40 to 200 km.

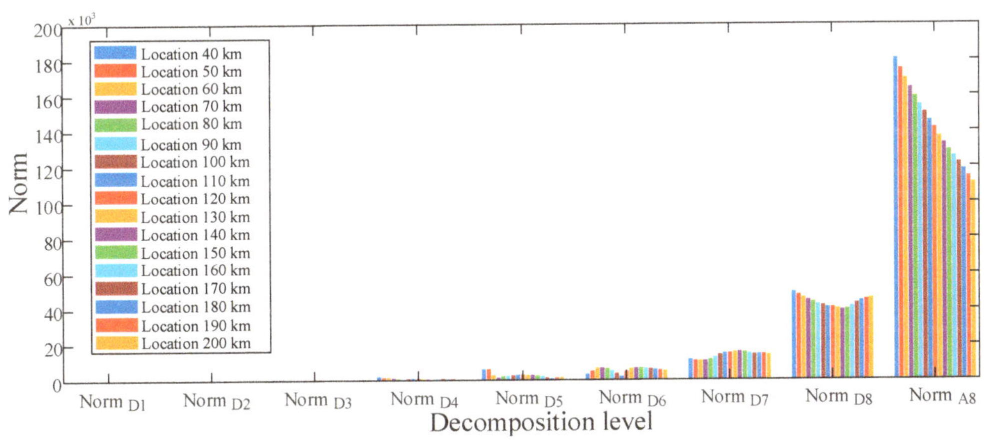

Figure 8. Feature vector extracted for ground fault at various locations of the current signal.

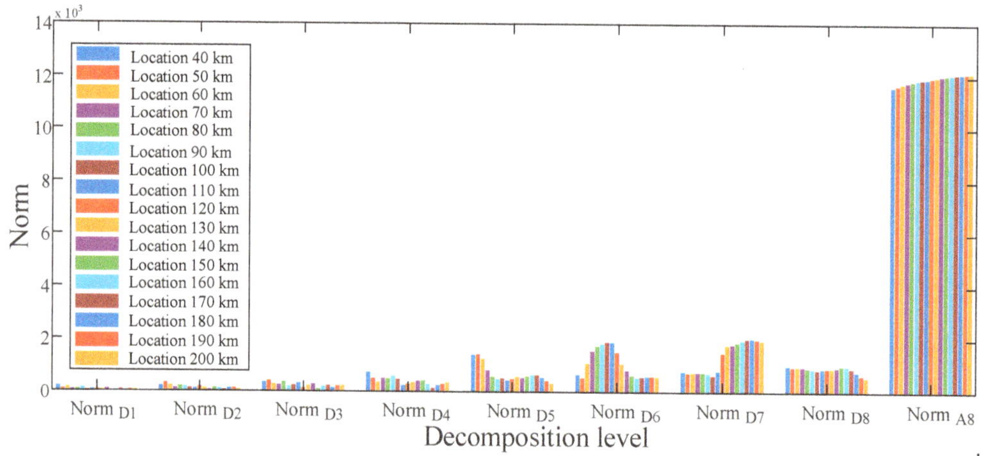

Figure 9. Feature vector extracted for ground fault at various locations of the voltage signal.

In Figure 9, the norm values for each location are significantly different in the dominant frequency band between D5 and D8 and can be used as input vectors to establish fault estimation rules. Similarly, as illustrated in Figure 10, a unique signature of the pole-to-pole fault may be derived at different frequency bands. A schematic diagram for the feature vector development process is shown in Figure 11.

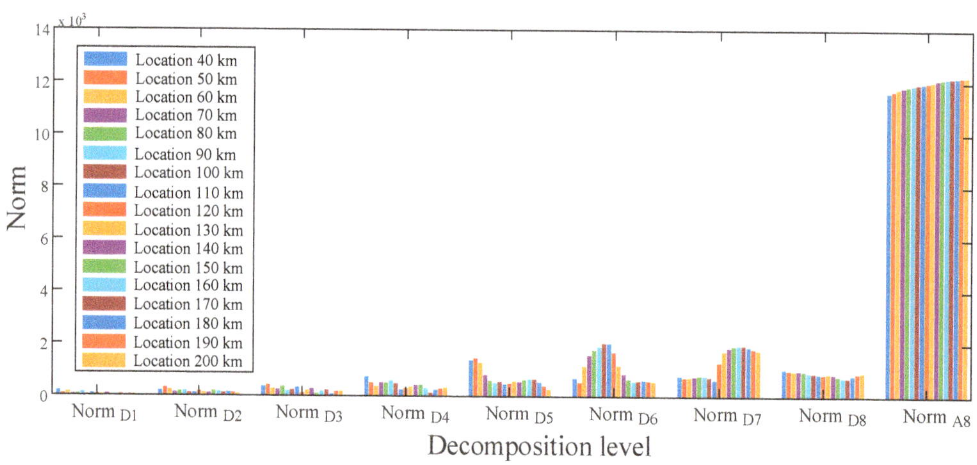

Figure 10. Feature vector extracted for the PTP fault at various locations of the voltage signal.

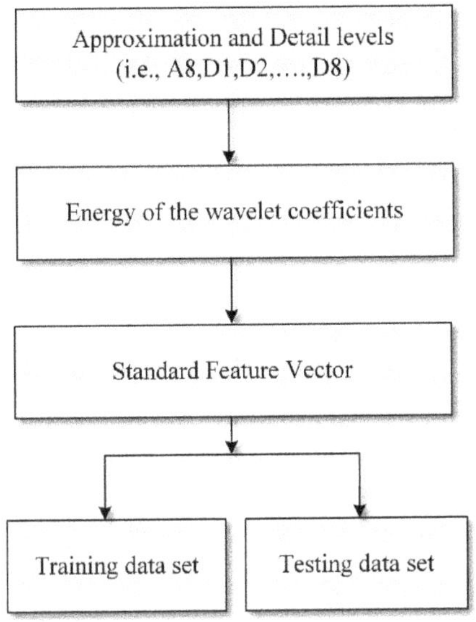

Figure 11. Flow chart for the development of the feature vector.

4.2.2. Training Set-Up

Following preprocessing strategies, these extracted features are standardized for computational simplification. The decluttered training dataset is then applied to the BO-based AI model to find the appropriate hyperparameters for the FFNN once the feature vectors have been determined. The input vector $\mathbf{p} = (x_1, x_2, x_3, x_4)$ of 10 ms is designed for the FFNN input; two inputs (x_1, x_2) represent the transient dc current second norm from positive and negative poles, while the rest (x_3, x_4) indicate the dc voltage second norm from positive and negative poles. This corresponds to 36 inputs for each training sample (total training samples = k = 714). In doing so, BO's AI model is modified each time until the maximum number of iterations is reached. BO then selects the ideal FFNN hyperparameters that result in the lowest RMSE, and the FFNN is trained using the LMBP algorithm. The final RMSE obtained is 0.0132, with a total evaluation time of 39.3428 s for 30 iterations. Some key hyperparameters of the multilayer FFNN model obtained via BO are presented in Table 6.

Table 6. Optimized parameters.

Hyperparameters	Range	Fault Location Model
Learning Rate	$[1 \times 10^{-2}–1]$	0.010037
Hidden Layers/Neurons (NHL)	[1–40]	28
Momentum	[0.001–0.005]	0.0028608
Epochs	[20–1000]	994
Gradient	$[1 \times 10^{-7}–10^{-6}]$	1.2925×10^{-7}
Validation	[0–6]	4

5. Simulation Results and Discussions

A. Metric for Evaluation and Testing Set-Up

Although, during validation, the selected models' average estimation accuracy was 98.94%. However, we tested our method for further investigation using case studies given in Table 7. For verification and more in-depth analysis, a performance index based on percentage error was used as follows:

$$Percentage\ error = \frac{Actual\ Location - Prediction\ location}{Total\ lenght\ of\ transmission\ line} \times 100 \quad (61)$$

Table 7. Testing fault scenarios for testing data.

Transient Period [10 ms]	Testing Samples	Fault Resistance (Ω)	Fault Distance (km)	Noise (dB)
10 ms	400	10, 35, 60, 85, ... , 435, 460, 485	5, 15, 25, ... , 175, 185, 195	20, 25, 45
Total faulty sample = 400/each fault type, Total testing samples = [(400) ∗ 2] = 800, Refer Table 6 for fault distance				

5.1. Case 1 (Fault Location)

In Case 1 (under varying fault locations and fault resistance), the functionality of the proposed technique was tested using the scenarios given in Table 7. After thorough training, fault analyses were carried out with varying fault distances and resistances. Table 8 shows the 800 test samples, absolute and percentage errors for two types of dc-link faults: PTP and PTG. It can be observed that the percentage error for the testing dataset was found to be 0.4927% and 0.5361% for the PTP fault and PTG fault, respectively. The proposed technique's total percentage error was found to be 0.5144 percent, which demonstrated that the misclassification was well within acceptable bounds.

Table 8. Fault location estimation errors.

Fault Type	Total Faults	Max Absolute Error (km)	Max Percentage Error (%)	Overall Absolute Error (km)	Overall Percentage Error (%)
PTP	400	2.6350	1.3174	0.9853	0.4927
PTG	400	2.6412	1.3206	1.0723	0.5361
Average Error	NA	NA	NA	1.0288	0.5144

In addition, Figure 12 depicts the percentage inaccuracy for the proposed technique in locating PTP faults on line 13 PTP faults with fault distances ranging from 5 km to 200 km. With a maximum percentage error of 1.3174% at 175 km and a minimum value of 0.00103% at 15 km, the findings revealed that the proposed algorithm had no major impact on the variance of fault distance. Therefore, the proposed approach is suitable for locating close-in and far-away faults.

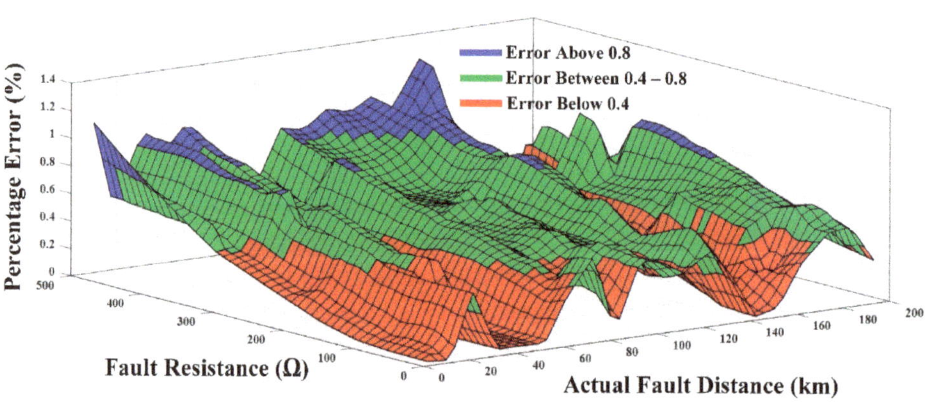

Figure 12. Accuracy of the proposed technique.

5.2. Case 2 (Fint)

Apart from fault location, it is important to note that the characteristics and amplitude of faulty signals, such as voltage and current measured at the local terminal, are also determined by fault parameters such as fault resistance. Therefore, it is crucial to highlight the proposed approach's performance under diverse fault resistances. This section analyzes the proposed algorithm's performance for in-depth fault resistance validity ranging from 10 to 385 Ω, and the results are given in Table 9. Notably, in the event of high fault resistance, such as 385 Ω, with an actual fault distance of 185 km, the energy of the travelling waves tended to be on the lower side, bringing the system closer to the steady state. However, the proposed algorithm with selected features extracted even the most minute voltage and current information. For example, the predicted fault distances for PTP and PTG at 385 Ω were 183.63147 km and 186.93141 km, respectively. The associated misclassification of 0.68427% and 0.96571% for each fault type was well within acceptable limits.

Table 9. Fault resistance estimation errors.

Fault Location	**Fault Resistance (Ω)**	**Fault Type**		**dc-Link Fault Location Results**					
				Predicted Location		**Absolute Error**		**Percentage Error (%)**	
				PTP	**PTG**	**PTP**	**PTG**	**PTP**	**PTG**
5 km of dc link	10	PTP	PTG	5.02021	5.03022	0.02021	0.03022	0.01011	0.01511
	110	PTP	PTG	5.07141	5.08142	0.07141	0.08142	0.03571	0.04071
	260	PTP	PTG	5.63413	5.76481	0.63413	0.76481	0.31707	0.38241
35 km of dc link	35	PTP	PTG	35.05123	35.07134	0.05123	0.07134	0.025615	0.03567
	235	PTP	PTG	35.62858	36.10184	0.62858	1.10184	0.31429	0.55092
	285	PTP	PTG	35.86144	36.31471	0.86144	1.31471	0.43072	0.65736
125 km of dc link	260	PTP	PTG	126.67141	126.81487	1.67141	1.81487	0.83571	0.90744
	385	PTP	PTG	126.76175	126.91231	1.76175	1.91231	0.88088	0.956155
	110	PTP	PTG	126.01522	126.52812	1.01522	1.52812	0.50761	0.76406
185 km of dc link	260	PTP	PTG	186.94571	186.75387	1.94571	1.75387	0.972855	0.87694
	385	PTP	PTG	183.63147	186.93141	1.36853	1.93141	0.68427	0.96571
	110	PTP	PTG	186.34578	186.53681	1.34578	1.53681	0.67289	0.76841
Normal operation	X	X X	X	NOT APPLICABLE		NOT APPLICABLE		NOT APPLICABLE	

5.3. Case 3 (Noisy Events)

In this case, a white Gaussian was added to the testing signals to examine the proposed fault-locating scheme under various noisy occurrences. Original signals with SNRs ranging from 20 to 45 dB were employed to assess fault location performance. Table 10 indicates that the proposed scheme could locate all sorts of faults with a reasonable mean percentage error rate for close-in, mid-point of line, and far-end of line. In the case of 45 dB noise additions at the far end of 155 km of the dc-link, the total mean percentage error was 0.72424% and 0.83147% for PTP and PTG faults. It is worth noting that the proposed method was noise-resistant because of the denoising process with better threshold settings and functions. This improved the estimation accuracy despite the high noise level of 20 dB with an overall mean percentage error of 0.9411% and 0.8561% for PTP and PTG faults, respectively.

Table 10. Results under the different noisy event.

Noise (dB)	Fault Location	Fault Resistance (Ω)	Fault Type		dc-Link Fault Location Results					
					Predicted Location		Absolute Error		Percentage Error (%)	
					PTP	PTG	PTP	PTG	PTP	PTG
25	5 km of dc link	10	PTP	PTG	5.04512	5.06727	0.04512	0.06727	0.02256	0.03364
		110	PTP	PTG	6.01202	6.03567	1.01202	1.03567	0.50601	0.51784
		260	PTP	PTG	6.26783	6.15872	1.26783	1.15872	0.63392	0.57936
20	45 km of dc link	35	PTP	PTG	46.06982	46.23672	1.06982	1.23672	0.53491	0.61836
		235	PTP	PTG	46.84612	47.03452	1.84612	2.03452	0.92306	1.01726
		285	PTP	PTG	47.03487	46.76324	2.03487	1.76324	1.01744	0.88162
45	155 km of dc link	260	PTP	PTG	156.96342	154.06853	1.96342	0.93147	0.98171	0.46574
		385	PTP	PTG	154.13647	153.02356	0.86353	1.97644	0.43177	0.98822
		110	PTP	PTG	154.43628	154.36571	0.56372	0.63429	0.28186	0.31715

5.4. Case 4 (Comparison with Existing Methods)

To further validate the proposed scheme's robustness, Figure 13 replaces it with intelligent adversaries such as the conventional FFNN and BP-NN with an original current signal as the input under the testing conditions listed in Table 7.

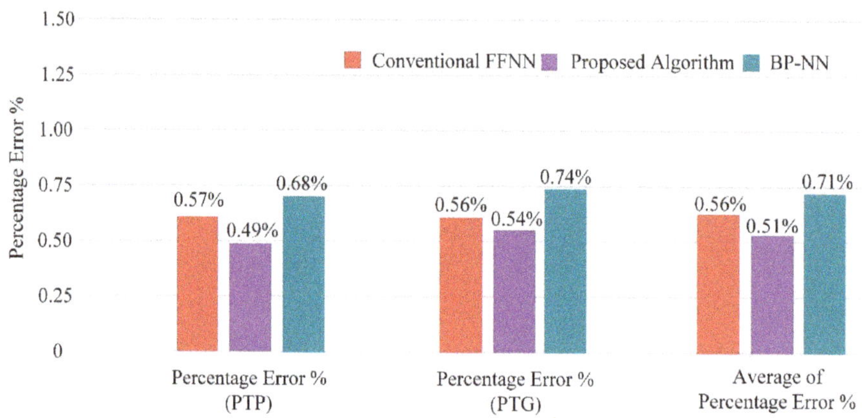

Figure 13. Comparative analysis.

On a dual 2.9 GHz, Intel Core i7 with 16 GB RAM, the current version of the algorithm implemented in Matlab® R2020a took 39.3428 s to run. Thirty ANN models were selected, trained, and validated with this runtime. It was approximately five times faster than a conventional FFNN configured manually with hyperparameters. The results showed that the proposed algorithm performed better than the BP-NN and had the lowest percentage error (i.e., 0.49%, 0.54% and 0.51%) for all fault types. In terms of percentage error, the conventional FFNN with hyperparameters such as 15 neurons in the hidden layer and a learning rate of 0.01 gave an average percentage error of 0.56%. This showed that efficient features and regulating parameters in the proposed algorithm helped to increase the interpretability of the spectrum generated by the wavelet.

6. Comparison and Analysis

This section compares the proposed methodology with existing fault estimation schemes for the MT-HVdc grid.

6.1. Non-AI-Based Methods

The proposed fault location method utilizes a continuous wavelet transform on dc line current signals in the MT-HVdc network [36]. The technique is quite efficient; however, a high sampling frequency of 200 kHz and time-synchronized measurements are required. Further, evaluation under high fault resistance has not been investigated thoroughly. Another work used time-stamped measurements to locate faults at a 200 kHz sampling frequency [37]. The proposed model is robust against noise measurement, but high sampling frequency and synchronized measurements could be a barrier to practical applications. The single-ended TW-based fault location model has no synchronized measurement issue [38] but has a high sampling frequency (100 kHz) [39]. In another example, modal voltage and current measurements are sampled at 1 MHz to develop a single-end fault location model [40]. However, it has only been tested for 100 Ω fault resistance. All the aforementioned TW-based fault location models require a high sampling frequency for good accuracy. Such a requirement is frequently considered a drawback. In comparison, the proposed single-end fault location approach operates with reasonable sampling frequency and tests against fault resistance as high as 485 Ω.

6.2. AI-Based Methods

Among the fault location approaches, learning-based techniques fall into a distinct category. Even though such practices are commonly utilized in AC systems for fault localization, few papers discuss their relevance to MT-HVdc networks. For example, an extreme learning machine was proposed to locate the fault in the MT-HVdc network [41]. Voltage and current measurements were captured at a 500 kHz sampling frequency during the learning phase to perform the wavelet transform and s-transform for feature extraction. However, the entire scheme has been tested for fault resistance up to 100 Ω. Similarly, the high voltage and current measurements sampled at 200 kHz and the investigation of highly resistive faults are missing [42]. Another method applied a traditional two-ended TW-based fault location algorithm to current measurements sampled at 5 kHz [43]. The distance inaccuracy caused by the moderate sampling frequency was subsequently reduced using a machine-learning approach. However, utilizing multiple distributed sensors on long transmission added cost to the method. With the help of the ANN, the real-time implementation of the proposed method is quite efficient. It has been proven to have a low execution time on low-spec machines [44]. Further, all the aforementioned models do not discuss the optimization of the machine-learning model. The proposed approach optimizes the pre-training set-up with the help of Bayesian optimization.

7. Conclusions

At first, a novel dc fault location scheme based on AI for a meshed dc grid is proposed. The BO-based FFNN model with DWT application is used to determine the best hyperparameters that improve the selected model's performance while keeping the RMSE low. Levenberg–Marquardt backpropagation is used to adjust weights and biases during training for the chosen multilayer FFNN model. The contribution of this work is summarized as follows:

1. The wavelet coefficient energies of voltage and current over 10 ms are calculated and denoised during the learning phase for feature extraction. This leads to fewer features yet is robust for the learning model.
2. A comprehensive training dataset is collected to train the multilayer FFNN model for different fault locations by varying fault impedance.

3. The performance of this model is then evaluated on data points that are not included in the training dataset. The study results show that the fault location can be calculated using the FFNN for fault resistance up to 485 Ω.
4. Because the signal and Gaussian noise are integrated into the FFNN training sets, the influence of the noise-contained environment is reduced.
5. Due to plug-and-play capability, the suggested intelligent algorithm is tailored for a multi-vendor-based fault location estimation strategy in meshed MT-HVdc grids.
6. The case studies show that the proposed scheme performs well against many variables, such as different fault resistances, transmission line lengths, and non-ideal noise events. Thus, that makes it feasible for practical application in the MT-HVdc grid.

In future work, variable time windows will be used to consider the effect of the fault location, fault resistance, and computational burden. This work provides an analysis of the fault location estimation method for HVdc cable grids that can be applied to hybrid cable–overhead line systems as well.

Author Contributions: Conceptualization, M.Z.Y.; methodology, M.Z.Y. and M.F.T.; software, M.Z.Y.; validation, M.A.K. and M.Z.Y.; mathematical analysis, M.A.K. and M.Z.Y.; investigation, M.F.T.; resources, M.Z.Y. and A.R.; writing original draft preparation, M.Z.Y.; writing review and editing, M.Z.Y. and A.R.; visualization, M.Z.Y. and A.R.; supervision, A.R.; project administration, M.A.K., F.B., M.F.T. and M.Z.Y.; funding acquisition, M.Z.Y. All authors have read and agreed to the published version of the manuscript.

Funding: This research received no external funding.

Institutional Review Board Statement: Not applicable.

Informed Consent Statement: Not applicable.

Data Availability Statement: Not applicable.

Acknowledgments: The authors extend their appreciation for the support provided by the Hubei University of Automotive Technology (Shiyan, China) under a long-term innovation project (Project name: BK202211).

Conflicts of Interest: The authors declare no conflict of interest.

References

1. Fairley, P. China's ambitious plan to build the world's biggest supergrid. *IEEE Spectr.* **2019**, *1*, 35–41.
2. He, Z.-Y.; Liao, K.; Li, X.-P.; Lin, S.; Yang, J.-W.; Mai, R.-K. Natural frequency-based line fault location in HVDC lines. *IEEE Trans. Power Deliv.* **2014**, *29*, 851–859. [CrossRef]
3. Suonan, J.; Gao, S.; Song, G.; Jiao, Z.; Kang, X. A novel fault-location method for HVDC transmission lines. *IEEE Trans. Power Deliv.* **2009**, *25*, 1203–1209. [CrossRef]
4. Zhang, M.; Li, J.; Li, Y.; Xu, R. Deep learning for short-term voltage stability assessment of power systems. *IEEE Access* **2021**, *9*, 29711–29718. [CrossRef]
5. Xiao, D.; Chen, H.; Wei, C.; Bai, X. Statistical Measure for Risk-Seeking Stochastic Wind Power Offering Strategies in Electricity Markets. *J. Mod. Power Syst. Clean Energy* **2021**, *10*, 1437–1442. [CrossRef]
6. Nsaif, Y.M.; Hossain Lipu, M.S.; Hussain, A.; Ayob, A.; Yusof, Y.; Zainuri, M.A.A. A Novel Fault Detection and Classification Strategy for Photovoltaic Distribution Network Using Improved Hilbert–Huang Transform and Ensemble Learning Technique. *Sustainability* **2022**, *14*, 11749. [CrossRef]
7. Zhang, C.; Song, G.; Wang, T.; Yang, L. Single-ended traveling wave fault location method in DC transmission line based on wave front information. *IEEE Trans. Power Deliv.* **2019**, *34*, 2028–2038. [CrossRef]
8. Hamidi, R.J.; Livani, H. Traveling-wave based fault-location algorithm for hybrid multiterminal circuits. *IEEE Trans. Power Deliv.* **2016**, *32*, 135–144. [CrossRef]
9. Ahmadimanesh, A.; Shahrtash, S.M. Transient-based fault-location method for multiterminal lines employing S-transform. *IEEE Trans. Power Deliv.* **2013**, *28*, 1373–1380. [CrossRef]
10. Perveen, R.; Mohanty, S.R.; Kishor, N. Fault location in VSC-HVDC section for grid integrated offshore wind farm by EMD. In Proceedings of the 18th Mediterranean Electrotechnical Conference (MELECON), Lemesos, Cyprus, 18–20 April 2016; IEEE: Lemesos, Cyprus, 2016; pp. 1–5.
11. Farshad, M.; Sadeh, J. Accurate single-phase fault-location method for transmission lines based on k-nearest neighbor algorithm using one-end voltage. *IEEE Trans. Power Deliv.* **2012**, *27*, 2360–2367. [CrossRef]

12. Samantaray, S. A systematic fuzzy rule based approach for fault classification in transmission lines. *Appl. Soft Comput.* **2013**, *13*, 928–938. [CrossRef]
13. Ola, S.R.; Saraswat, A.; Goyal, S.K.; Jhajharia, S.; Rathore, B.; Mahela, O.P. Wigner distribution function and alienation coefficient-based transmission line protection scheme. *IET Gener. Transm. Distrib.* **2020**, *14*, 1842–1853. [CrossRef]
14. Ram Ola, S.; Saraswat, A.; Goyal, S.K.; Jhajharia, S.; Khan, B.; Mahela, O.P.; Haes Alhelou, H.; Siano, P. A protection scheme for a power system with solar energy penetration. *Appl. Sci.* **2020**, *10*, 1516. [CrossRef]
15. Samantaray, S.; Dash, P.; Panda, G. Fault classification and location using HS-transform and radial basis function neural network. *Electr. Power Syst. Res.* **2006**, *76*, 897–905. [CrossRef]
16. Salat, R.; Osowski, S. Accurate fault location in the power transmission line using support vector machine approach. *IEEE Trans. Power Syst.* **2004**, *19*, 979–986. [CrossRef]
17. Luo, G.; Yao, C.; Tan, Y.; Liu, Y. Transient signal identification of HVDC transmission lines based on wavelet entropy and SVM. *J. Eng.* **2019**, *2019*, 2414–2419. [CrossRef]
18. Malathi, V.; Marimuthu, N.; Baskar, S. Intelligent approaches using support vector machine and extreme learning machine for transmission line protection. *Neurocomputing* **2010**, *73*, 2160–2167. [CrossRef]
19. Farshad, M. Detection and classification of internal faults in bipolar HVDC transmission lines based on K-means data description method. *Int. J. Electr. Power Energy Syst.* **2019**, *104*, 615–625. [CrossRef]
20. Ekici, S.; Yildirim, S.; Poyraz, M. Energy and entropy-based feature extraction for locating fault on transmission lines by using neural network and wavelet packet decomposition. *Expert Syst. Appl.* **2008**, *34*, 2937–2944. [CrossRef]
21. Jayamaha, D.; Lidula, N.; Rajapakse, A.D. Wavelet-multi resolution analysis based ANN architecture for fault detection and localization in DC microgrids. *IEEE Access* **2019**, *7*, 145371–145384. [CrossRef]
22. Merlin, V.L.; dos Santos, R.C.; Le Blond, S.; Coury, D.V. Efficient and robust ANN-based method for an improved protection of VSC-HVDC systems. *IET Renew. Power Gener.* **2018**, *12*, 1555–1562. [CrossRef]
23. Karmacharya, I.M.; Gokaraju, R. Fault location in ungrounded photovoltaic system using wavelets and ANN. *IEEE Trans. Power Deliv.* **2017**, *33*, 549–559. [CrossRef]
24. Torun, H.M.; Swaminathan, M.; Davis, A.K.; Bellaredj, M.L.F. A global Bayesian optimization algorithm and its application to integrated system design. *IEEE Trans. Very Large Scale Integr. Syst.* **2018**, *26*, 792–802. [CrossRef]
25. Chen, P.; Merrick, B.M.; Brazil, T.J. Bayesian optimization for broadband high-efficiency power amplifier designs. *IEEE Trans. Microw. Theory Tech.* **2015**, *63*, 4263–4272. [CrossRef]
26. Park, S.J.; Bae, B.; Kim, J.; Swaminathan, M. Application of machine learning for optimization of 3-D integrated circuits and systems. *IEEE Trans. Very Large Scale Integr. Syst.* **2017**, *25*, 1856–1865. [CrossRef]
27. Lv, C.; Xing, Y.; Zhang, J.; Na, X.; Li, Y.; Liu, T.; Cao, D.; Wang, F.-Y. Levenberg–Marquardt backpropagation training of multilayer neural networks for state estimation of a safety-critical cyber-physical system. *IEEE Trans. Ind. Inform.* **2017**, *14*, 3436–3446. [CrossRef]
28. Soualhi, A.; Makdessi, M.; German, R.; Echeverría, F.R.; Razik, H.; Sari, A.; Venet, P.; Clerc, G. Heath monitoring of capacitors and supercapacitors using the neo-fuzzy neural approach. *IEEE Trans. Ind. Inform.* **2017**, *14*, 24–34. [CrossRef]
29. Hagan, M.T.; Menhaj, M.B. Training feedforward networks with the Marquardt algorithm. *IEEE Trans. Neural Netw.* **1994**, *5*, 989–993. [CrossRef]
30. Sadiq, M.T.; Yu, X.; Yuan, Z.; Aziz, M.Z.; Siuly, S.; Ding, W. Toward the development of versatile brain–computer interfaces. *IEEE Trans. Artif. Intell.* **2021**, *2*, 314–328. [CrossRef]
31. Sharifzadeh, M.; Sikinioti-Lock, A.; Shah, N. Machine-learning methods for integrated renewable power generation: A comparative study of artificial neural networks, support vector regression, and Gaussian Process Regression. *Renew. Sustain. Energy Rev.* **2019**, *108*, 513–538. [CrossRef]
32. Bull, A.D. Convergence rates of efficient global optimization algorithms. *J. Mach. Learn. Res.* **2011**, *12*, 2879–2904.
33. Leterme, W.; Ahmed, N.; Beerten, J.; Ängquist, L.; Van Hertem, D.; Norrga, S. A New HVDC Grid Test System for HVDC Grid Dynamics and Protection Studies in EMT-Type Software. In Proceedings of the 11th IET International Conference on AC and DC Power Transmission, Birmingham, AL, USA, 10–12 February 2015; IET, 2015; pp. 1–7.
34. Wang, M.-H.; Lu, S.-D.; Liao, R.-M. Fault diagnosis for power cables based on convolutional neural network with chaotic system and discrete wavelet transform. *IEEE Trans. Power Deliv.* **2021**, *21*, 2285–2294. [CrossRef]
35. Ukil, A.; Yeap, Y.M.; Satpathi, K. *Fault Analysis and Protection System Design for DC Grids*; Springer: Berlin/Heidelberg, Germany, 2020.
36. Nanayakkara, O.K.; Rajapakse, A.D.; Wachal, R. Traveling-wave-based line fault location in star-connected multiterminal HVDC systems. *IEEE Trans. Power Deliv.* **2012**, *27*, 2286–2294. [CrossRef]
37. Nanayakkara, O.K.; Rajapakse, A.D.; Wachal, R. Location of DC line faults in conventional HVDC systems with segments of cables and overhead lines using terminal measurements. *IEEE Trans. Power Deliv.* **2011**, *27*, 279–288. [CrossRef]
38. Azizi, S.; Sanaye-Pasand, M.; Abedini, M.; Hasani, A. A traveling-wave-based methodology for wide-area fault location in multiterminal DC systems. *IEEE Trans. Power Deliv.* **2014**, *29*, 2552–2560. [CrossRef]
39. Tzelepis, D.; Psaras, V.; Tsotsopoulou, E.; Mirsaeidi, S.; Dyśko, A.; Hong, Q.; Dong, X.; Blair, S.M.; Nikolaidis, V.C.; Papaspiliotopoulos, V. Voltage and current measuring technologies for high voltage direct current supergrids: A technology review identifying the options for protection, fault location and automation applications. *IEEE Access* **2020**, *8*, 203398–203428. [CrossRef]

40. Ashouri, M.; da Silva, F.F.; Bak, C.L. On the application of modal transient analysis for online fault localization in HVDC cable bundles. *IEEE Trans. Power Deliv.* **2019**, *35*, 1365–1378. [CrossRef]
41. Hadaeghi, A.; Samet, H.; Ghanbari, T. Multi extreme learning machine approach for fault location in multi-terminal high-voltage direct current systems. *Comput. Electr. Eng.* **2019**, *78*, 313–327. [CrossRef]
42. Livani, H.; Evrenosoglu, C.Y. A single-ended fault location method for segmented HVDC transmission line. *Electr. Power Syst. Res.* **2014**, *107*, 190–198. [CrossRef]
43. Tzelepis, D.; Dyśko, A.; Fusiek, G.; Niewczas, P.; Mirsaeidi, S.; Booth, C.; Dong, X. Advanced fault location in MTDC networks utilising optically-multiplexed current measurements and machine learning approach. *Int. J. Electr. Power Energy Syst.* **2018**, *97*, 319–333. [CrossRef]
44. Tsotsopoulou, E.; Karagiannis, X.; Papadopoulos, P.; Dyśko, A.; Yazdani-Asrami, M.; Booth, C.; Tzelepis, D. Time-domain protection of superconducting cables based on artificial intelligence classifiers. *IEEE Access* **2022**, *10*, 10124–10138. [CrossRef]

sensors

Article

Hybrid CNN–Transformer Network for Electricity Theft Detection in Smart Grids

Yu Bai, Haitong Sun *, Lili Zhang and Haoqi Wu

School of Electronical and Information Engineering, Shenyang Aerospace University, Shenyang 110136, China; yubai@sau.edu.cn (Y.B.); 20052727@sau.edu.cn (L.Z.); wuhaoqi@stu.sau.edu.cn (H.W.)
* Correspondence: sunhaitong@stu.sau.edu.cn

Abstract: Illicitly obtaining electricity, commonly referred to as electricity theft, is a prominent contributor to power loss. In recent years, there has been growing recognition of the significance of neural network models in electrical theft detection (ETD). Nevertheless, the existing approaches have a restricted capacity to acquire profound characteristics, posing a persistent challenge in reliably and effectively detecting anomalies in power consumption data. Hence, the present study puts forth a hybrid model that amalgamates a convolutional neural network (CNN) and a transformer network as a means to tackle this concern. The CNN model with a dual-scale dual-branch (DSDB) structure incorporates inter- and intra-periodic convolutional blocks to conduct shallow feature extraction of sequences from varying dimensions. This enables the model to capture multi-scale features in a local-to-global fashion. The transformer module with Gaussian weighting (GWT) effectively captures the overall temporal dependencies present in the electricity consumption data, enabling the extraction of sequence features at a deep level. Numerous studies have demonstrated that the proposed method exhibits enhanced efficiency in feature extraction, yielding high F1 scores and AUC values, while also exhibiting notable robustness.

Keywords: electricity theft detection; transformer neural network; convolutional neural network; smart grids

Citation: Bai, Y.; Sun, H.; Zhang, L.; Wu, H. Hybrid CNN–Transformer Network for Electricity Theft Detection in Smart Grids. *Sensors* **2023**, *23*, 8405. https://doi.org/10.3390/s23208405

Academic Editor: Arshad Arshad

Received: 31 August 2023
Revised: 9 October 2023
Accepted: 10 October 2023
Published: 12 October 2023

1. Introduction

With the advancement of smart grid technology and the ongoing expansion of power system infrastructure, the power industry, as a fundamental sector facilitating national economic growth, has increasingly emphasized the need to enhance the economic efficiency and ensure the stable operation of power companies [1]. The categorization of electricity losses can be divided into two main types: technical losses (TLs) and non-technical losses (NTLs) [2]. Technical losses are a result of disparities in infrastructure and energy dissipation, whereas non-technical losses emerge from the disparity between the total power transferred over distribution lines and the power consumed by customers. Electricity theft is the predominant type of non-technical loss, encompassing a range of techniques including private cables, physical manipulation of meter counting components, and destructive modification of meter facilities resulting in inconsistent meter readings [3]. Electricity theft not only carries significant economic consequences for the nation but also poses a threat to public safety, since it heightens the risk of mishaps such as fires and electric shocks. According to the source cited as [4], the aggregate financial impact of power theft on a global scale is estimated to be around CAD 100 million per year. This substantial amount of money, if not lost to theft, might instead be utilized to supply electricity to around 77,000 households for a duration of 1 year.

Numerous potential resolutions to the issue of power theft have been put out in the existing body of scholarly work [5–7]. The existing body of literature classifies these solutions into two primary categories: hardware-based solutions and data-driven solutions.

Hardware-based solutions primarily center around the development of intelligent devices and sensors with the capability to identify and detect irregularities. Nevertheless, it should be noted that the aforementioned solutions incur significant maintenance expenses, exhibit lower levels of efficiency, and require a substantial amount of time [8]. Furthermore, they demonstrate an elevated false-positive rate (FPR). On the other hand, there exists a plethora of data-driven methodologies aimed at detecting instances of electricity theft [9]. These solutions utilize methodologies rooted in artificial intelligence (AI) [10], game theory (GT) [11], and machine learning (ML), which are extensively applied in various fields such as healthcare, education, and transportation. According to cited source [12], solutions that are driven by data have enhanced resilience, efficiency, and comprehensibility. Furthermore, the scholarly literature [13] presents a methodology centered on grid analysis as a means to examine the identification of abnormal power consumption patterns. This methodology involves scrutinizing several parameters of the grid, such as current, voltage, and others, in order to find any atypical usage behavior. Anomaly detection encompasses the utilization of diverse data types, encompassing network-related data such as the operational state of switches and circuit breakers, alongside sensor data like voltage and current magnitudes captured by remote terminal units.

During the early phases of classification research, conventional machine learning techniques [14,15] were employed for the purposes of feature extraction and classification. The approaches employed in this study encompassed support vector machines (SVMs) [16,17], decision trees (DTs) [18,19], and nearest neighbors [20,21]. With the advancement of machine learning algorithms, there has been an increasing adoption of integrated learning algorithms that consist of several individual learners for the purpose of power theft detection. Several studies have presented several strategies for detecting instances of electricity theft, utilizing integrated learning algorithms such as random forest (RF), Adaboost, and XGBoost [22–25]. The experimental findings provided evidence that the integrated learning algorithms exhibit superior performance compared to conventional approaches. In a specific research investigation [26], deep learning models [27] were utilized as binary classifiers, with the purpose of detecting instances of energy theft. The researchers examined various deep learning architectures, such as CNN, Multi-Layer Perceptron (MLP), Long–Short-Term Memory (LSTM), and Gated Recurrent Unit (GRU) networks. Pereira et al. [28] employed a CNN for the purpose of detecting instances of power theft. Additionally, they conducted a comparative analysis of different oversampling approaches to investigate the potential effects of dataset imbalance. Zheng et al. [29] utilized a CNN to extract periodic features from load data that were transformed into a two-dimensional format. These extracted features were subsequently combined with global characteristics acquired from one-dimensional load data, which were captured using a fully connected network. The purpose of this approach was to detect instances of power theft. In a separate investigation, the authors of study [30] employed a fusion of clustering algorithms and Long–Short-Term Memory networks in order to identify instances of electricity theft. The methodology employed entailed forecasting the subsequent electricity usage of a client at each given time and afterwards evaluating the disparity between the projected values and the actual data. Deep learning techniques provide the advantage of automated sequence feature extraction in comparison to conventional machine learning algorithms.

The issue of detecting electricity theft has been extensively explored in academic research, leading to the development and widespread adoption of several hybrid neural network models that incorporate deep learning techniques. The study conducted by [2] introduced a hybrid neural network that integrates Long–Short-Term Memory (LSTM) and Multilayer Perceptron (MLP) models. This hybrid network demonstrates the ability to extract characteristics from diverse data sources. The authors Ismail et al. [31] proposed a hybrid neural network model that combines CNN and GRU to tackle the issue of electricity theft in distributed generation systems. The researchers in [32] devised a novel hybrid neural network architecture that integrates the GRU, CNN, and Particle Swarm Optimization (PSO) algorithms. This model was trained and evaluated using real-time data on electricity use. The utilization of the CNN facilitates the reduction of dimensionality and redundancy within time series data. The classification of consumption patterns into normal and fraudulent categories is achieved by the utilization of the GRU network and particle swarm algorithm. The integration of the long- and short-term memory strategies into CNN technology was found to boost e-fraud detection, as demonstrated in a study conducted by [33]. The optimal values of the hyperparameters for the CNN–LSTM were computed using meta-heuristic techniques, namely Black-Widow Optimization (BWO) and Blue-Monkey Optimization (BMO). The aforementioned works [30–33] have introduced detectors that function as hybrid deep-learning models, specifically designed for the purpose of feature extraction.

The aforementioned models have demonstrated favorable outcomes in the domain of electricity theft detection, yet certain concerns persist. The initial approach in many electrical theft detection models relies on CNNs. However, CNNs have limitations in properly capturing the global characteristics of time series data and calculating the relative correlations among the retrieved features. The excessive dependence on the initial input data presents a notable limitation. Furthermore, it is possible for the model to experience overfitting as a result of the disparity between the amount of data available in the training set and the intricacy of the model. As a result, the model's capacity to generalize to real-world scenarios is constrained. Therefore, it is imperative to consider the importance of mitigating model overfitting and improving feature extraction capabilities. Ding et al. [34] introduced a multivariate-branching block (DBB) as a means to extract feature information. The DBB accomplishes this by integrating several branches with diverse widths and complexities. The Gaussian-weighted feature-tokenization transformer module (FTT) was introduced by Sun et al. [35]. The FTT module aims to investigate the transformer's ability to capture local spatial semantic information and effectively represents the links between adjacent sequences. Moreover, Shi et al. [36] introduced a novel methodology for detecting power theft through an end-to-end approach by utilizing the transformer neural network. This study presents a novel hybrid model named the DSDB CNN and the Gaussian-weighted transformer network (DSDBGWT), which integrates a CNN with a DSDB structure and a GWT network. In contrast to a CNN, the DSDBGWT model demonstrates enhanced proficiency in extracting global features and determining the relative relationships among various characteristics. As a result, it diminishes its dependence on the initial input data when performing classification tasks. In order to augment the model's ability to extract features, a GWT module is utilized, which is particularly well-suited for processing sequences of extended duration. The present module effectively captures the characteristics of extended temporal sequences through the computation of attention coefficients, which are determined by the positional information of the input sequences. As a result, the model demonstrates enhanced efficacy in the detection of electricity theft. In order to address the issue of overfitting in the model, the initial step involves incorporating suitable normalization layers (LN) into both the regular block and transformer block. Furthermore, the dropout regularization technique is utilized to stochastically deactivate a certain proportion of neurons throughout the training process.

The main contributions of this article are summarized as follows:

(1) We propose a simple and efficient DSDB convolutional module in our network to extract inter- and intra-periodic features from sequences. This module replaces the traditional CNN structure, resulting in a lightweight model while improving model accuracy;

(2) We employ a transformer network with Gaussian weighting. The attention weights in this network can be attenuated based on the distance between related symbols. This allows for a more rational allocation of the attention mechanism, leading to more efficient extraction of sequence features and improved model accuracy;

(3) The systematic combination of CNN network and GWT network can fully extract the electricity consumption information in the sequences and accurately and efficiently recognize the semantic features, thus significantly improving the classification accuracy.

Extensive experiments on the China National Grid dataset show that our DSDBGWT model outperforms other existing methods.

The remainder of this paper is structured as follows. Section 2 presents the framework of the proposed model and provides specific details on the implementation of its constituent modules. In Section 3, we elaborate on the dataset processing, conduct comparative experiments to assess the effectiveness of our framework, and discuss the experimental results. Finally, Section 4 concludes the paper.

2. Materials and Methods

The overall architecture of the hybrid model for electricity theft detection (DSDBGWT) based on a CNN with DSDB and a GWT network is shown in Figure 1. The framework has three distinct modules: a CNN that incorporates a DSDB structure to facilitate shallow feature extraction, a GWT network designed specifically for long-distance feature extraction, and a classification module. Initially, the original sequence is segmented on a weekly basis using patch [37] to effectively capture the overall characteristics and minimize computing workload, while still retaining the information from the original sequence. Following the implementation of the patch, two DSDB structures are employed, possessing identical structures. This approach enables the extraction of inter- and intra-week features of electricity consumption information with enhanced accuracy and efficiency. Additionally, this significantly reduces the computational burden associated with the convolutional operation. Subsequently, the output data generated by the CNN are once again divided into discrete four-week intervals, employing patches as the input for the transformer model. The GWT network is capable of extracting a sequence's global features, which can produce varying weighting weights based on the input data's distance. Thus, enhancement in feature extraction accuracy is achieved. It is important to highlight that this approach differs from the standard transformer in that it does not incorporate a class token and position embedding into the transformer's tokens. Consequently, it does not engage in MLP processing within the tokens, but instead prioritizes the extraction of deep features from the tokens. Ultimately, the outcomes of the encoding process for each token are fed into the classification module.

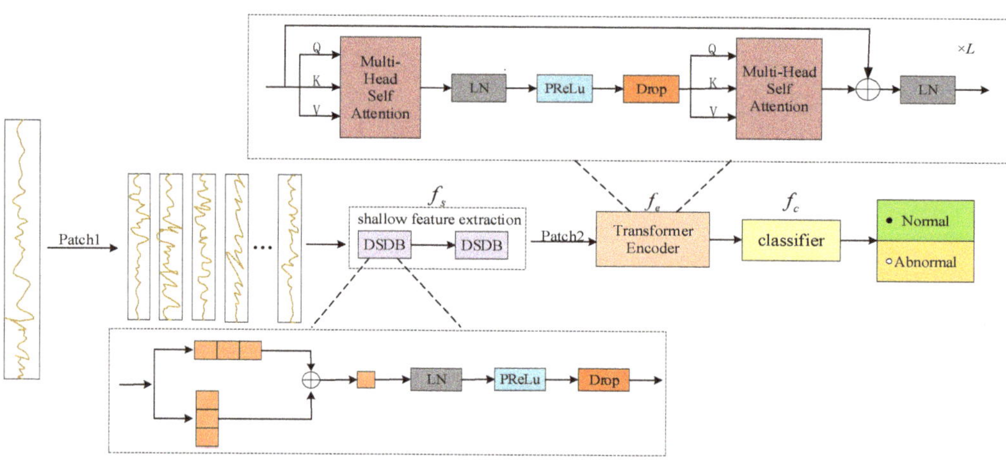

Figure 1. The overall architecture of the proposed hybrid neural network, combining DSDB convolutional neural network and GWT network (DSDBGWT).

The DSDBGWT network model proposed in this paper can be denoted as $f = f_c \odot f_e \odot f_s$, with parameters $\omega = \{\omega_s, \omega_e, \omega_c\}$. Here, f_s is a convolutional neural network, used for shallow feature extraction, and its output is $v = f_e(x;\omega_e)$; f_e is a transformer network, used for long-distance feature extraction, and its output is $z = f_e[f_s(x;\omega_s);\omega_e]$;

and f_c is a classifier, used for result categorization, which maps instances from the representations to the corresponding logics, which can be transformed into similar classes by $p(y|z;\omega) = sigmoid(f_c(z;\omega_c))$. We optimize the end-to-end parameters by minimizing the cross-entropy loss on the set of markers denoted as $\sum_{(x,y)\sim D_{train}}[l(f_c \odot f_e \odot f_s(x;\omega), y)]$. We define x to represent the input data, $p(;)$ to represent the derived probability value, and $\sum_{(x,y)}$ to represent the sum of the loss functions.

2.1. Data Preprocessing

The suggested approach is used for the smart meter data of consumers' daily electricity consumption, which is sourced from the State Grid Corporation of China (SGCC) [38]. The dataset provided includes authentic power consumers as well as those engaged in electricity theft, with more information about the dataset available in Table 1. Figure 2 illustrates the electricity consumption patterns of two users within the dataset. User 1 exhibits the highest electricity usage, with a daily consumption reaching close to 2000. In contrast, User 2 represents the majority of electricity users, ranging from a few kWh to a dozen kWh per day. This discrepancy highlights the significant variation in electricity consumption among users. To address this, it is necessary to normalize the data. Normalization not only stabilizes the dataset but also enhances the convergence speed and overall efficiency of the model. Furthermore, it is evident that the data from User 2 exhibits discontinuity in certain instances. This can be attributed to various intricate factors encountered during the meter collection process, such as unreliable transmission of data due to smart meter faults, irregular system maintenance, occurrence of special events, and other multifaceted elements. Consequently, these factors contribute to the absence of electricity consumption data. In order to mitigate the impact of data variations on the neural network model, it is imperative to employ appropriate data preprocessing techniques. This study undertakes the normalization of raw data and addresses the issue of missing values through appropriate processing techniques.

Table 1. Raw data status.

Description	Value
Total number of electricity consumers	42,372
Number of abnormal electricity consumers	3615
Time span	1 January 2014–31 October 2016
Proportion of missing data	25.7%
Maximum daily consumption of electricity by customers	2782.2

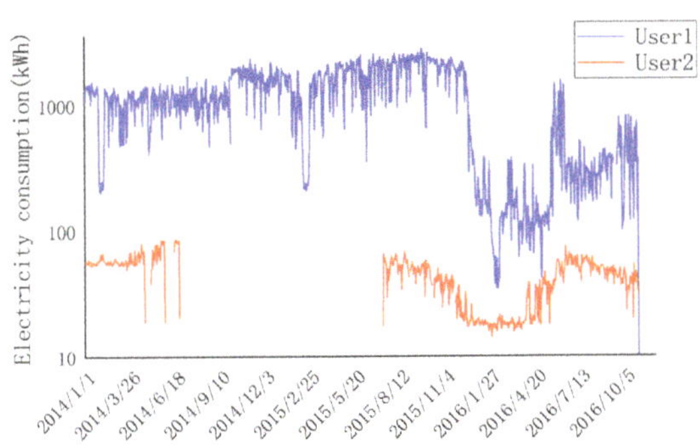

Figure 2. Display of electricity consumption by the largest user in the dataset and by an average user.

(1) The process of normalization.

The act of normalizing the dataset has the effect of increasing the numerical conditions of the dataset, which in turn enhances the stability of the optimization method.

Consequently, this phenomenon enhances the speed of model training and augments the efficiency of the algorithm. In addition, the process of normalization serves to standardize the distribution of data and reduce the influence of outliers on the model, improving its resilience. We choose the scaling method of $MAX - MIN$ to normalize the data according to the following equation. In the normalization process, we leave the missing values untouched first:

$$n(x) = \frac{x - min(x)}{max(x) - min(x)} \tag{1}$$

Here, x represents the user's electricity consumption on a specific day, while $min(x)$ and $max(x)$ represent the minimum and maximum values, respectively, across the entire dataset.

(2) Missing value processing.

Missing values are predominantly observed when there is a lack of data at a particular point in time, typically resulting from mistakes in the measuring instrument. The inclusion of these omitted values serves to improve the overall quality of the data, enhancing its trustworthiness and suitability for analytical and modeling purposes. The zero-replacement approach is employed to address the presence of missing data that meet the specified requirements:

$$f(x_t) = \begin{cases} 0 x_t \in NAN \\ x_t x_t \notin NAN \end{cases} \tag{2}$$

where x_t indicates the user's electricity consumption at a given time and $x_t \in NAN$ indicates that x_t is a null value.

The network encountered difficulty distinguishing between the original value being zero and the missing value being imputed as zero, due to the preexistence of zero values in the samples. In order to tackle this matter, we implemented an additional input channel by using a binary mask [39]. Within the mask matrix, the original data's missing value is designated as 0, whereas the normal value of 0 is designated as 1. By employing this approach, the neural network is capable of differentiating between these two situations, thereby improving the resilience of the model.

The initial dataset, denoted as X, comprises the electricity consumption data for a specific electricity user (referred to as M) over a time period of L days in the past. Therefore, we can represent the original dataset as $X \in \mathbb{R}^{M \times L}$. The dataset undergoes preprocessing, which involves normalizing the raw data, processing missing values, and adding binary masks. These processes transform the dataset from a two-dimensional structure to a three-dimensional structure for variable $X' \in \mathbb{R}^{M \times L \times 2}$.

2.2. Patch

Due to the considerable length of the sample sequence, it is necessary to employ the patch technique to partition the data into several subsequences at specific intervals. This strategy not only maintains the intrinsic properties of sequence but also enables more effective management and processing of the data for a range of activities, such as model training, feature extraction, and predictive analytics. The length of the patch is represented by the variable P. The sampling step is marked as S. The total number of patches is indicated by the variable N. The electricity usage per user over a period of L days is symbolized by the variable L. The calculation formula can be expressed as follows:

$$N = \left\lfloor \frac{(L - P)}{S} \right\rfloor + 1 \tag{3}$$

Electricity consumption data for normal users are usually more cyclical than for abnormal users [29,40,41] (detailed analysis in Section 3.3). To effectively process such periodic data using CNN, we employ the Patch architecture. Taking a specific sample as an example, the preprocessed data has a spatial size of $H \times W$. Utilizing the Patch architecture and considering the weekly periodicity of the data, we set the parameters

$P = 7$ and $S = 7$, thus transforming the data into a three-dimensional space represented as $H \times \left(\left\lfloor \frac{(L-P)}{S} \right\rfloor + 1 \right) \times S$ after Patch processing, as illustrated in Figure 3. Similarly, as shown in Figure 4, the inputs to the transformer network undergo processing using Patch. The decision to employ Patch processing on a four-week cycle is motivated by the transformer network's exceptional feature extraction capabilities and its proficiency in capturing distant features. The parameters Patch_size = 28 and Stride_size = 28 were set to partition the data based on monthly time intervals. This Patch architecture transforms the dimensionality of the output data from $U \times K \times V$ to $U \times (K \times V)$, then to $U \times \left(\left\lfloor \left(\frac{K \times V - P)}{S} \right\rfloor + 1 \right) \times S$. Moreover, the Patch operation reduces the number of input channels from L to approximately $\frac{L}{S}$, resulting in a reduction in computational complexity by a factor of S. Additionally, the Patch operation enables the model to have a stronger ability to refer back to earlier data, enhancing the network's learning capability and leading to significant improvements in prediction performance.

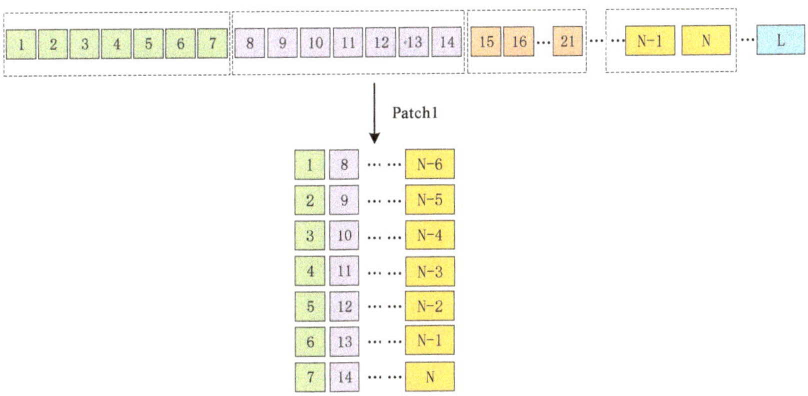

Figure 3. Patch architecture for CNN (N followed by data that are not divisible).

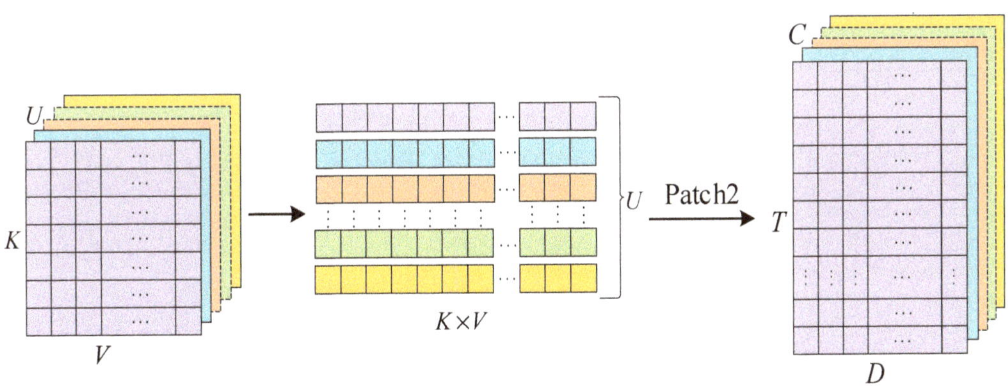

Figure 4. Patch architecture for transformer networks (each channel is patched separately).

2.3. Shallow Feature Extraction for DSDB Structures

After performing data preprocessing, the sequence features of the samples are extracted using 2D convolution. Figure 5 depicts the implementation of a DSDB structure during this phase. Each branch within the structure incorporates a convolution kernel of different scales, enabling the extraction of more complete feature information compared to a single convolutional network. Taking inspiration from work with ACNet [42], we propose the incorporation of asymmetric convolution into our approach. Specifically, we construct the convolution kernels for each branch to have dimensions of 1 × s and s × 1, respectively. The convolution kernel with dimensions s × 1 is utilized for feature extraction within a singular cycle, whereas the 1 × s convolution kernel is employed for extracting features across different cycles. The process involves the linear combination of two convolutions that are applied to the same locations but on different channels. This results in the emphasis of the squared convolution kernel in both horizontal and vertical directions, thereby highlighting distinct locally prominent features from various orientations. After the integration of the outputs from both branches into a single DSDB output, a 1 × 1 convolutional kernel is employed to maintain the inherent structure of the original sequence. In order to accelerate the rate at which the model converges during training and improve the overall generalization ability of the network [43], a normalization layer is implemented following the convolutional layer in each branch. The normalization layer is responsible for ensuring the normalization of the feature mapping in each branch. This process results in the output features becoming nonlinear and effectively reduces data dispersion. Additionally, it serves as a preventive measure against problems such as gradient explosion or gradient vanishing. Following the normalizing procedure, the PReLu activation function is employed to counteract linearity inside the network, so enabling the network to acquire knowledge about nonlinear mappings in a hierarchical fashion. Ultimately, the utilization of Dropout serves as a means to change data in order to mitigate the occurrence of overfitting.

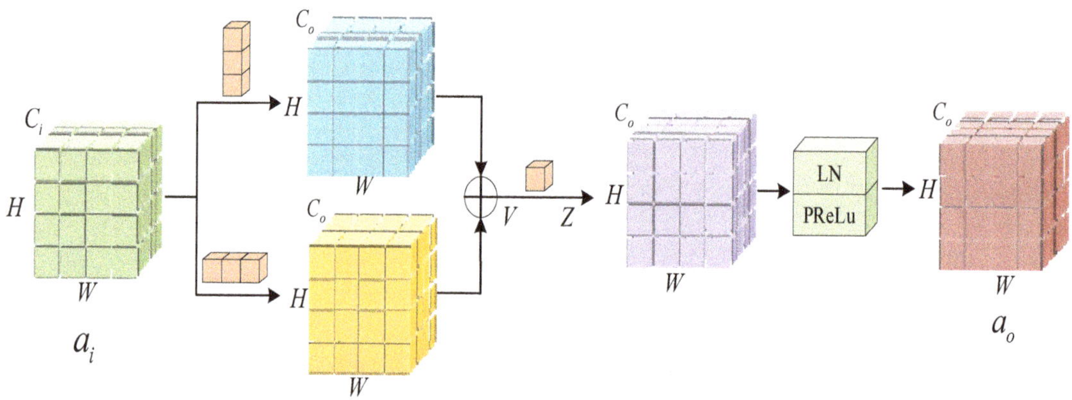

Figure 5. Convolutional neural network model with DSDB structure.

The model employs two distinct DSDB structures, and the subsequent description pertains to both of these DSDBs. The input data of the DSDB CNN are denoted as $a_i \in \mathbb{R}^{H \times W \times C_i}$, where $H \times W$ represents the spatial size and C_i represents the number of input channels. The mathematical expression for the DSDB convolution at point (h, w) of its jth channel can be represented by the following formula:

$$V_j^{h,w} = \sum_{k}^{C_i} \sum_{h'=0}^{S} \omega_{v_1,j,k}^{h',0} \cdot a_{i,k}^{h+h'-\lfloor \frac{S}{2} \rfloor, w} + \sum_{k}^{C_i} \sum_{w'=0}^{S} \omega_{v_2,j,k}^{0,w'} \cdot a_{i,k}^{h, w+w'-\lfloor \frac{S}{2} \rfloor} \quad (4)$$

Here, v_1 represents the s × 1 convolution kernel, v_2 represents the 1 × s convolution kernel, k represents the sum of C_i channels, h' represents the corresponding position ranging from 1 to s in the s × 1 convolution kernel, and, similarly, w' represents the corresponding position ranging from 1 to s in the 1 × s convolution kernel. To maximize data utilization, it is essential to employ the padding operation by adding zeros around the space $H \times W$. The padding size is determined by $\left\lfloor \frac{S}{2} \right\rfloor$; at this point, $V \in \mathbb{R}^{H \times W \times C_O}$.

Following the DSDB structure, a 1×1 convolution kernel is utilized to perform the convolution operation. Consequently, the value at the (h, w) position of the jth channel can be obtained as follows:

$$Z_j^{h,w} = \sum_{l}^{C_o} \omega_{z,j,l} \cdot V_l^{h,w} \tag{5}$$

where l represents the sum of C_O channels. At this point, $Z \in \mathbb{R}^{H \times W \times C_O}$.

Finally, we perform linear normalization processing and use PReLu activation function to obtain $a_o = \phi(\gamma Z + \beta)$.

2.4. Gaussian-Weighted Transformer Encoder Module

Regarding the detection of electricity theft, past research has predominantly concentrated on shallow feature extraction using convolutional neural networks, resulting in favorable outcomes. However, when addressing the issue of electricity theft, it is crucial to consider the correlation of data over an extended period. The data samples in this case consist of long time series. CNN has limitations in representing features for such long-time series data. The shallow feature extraction of CNN restricts their ability to capture long-term dependencies, as the extensive use of convolutional operations can only encompass a limited range of features. Moreover, the sample data contain a small number of anomalous samples, accounting for only 8.5% of the total. Relying solely on CNNs not only fails to extract more positive outcomes, but also runs the risk of gradient vanishing. To address the challenge at hand, this study introduces a transformer network into the framework. By incorporating the transformer network, the model is able to effectively capture global dependencies, enabling the extraction of long-distance characteristics. Additionally, the transformer network offers parallel computing capabilities, enhancing the overall efficiency of the network. This paper aims to enhance the precision of the model by enhancing the transformer network's Gaussian-weighted attention mechanism. The proposed improvement involves incorporating a Gaussian-weighted self-attention mechanism into the original network. This mechanism combines features extracted from W^Q, W^K, and W^V using a Gaussian-weighted matrix, thereby eliminating the reliance on attention weights for feature utilization. The weights undergo attenuation based on the proximity of tokens, with the degree of attenuation being defined by the Gaussian variance. This variance is acquired through the training process. The proposed method has the capability to comprehensively and precisely capture the temporal dependencies on a worldwide scale inside electricity consumption data. Consequently, this approach has the potential to enhance the effectiveness of power theft detection to a greater extent.

The module comprises two blocks of multi-head self-attention mechanism (MSA), as depicted in Figure 1. The residual operation is iterated by using the input channels as the heads of the first MSA, and mapping the output of the first MSA to the second MSA as the input heads of the second MSA, with the same dimensions for A^0 and A^2. The matrix dimensions of the input and output features are shown in Figure 6. To encompass the global relationship, the multi-head attention mechanism incorporates three learnable weight matrices, namely, W^Q, W^K, and W^V. The ith self-attention (SA) is chosen using the three learnable weight matrices mentioned above. It is then linearly normalized to obtain Scaled Dot-Product Attention, as depicted in Figure 7. This section introduces the concept of Gaussian-weighted self-attention, which allows for the utilization of varying weights based on the proximity of tokens. This feature enhances the accuracy of the findings obtained.

The formula for the SA mechanism is as follows:

$$SA = Attention(Q, K, V) = softmax(\frac{QK^T}{\sqrt{d_K}})V \tag{6}$$

where d_K is the dimension of K.

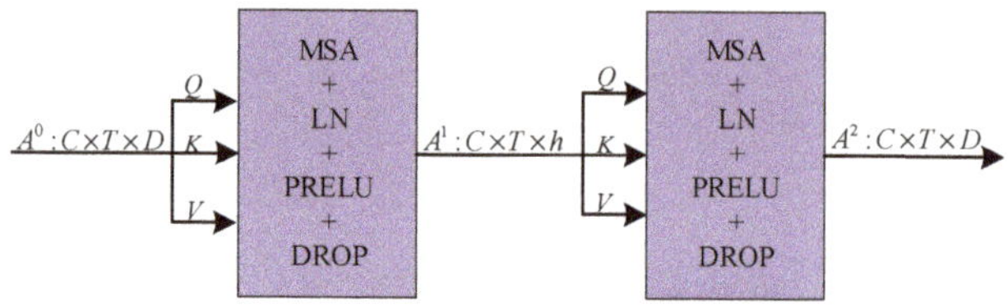

Figure 6. The matrix dimension of input and output features.

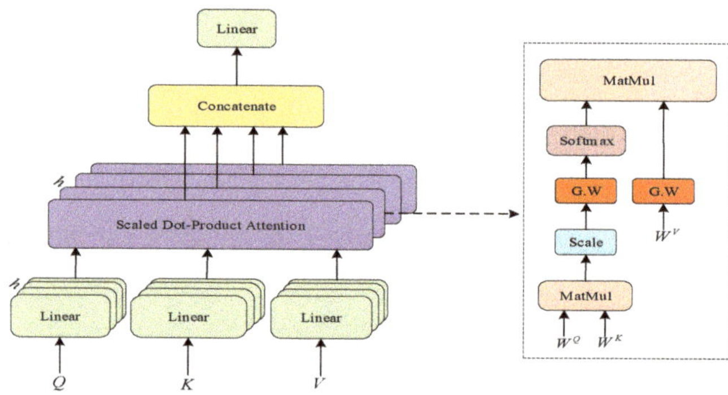

Figure 7. Multi-head self-attention module and the internal structure of a single self-attention.

The diagram illustrating the internal architecture of Gaussian-weighted self-attention is depicted in Figure 8. In this context, B represents the size of the batch, T defines the length of the sequence, D marks the dimension of the input, and E relates to the number of units in the self-attention mechanism. The matrices for the query, key, and value are defined in the following manner:

$$\begin{aligned} Q_i^W &= W^Q A^{l-1} \\ K_i^W &= W^K A^{l-1} \\ V_i^W &= W^V A^{l-1} \end{aligned} \tag{7}$$

where A^{l-1} is the input to the lth hidden layer (l = 0, 1, 2). W^Q, W^K, and W^V are network parameters. The score matrix in our proposed method is scaled by utilizing a Gaussian weighting matrix. This matrix is computed through the multiplication of key and query matrices, as described below:

$$S_i = G_S^l \circ \left(\frac{Q_i^\omega (K_i^\omega)^T}{\sqrt{d}} \right) \tag{8}$$

$$V_i = G_v^l \circ V_i^\omega \tag{9}$$

$$O_i = softmax(S_i) \circ V_i \tag{10}$$

G_s^l is the Gaussian weight matrix.

Within the MSA block, a series of weight matrices in variables Q, K, and V are subjected to the same operating technique. This results in the calculation of multiple head-attention values. Afterwards, the outcomes of each individual head attention are combined. The mathematical representation of this process can be expressed by the following equation:

$$MSA(Q, K, V) = Concat(O_0, O_1, ..., O_{h-1}) \tag{11}$$

where h represents the number of heads.

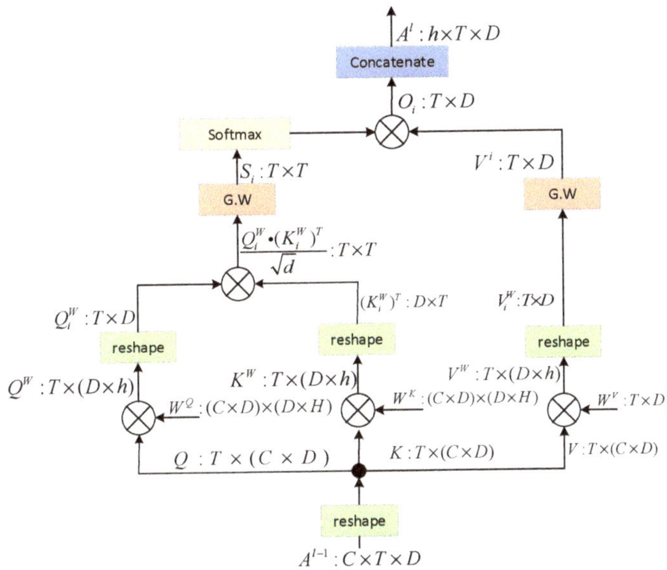

Figure 8. Proposed multi-head self-attention block diagram. The G.W. block performs element-wise multiplication of the Gaussian-weight matrix with the generated score matrix. The matrix dimensions are noted beside each signal.

Ultimately, the characteristics acquired within the transformer are afterward inputted into the classifier for the purpose of categorization. This classifier comprises two fully linked layers, with the Sigmoid activation function being employed for the output of the final layer. The function maps the output values within the range of 0 and 1. Individuals with a value equal to or beyond a threshold of 0.5 are classified as engaging in electro-pilfering, whilst individuals falling below this threshold are categorized as regular users.

2.5. Overall Algorithm Steps

The overall process of the proposed DSDBGWT is shown in Algorithm 1.

Algorithm 1 DSDBGWT Model

Input: Input a dataset $X \in \mathbb{R}^{1035 \times 2}$; patch size s1 = 7; patch size s2 = 28; training sample rate = 80%.
Output: Normal and abnormal prediction of test sets
1: Set batch size to 100, optimizer Adam (learning rate: 10^{-4}), epochs number e to 80.
2: Perform patch1 in the X, available to $X \in \mathbb{R}^{7 \times 147 \times 2}$ and divide them into training dataset and test dataset.
3: Generate training loader and test loader.
4: **for** i = 1 to e **do**
5: Perform DSDB convolution layer.
6: Perform patch 2 to change $X \in \mathbb{R}^{7 \times 147 \times 16}$ to $X \in \mathbb{R}^{36 \times 28 \times 16}$.
7: Perform a transformer network using Gaussian weighting.
8: Spread the transformer output to pass into the classifier.
9: Use the sigmoid function to identify the labels.
10: **end for**
11: Use test dataset with the trained model to get predicted labels.

---

3. Experimental Results and Analysis

3.1. Raw Electricity Consumption Dataset

The methodology was evaluated using a genuine dataset acquired from the State Grid Corporation of China. The dataset consists of a collection of daily power usage data series spanning from January 2014 to October 2016. This dataset encompasses a total of 43,272 customers. Approximately 8.55% of the aforementioned consumers were detected by the data source as participating in electricity theft operations and, as a result, were categorized as anomalous. We preprocessed the dataset according to the method in Section 2.1. It was randomly divided into five separate subsets of equal size while maintaining the original ratio of abnormal samples to normal samples. Four of these subsets were used as the training set to train the models, while the remaining subset was used as the test set to evaluate the models. The aforementioned procedure was iterated for the five potential choices, wherein a distinct subset was selected as the test set on each occasion. Consequently, five models are trained, with each model being evaluated on its respective test set to determine the test error. This process yields five test results, which are subsequently averaged. By repeating the aforementioned steps three times, the final results are obtained.

3.2. Experimental Setting

The experiments conducted in this paper were carried out on a server equipped with an Intel(R) Core (TM) i5-1035G1 CPU operating at a frequency of 1.7 GHz, with a maximum turbo frequency of 2.19 GHz. The server also had a total of 128 GB of RAM and was equipped with an NVIDIA GeForce RTX 3090 Ti GPU. The PyTorch 1.10.0 deep learning framework and Python 3.9 compiler were utilized on an Ubuntu machine to create the specific software. In the experiments, the batch size was set to 100, the learning rate was set to 0.001, the epoch was set to 80, and the Adam optimizer was used to make the model converge quickly.

3.3. Data Description

The chosen dataset is published by the State Grid of China and contains electricity consumption data of 43,272 electricity users over a period of 1035 days. The dataset contains electricity consumption data of 42,372 customers over a total of 1034 days from 1 January 2014 to 31 October 2016, of which 38,757 customers are normal electricity users (marked as 0) and the remaining 3615 customers are identified as electricity theft users (marked as 1). The details of the dataset are shown in Table 1.

The anomalous manifestations of electricity theft are not only shown on the surface of the data, but their implied patterns and trends are equally characterized. In particular, Figure 9a gives an example of the electricity consumption data of a normal electricity user in one year (i.e., 2016), and Figure 9b represents an example of the electricity consumption data of an electricity theft user in one year. As can be seen from Figure 9, the electricity consumption data of normal users in July, August, and September are higher than that in other months (high air conditioning usage in summer), but overall are relatively stable. The overall data in other months are generally consistent with little fluctuation; the data of the electricity theft user appear to be abnormally chaotic, and the decline in electricity consumption in a certain month is particularly high, which is not in line with the normal pattern of electricity consumption. As shown in Figure 10, the electricity consumption data for four weeks (February 2015) of normal users and electricity theft users are extracted for further analysis. Figure 10a shows that, under normal circumstances, normal electricity users can exhibit significant periodicity, with weekly electricity consumption usually peaking on day 2 or 3, often reaching a low on day 4, and then starting to rise again, whereas the electricity consumption data for those defined as stealing (Figure 10b) fluctuates cyclically for the first two weeks (i.e., week 1 and week 2). However, from the second week onwards, electricity consumption decreases significantly and, thereafter, electricity consumption remains at a low level.

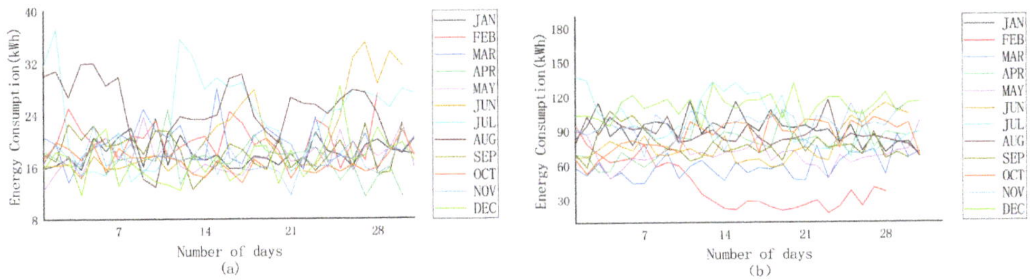

Figure 9. Average monthly electricity consumption in 2015. (**a**) Normal energy users. (**b**) Energy theft users.

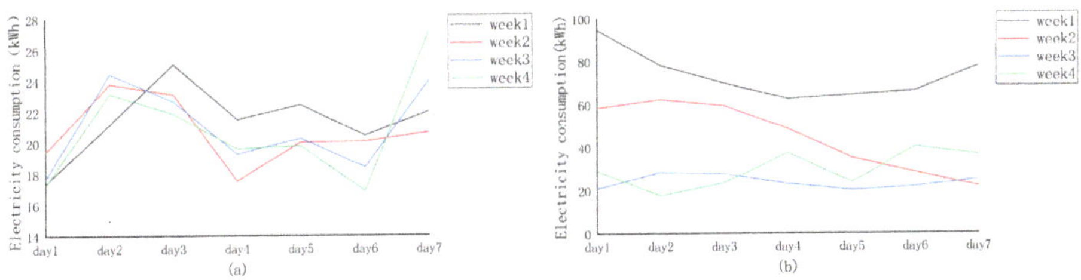

Figure 10. Average daily electricity consumption every four weeks (February 2015). (**a**) Normal energy users. (**b**) Energy theft users.

In order to better analyze the periodicity of normal customers and the non-periodicity of electricity theft users, we performed a correlation analysis on the electricity consumption data. Figure 3 shows the Pearson correlation coefficient (PCC) of the electricity consumption of the above two users over a four-week period. In this case, Figure 11a shows the PCC values for normal users and Figure 11b shows the PCC values for electricity theft users. From Figure 11a, we can find that the electricity consumption data of normal users have a strong positive correlation. Most of their PCC values are around 0.5, and some even reach 0.9 (a closer PCC value to 1 means that a stronger correlation [30]), whereas the PCC value of the electricity consumption data of abnormal users is not more than 0.4 (Figure 11b), and even the phenomenon of negative PCC values occurs, which means that they show a negative correlation.

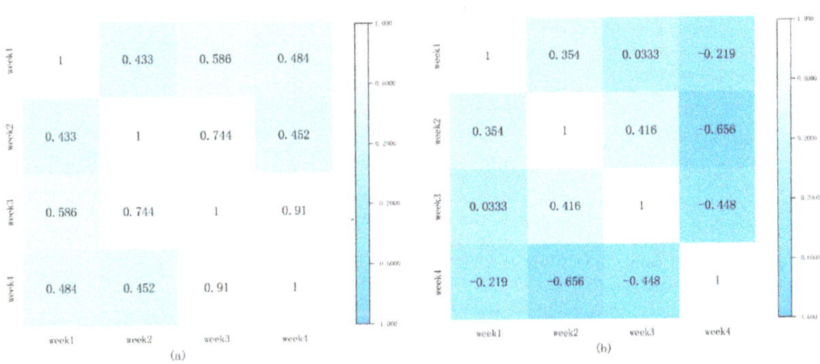

Figure 11. PCC of electricity consumption by week (February 2015). (**a**) Normal energy users. (**b**) Energy theft users.

By statistically analyzing the electricity consumption data of normal users and electricity theft users, we can find that the electricity consumption data of electricity theft users are usually not periodic or non-periodic compared to normal users. Therefore, weekly, monthly, quarterly, and annual electricity consumption data can be used as benchmarks for feature extraction.

3.4. Evaluation Indicators

In order to evaluate the efficacy of the model, its performance was assessed using various metrics, including precision, recall, F1 score (F1), Area Under the Curve (AUC), and Mean Average Precision (MAP). The measurements encompass four primary error rates, namely, false positive (FP), false negative (FN), true positive (TP), and true negative (TN) [2,31].

The recall metric is defined as the ratio of accurately recognized instances of electricity theft by the model to the total number of real electricity theft samples:

$$recall = \frac{TP}{TP + FN} \tag{12}$$

Precision is a metric that quantifies the proportion of samples accurately identified by the model as instances of power theft relative to the overall number of samples categorized as instances of electricity theft across all detection tests:

$$precision = \frac{TP}{TP + FP} \tag{13}$$

The F1 score, also known as the balanced score, is a statistical measure used to assess the precision of a binary classification model. The evaluation metric takes into account both the precision and recall of the classification model:

$$F1 = \frac{2 \times precision \times recall}{precision + recall} \tag{14}$$

AUC is defined as the area under the ROC curve and is used to measure the overall quality of the classifier. The larger the value of AUC, the better the performance of the classifier:

$$AUC = \frac{\sum_{i \in positiveClass} Rank_i - \frac{M(1+M)}{2}}{M \times N} \tag{15}$$

where $Rank_i$ denotes the rank value of sample i, M is the number of normal samples, and N is the number of electricity theft samples.

MAP is a position sensitive indicator; if the abnormal samples are ranked higher than the normal samples, the higher the value of MAP. It can be calculated as follows:

$$MAP@K = \frac{1}{m} \sum_{i=1}^{m} \frac{i}{p_i} \tag{16}$$

Considering the top K users in the sorted list, m is the number of selected users who have actually performed a power theft operation and $p_i (i = 1, 2, 3, \ldots, m)$ denotes the position of each anomaly in the sorted list. In our experiments, we compute this metric for all samples in a given list and abbreviate the metric as MAP@ALL.

3.5. Comparison with Advanced Methods

In order to demonstrate the efficacy of the suggested model, a selection of representative methodologies has been chosen to perform comparative tests using the DSDBGWT model. These methods integrate both representative and high-level scholarly publications with publicly accessible source code, spanning the period from 2001 to 2022. It is noteworthy to emphasize that the aforementioned methods were applied to a preprocessed dataset in order to ensure a fair comparison:

(1) Random forest (RF) [44]: The RF classifier, also known as random forest, is a machine learning algorithm composed of several decision trees;

(2) MiniRocket [45]: The MiniRocket model is a time series classification model that operates at rapid speeds. It utilizes a concise collection of predetermined convolutional kernels to convert the input time series data. The extracted features are subsequently employed in the training of a linear classifier;
(3) Wide and Deep CNN (Wide and Deep) [29]: The Wide and Deep model, which has a wide component and a deep CNN component, has gained significant traction as a fundamental approach in various domains;
(4) Hybrid-Order Representation Learning [40]: The electrical behavior classifier employs a comprehensive representation that combines first-order and second-order variables to detect occurrences of electricity theft;
(5) Hybrid Attention (HyAttn) [39]: The extraction of features is performed using a convolutional module that is enhanced by an MSA technique. Subsequently, the classification of these features is carried out evenly by concatenating convolutional layers with a kernel size of 1.

To ensure the integrity of the experimental findings, the network architecture and associated parameters from both classical and contemporary methodologies in the existing literature are employed to replicate the models for comparison studies. All tests were conducted using identical hardware configurations and maintained a consistent ratio of training to testing samples. The empirical findings are shown in Table 2.

Table 2. Performance comparison of different methods.

Methods	F1	AUC	MAP@ALL
RF [44]	0.386 ± 0.011	0.804 ± 0.018	0.603 ± 0.011
MiniRocket [45]	0.427 ± 0.008	0.829 ± 0.013	0.683 ± 0.009
Wide and Deep CNN [29]	0.468 ± 0.004	0.862 ± 0.011	0.751 ± 0.007
Hybrid-Order Representation Learning [40]	0.594 ± 0.004	0.895 ± 0.007	0.807 ± 0.006
HyAttn [39]	0.609 ± 0.003	0.907 ± 0.006	0.831 ± 0.006
DSDBGWT (proposed)	**0.629 ± 0.002**	**0.923 ± 0.004**	**0.834 ± 0.004**

Table 2 presents a comprehensive overview of the performance exhibited by all the approaches that were compared. The classification methods RF and MiniRocket, although known for their strong performance, are not specifically tailored for the purpose of power theft detection. The utilization of a Wide and Deep CNN in a CNN-based framework yields forecasts that are more dependable. The Wide and Deep CNN exhibits the capability to capture periodicity in weekly patterns through the utilization of deep CNN models and the integration of global knowledge from wide components. However, the performance of the model is constrained by the simplistic approach of stacking convolutional and fully connected layers, resulting in limited effectiveness for long-distance feature extraction and consequently leading to its poor accuracy. The HORLN model leverages first-order information to conduct shallow feature extraction on the sample sequence. Subsequently, the recovered features from the first-order information are employed as input for second-order processing. Despite the implementation of shallow feature extraction and long-distance feature extraction, the current model lacks the necessary level of granularity. HyAttn significantly enhances performance by integrating extended convolutional layers and including a self-attention mechanism. This approach effectively leverages both CNN and SA to extract shallow features and long-distance features from the input data simultaneously. However, it lacks selectivity in extracting features across long distances and does not dynamically adjust the weights of feature extraction across tokens while considering temporal considerations. The extracted features are not sufficiently complete, leaving potential for further improvement in accuracy. The model proposed in this article, known as the DSDBGWT model, incorporates a DSDB structure to enhance the extraction of comprehensive feature information during shallow feature extraction. Additionally, it incorporates Gaussian weighting processing on the token during training, enabling accurate and efficient feature extraction and F1 score calculation. The AUC and MAP@ALL metrics exhibit increases of 3.28%, 1.76%, and 0.36% compared to the highest values achieved by the aforementioned methods.

3.6. Parametric Analysis

The examination of parameters examines several elements that impact both the performance of classification and the process of training. The factors encompassed in this analysis consist of the number of output channels inside the convolutional network, the count of tokens, and the number of heads involved in the multi-head attention mechanism.

The augmentation of channels within the convolutional kernel improves the model's ability to extract features. However, this augmentation also introduces greater complexity to the model, which can potentially result in overfitting issues. The discussion revolves around the number of output channels in the two convolutional layers of the convolutional neural network. The impact of this parameter on F1, AUC, and MAP@ALL metrics is illustrated in Figure 12. The number of output channels for the first convolutional layer is denoted as output_DSDB1, whereas the number of output channels for the second convolutional layer is denoted as output_DSDB2. Based on the data presented in Figure 12, it can be observed that F1 achieves optimal performance when the values of output_DSDB1 and output_DSDB2 are set to 32 and 16, respectively, resulting in a performance metric of 0.629. Additionally, this configuration corresponds to the largest AUC value of 0.923. In the context of MAP@ALL, the maximum value is observed at output_DSDB1 = 48 and output_DSDB2 = 32, with a corresponding value of 0.848. In terms of the parameters, if the number of output channels of output_DSDB1 is doubled, it will lead to a doubling of the number of input channels of output_DSDB2. Consequently, this will not only increase the parameters of output_DSDB1, but will also increase the parameters of output_DSDB2. The excessive number of parameters can negatively impact the efficiency of the model. Therefore, in order to maintain model accuracy, measures need to be taken. Simultaneously, it is imperative to minimize the selection of output channels. After considering all relevant factors and analyzing the experimental findings, we have determined that the optimal number of output channels is output_DSDB1 = 32 and output_DSDB2 = 16. At this configuration, the corresponding values for F1, AUC, and MAP@ALL metrics are 0.629, 0.923, and 0.834, respectively.

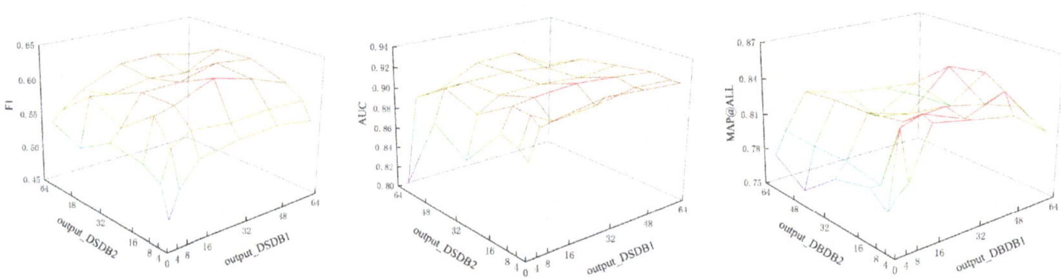

Figure 12. The impact of the number of output channels in CNN on various evaluation metrics.

The computational cost is directly influenced by the quantity of tokens in MSA. To regulate the CNN output features at various scales, we employ patching, which ultimately controls the quantity and dimensions of tokens. The fine-grained characteristics are influenced by the number of tokens, while the receptive field of the token features is determined by the dimension. The findings shown in Table 3 demonstrate the impact of token count on F1, AUC, and MAP@ALL within the context of MSA. The observed sample sequence exhibits periodicity not just on a weekly basis, but also on monthly and quarterly time scales. When the number of P is 7, 28, or 91, these correspond to the studies conducted in weekly, monthly, and quarterly patches, respectively. The table presents the performance metrics of F1, AUC, and MAP@ALL for different token values. It is seen that, when P = 28, F1 achieves a value of 0.629 and AUC achieves a value of 0.923. Comparatively, the impact of P = 7 and P = 28 on MAP@ALL is similar. Therefore, P = 28 (token = 36) is selected as the input for the transformer network. Based on the findings, it can be inferred that, while the transformer network exhibits strong capability in handling long-distance dependencies, its effectiveness is not only determined by the length of the sequence; rather, there exists a specific range within which the network performs optimally.

Table 3. The impact of the number of tokens in transformer networks on various evaluation metrics.

P, Token	F1	AUC	MAP@ALL
P = 7, Token = 147	0.570	0.915	0.835
P = 28, Token = 36	0.629	0.923	0.834
P = 91, Token = 11	0.576	0.903	0.816

The primary purpose of employing multiple heads is to concurrently execute numerous independent attention computations, while also connecting their respective outputs. The use of multi-head attention in neural networks enhances the capacity to capture more comprehensive feature information. Similar to how raising the number of channels in a convolutional kernel in a CNN amplifies model complexity, augmenting the number of attention heads in multi-head attention similarly substantially elevates model complexity. The impact of the number of heads in the multi-head attention mechanism on each evaluation parameter is depicted in Table 4. Based on the data presented in the table, it is evident that the F1 score exhibits an upward trend as the number of heads increases, particularly when the number of heads is relatively small. Notably, the F1 score reaches its peak value of 0.629 when the number of heads reaches 48. However, a gradual decline in the F1 score is observed as the number of heads further increases to 64 and 80. This observation demonstrates that an excessive number of heads is not essential. When a sufficient number of heads are present, this enables comprehensive utilization of all aspects of the feature information. However, as the number of heads increases, so does the number of parameters and the computational load. Consequently, this leads to a decrease in the efficiency of the model. In conclusion, the value of h = 48 was selected as the designated quantity of heads for the MSA.

Table 4. The impact of the number of heads in multi-head attention mechanism on various evaluation metrics.

h	F1	AUC	MAP@ALL
16	0.616	0.922	0.813
32	0.618	0.917	0.833
48	0.629	0.923	0.834
64	0.620	0.921	0.827
80	0.612	0.919	0.816

3.7. Ablation Experiments

To assess the efficacy of the multi-branch component, we substitute it with a conventional two-dimensional convolution kernel for verification purposes. In this particular case, the substitution of a 1 × 3 and 3 × 1 convolution kernel is made with a 3 × 3 convolution kernel, while leaving other structures unaltered. The classification results obtained from the SGCC dataset are depicted in Figure 13. The figure demonstrates that the suggested model exhibits enhancements in the F1 score, AUC, and MAP@ALL by 4.31%, 1.76%, and 1.33%, correspondingly, in comparison to the model ordinary convolution. This is because the 1 × 3 convolution kernel in the dual-branch structure we designed efficiently extracts the intra-week features in the power data, and the 3 × 1 convolution kernel efficiently extracts the intra-week features in the power data. The experimental results show that the proposed dual-branching part can enhance the feature extraction ability of the network model and that the scheme is feasible.

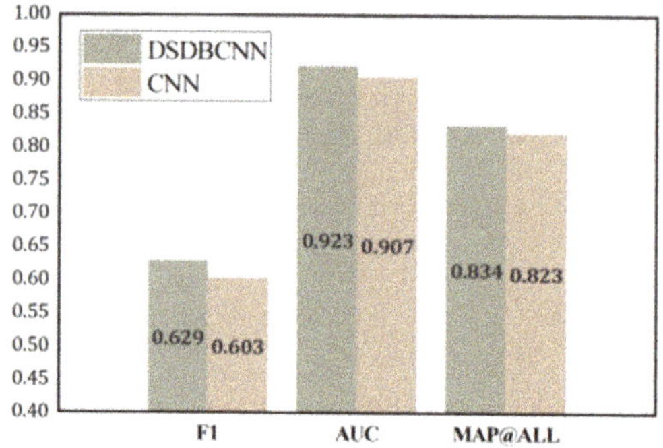

Figure 13. Performance comparison results between CNN with DSDB architecture and traditional CNN.

The proposed model utilizes a fusion of DSDB CNN and GWT techniques for the purpose of identifying instances of electricity theft among customers. A series of ablation experiments were performed on the SGCC dataset to comprehensively evaluate the efficacy of the approach. These experiments involved testing various combinations of components. Table 5 examines five combinations and evaluates the influence of various components on the overall model in terms of classification accuracy. "$\surd$" indicates that the structure is added to the model, and "$\times$" indicates that the structure is not used in the model.In this context, DSDB refers to a CNN with a DSDB structure. The term "conv" denotes the utilization of a regular 2D convolutional kernel. G.W. symbolizes the incorporation of Gaussian weighting treatment into the transformer model. Lastly, "tran" refers to the transformer network without Gaussianization. In Variant (1), the utilization of solely CNN is limited due to the absence of transformers. Consequently, the receptive field is restricted, leading to the extraction of primarily local information. As a result, the achieved F1 score is quite low. Variant (2) refers to the utilization of the tran network exclusively for long-distance feature extraction, while neglecting the use of CNN for shallow extraction of samples. Consequently, this approach exhibits limited capability in capturing local information and is susceptible to the issue of gradient vanishing. As a result, its F1 score is notably low, measuring only 0.426. Variant (3) entails the fusion of CNN with transformer. It is evident that the combination of these two models yields significantly improved accuracy compared to their individual implementations. This finding underscores the importance of incorporating both local and global temporal dependencies in the context of power theft detection. Notably, the F1 score of this combined approach reaches a value of 0.597. In (4), we use CNN with DSDB structure and transformer for combination. One can see that its accuracy is a little better than (3), which perfectly proves the effectiveness of the DSDB structure. In accordance with premise (3), we applied Gaussian weighting to the transformer, as described in (5). This approach considers both shallow and long-distance feature extraction, while also incorporating Gaussian weighting based on token distance closeness. As a result, the F1 score exhibits a 1% improvement compared to the approach outlined in (3). In (6), we once again integrate the CNN with DSDB structure using GWT. We replace the $k \times k$ convolution kernel with $1 \times k$ and $k \times 1$, allowing for simultaneous extraction of both inter-periodic and intra-periodic features. This modification not only enhances the model's efficiency by reducing parameter usage, but also improves its accuracy. The F1 score exhibits a significant increase of 5.36% when compared to the value obtained in (3). In conclusion, the examination of the amalgamated experimental findings serves to reinforce the soundness and credibility of our theoretical framework.

Table 5. Performance of different variants of DSDBCGW.

	DSDB	conv	G.W.	tran	F1	AUC	MAP@ALL
(1)	×	√	×	×	0.562	0.874	0.827
(2)	×	×	×	√	0.426	0.830	0.737
(3)	×	√	×	√	0.597	0.909	0.816
(4)	√	×	×	√	0.599	0.897	0.815
(5)	×	√	√	√	0.603	0.907	0.823
(6)	√	×	√	√	0.629	0.923	0.834

4. Conclusions

This research presents a novel approach for power usage anomaly identification by proposing a hybrid network that combines a DSDB CNN with a GWT network. The proposed model incorporates a DSDB to perform shallow feature extraction on the sample sequence. This approach not only enables the extraction of more comprehensive features but also efficiently decreases parameter usage and enhances efficiency. The GWT network is capable of extracting characteristics from long-distance sequences in a more reasoned manner by utilizing the Gaussian-weighted technique. To assess the efficacy of the approach, a comparative experiment was undertaken, employing DSDBGWT alongside other classification methods. The experiment was performed on the publicly available dataset of SGCC. The experimental findings demonstrate that the approach described in this research study is capable of effectively extracting the abnormal characteristics of power consumption from the provided training samples. Moreover, the method exhibits a notable enhancement in F1 performance, surpassing the current state-of-the-art method by a margin of 3.28%. This improvement signifies a significant advancement over the existing advanced method. The technique described in this study is limited to feature extraction from data on electricity consumption. In actuality, a variety of complex factors, like the weather, holidays, the economy, etc., also influence how much power people use. The proposed DSDBGWT has good scalability in the high-level semantic feature extraction of multimodal data. In the future, we will build on the DSDBGWT model by fusing the model with more modes of data to extract high-level features of electricity consumption sequences, thus further improving the classification accuracy.

Author Contributions: Conceptualization, Y.B.; methodology, Y.B.; software, Y.B.; validation, Y.B. and H.S.; formal analysis, Y.B. and H.S.; investigation, Y.B.; resources, Y.B. and H.S.; data curation, Y.B.; writing—original draft preparation, H.S.; writing—review and editing, Y.B., H.S., L.Z. and H.W.; visualization, Y.B.; supervision, Y.B.; project administration, Y.B.; funding acquisition, L.Z. All authors have read and agreed to the published version of the manuscript.

Funding: This research was funded by Liaoning Province Education Administration under Grant LJKZ0174.

Institutional Review Board Statement: Not applicable.

Informed Consent Statement: Not applicable.

Data Availability Statement: Not applicable.

Acknowledgments: We are very grateful to the editors and reviewers for their valuable comments, to the providers of all the data used in the paper, and to the people who helped to complete this study.

Conflicts of Interest: The authors declare no conflict of interest.

References

1. Dileep, G. A survey on smart grid technologies and applications. *Renew. Energy* **2020**, *146*, 2589–2625. [CrossRef]
2. Buzau, M.M.; Tejedor-Aguilera, J.; Cruz-Romero, P.; Gómez-Expósito, A. Hybrid deep neural networks for detection of non-technical losses in electricity smart meters. *IEEE Trans. Power Syst.* **2019**, *35*, 1254–1263. [CrossRef]
3. McLaughlin, S.; Holbert, B.; Fawaz, A.; Berthier, R.; Zonouz, S. A multi-sensor energy theft detection framework for advanced metering infrastructures. *IEEE J. Sel. Areas Commun.* **2013**, *31*, 1319–1330. [CrossRef]

4. *Smart Meters Help Reduce Electricity Theft, Increase Safety*; BCHydro, Inc.: Vancouver, BC, Canada, 2011; Available online: https://www.bchydro.com/news/conservation/2011/smart_meters_energy_theft.html (accessed on 20 March 2023).
5. Leite, D.; Pessanha, J.; Simões, P.; Calili, R.; Souza, R. A stochastic frontier model for definition of non-technical loss targets. *Energies* **2020**, *13*, 3227. [CrossRef]
6. Nabil, M.; Ismail, M.; Mahmoud, M.M.E.A.; Alasmary, W.; Serpedin, E. PPETD: Privacy-preserving electricity theft detection scheme with load monitoring and billing for AMI networks. *IEEE Access* **2019**, *7*, 96334–96348. [CrossRef]
7. Maamar, A.; Benahmed, K. A Hybrid Model for Anomalies Detection in AMI System Combining K-means Clustering and Deep Neural Network. *Comput. Mater. Contin.* **2019**, *60*, 15–39. [CrossRef]
8. Krysanov, V.; Danilov, A.; Burkovsky, V.; Gusev, P.; Gusev, K. Optimization of electric transmission lines (ETL) operation modes based on hardware solutions of process platform FACTS. In Proceedings of the 14th International Conference on Electromechanics and Robotics "Zavalishin's Readings", Kursk, Russia, 17–20 April 2019.
9. Saeed, M.S.; Mustafa, M.W.; Hamadneh, N.N.; Alshammari, N.A.; Sheikh, U.U.; Jumani, T.A.; Khalid, S.B.A.; Khan, I. Detection of non-technical losses in power utilities—A comprehensive systematic review. *Energies* **2020**, *13*, 4727. [CrossRef]
10. Winston, P.H. *Artificial Intelligence*; Addison-Wesley Longman Publishing Co., Inc.: Boston, MA, USA, 1984.
11. Lin, C.H.; Chen, S.J.; Kuo, C.L.; Chen, J.L. Non-cooperative game model applied to an advanced metering infrastructure for non-technical loss screening in micro-distribution systems. *IEEE Trans. Smart Grid* **2014**, *5*, 2468–2469. [CrossRef]
12. Liu, Y.; Liu, T.; Sun, H.; Zhang, K.; Liu, P. Hidden electricity theft by exploiting multiple-pricing scheme in smart grids. *IEEE Trans. Inf. Forensics Secur.* **2020**, *15*, 2453–2468. [CrossRef]
13. Yan, Z.; Wen, H. Performance analysis of electricity theft detection for the smart grid: An overview. *IEEE Trans. Instrum. Meas.* **2021**, *71*, 2502928. [CrossRef]
14. Mitchell, T.M. *Machine Learning*; McGraw-Hill: New York, NY, USA, 1997.
15. Jordan, M.I.; Mitchell, T.M. Machine learning: Trends, perspectives, and prospects. *Science* **2015**, *349*, 255–260. [CrossRef]
16. Toma, R.N.; Hasan, M.N.; Nahid, A.A.; Li, B. Electricity theft detection to reduce non-technical loss using support vector machine in smart grid. In Proceedings of the 2019 1st International Conference on Advances in Science, Engineering and Robotics Technology (ICASERT), Dhaka, Bangladesh, 3–5 May 2019; pp. 1–6.
17. Jindal, A.; Dua, A.; Kaur, K.; Singh, M.; Kumar, N.; Mishra, S. Decision tree and SVM-based data analytics for theft detection in smart grid. *IEEE Trans. Ind. Inf.* **2016**, *12*, 1005–1016. [CrossRef]
18. Monedero, I.; Biscarri, F.; León, C.; Guerrero, J.I.; Biscarri, J.; Millán, R. Detection of frauds and other non-technical losses in a power utility using Pearson coefficient, Bayesian networks and decision trees. *Int. J. Electr. Power Energy Syst.* **2012**, *34*, 90–98. [CrossRef]
19. Song, Y.Y.; Ying, L.U. Decision tree methods: Applications for classification and prediction. *Shanghai Arch. Psychiatry* **2015**, *27*, 130. [PubMed]
20. Aziz, S.; Naqvi, S.Z.H.; Khan, M.U.; Aslam, T. Electricity theft detection using empirical mode decomposition and K-nearest neighbors. In Proceedings of the IEEE International Conference on Emerging Trends in Smart Technologies (ICETST), Karachi, Pakistan, 26–27 March 2020; pp. 1–5.
21. Larose, D.T.; Larose, C.D. k-Nearest Neighbor Algorithm. 2014. Available online: onlinelibrary.wiley.com (accessed on 30 August 2023).
22. Meira, J.A.; Glauner, P.; State, R.; Valtchev, P.; Dolberg, L.; Bettinger, F.; Duarte, D. Distilling Provider-Independent Data for General Detection of Non-Technical Losses. In Proceedings of the 2017 IEEE Power and Energy Conference at Illinois (PECI), Champaign, IL, USA, 23–24 February 2017.
23. Avila, N.F.; Figueroa, G.; Chu, C.C. NTL detection in electric distribution systems using the maximal overlap discrete wavelet-packet transform and random undersampling boosting. *IEEE Trans. Power Syst.* **2018**, *33*, 7171–7180. [CrossRef]
24. Buzau, M.M.; Tejedor-Aguilera, J.; Cruz-Romero, P.; Gómez-Expósito, A. Detection of non-technical losses using smart meter data and supervised learning. *IEEE Trans. Smart Grid* **2018**, *10*, 2661–2670. [CrossRef]
25. Parmar, A.; Katariya, R.; Patel, V. A review on random forest: An ensemble classifier. In Proceedings of the International Conference on Intelligent Data Communication Technologies and Internet of Things (ICICI) 2018, Coimbatore, India, 7–8 August 2018; Springer International Publishing: Berlin/Heidelberg, Germany, 2019; pp. 758–763.
26. Li, S.; Han, Y.; Yao, X.; Song, Y.; Wang, J.; Zhao, Q. Electricity theft detection in power grids with deep learning and random forests. *J. Electr. Comput. Eng.* **2019**, *2019*, 4136874. [CrossRef]
27. Janiesch, C.; Zschech, P.; Heinrich, K. Machine learning and deep learning. *Electr. Mark.* **2021**, *31*, 685–695. [CrossRef]
28. Pereira, J.; Saraiva, F. Convolutional neural network applied to detect electricity theft: A comparative study on unbalanced data handling techniques. *Int. J. Electr. Power Energy Syst.* **2021**, *131*, 107085. [CrossRef]
29. Zheng, Z.; Yang, Y.; Niu, X.; Dai, H.; Zhou, Y. Wide and deep convolutional neural networks for electricity-theft detection to secure smart grids. *IEEE Trans. Ind. Inf.* **2017**, *14*, 1606–1615. [CrossRef]
30. Kocaman, B.; Tümen, V. Detection of electricity theft using data processing and LSTM method in distribution systems. *Sādhanā* **2020**, *45*, 286. [CrossRef]
31. Ismail, M.; Shaaban, M.F.; Naidu, M.; Serpedin, E. Deep learning detection of electricity theft cyber-attacks in renewable distributed generation. *IEEE Trans. Smart Grid* **2020**, *11*, 3428–3437. [CrossRef]

32. Ullah, A.; Javaid, N.; Samuel, O.; Imran, M.; Shoaib, M. CNN and GRU based deep neural network for electricity theft detection to secure smart grid. In Proceedings of the 2020 IEEE International Wireless Communications and Mobile Computing (IWCMC), Limassol, Cyprus, 15–19 June 2020; pp. 1598–1602.
33. Almazroi, A.A.; Ayub, N. A novel method CNN-LSTM ensembler based on Black Widow and Blue Monkey Optimizer for electricity theft detection. *IEEE Access* **2021**, *9*, 141154–141166. [CrossRef]
34. Ding, X.; Zhang, X.; Han, J.; Ding, G. Diverse branch block: Building a convolution as an inception-like unit. In Proceedings of the IEEE/CVF Conference on Computer Vision and Pattern Recognition, Nashville, TN, USA, 20–25 June 2021; pp. 10886–10895.
35. Sun, L.; Zhao, G.; Zheng, Y.; Wu, Z. Spectral–spatial feature tokenization transformer for hyperspectral image classification. *IEEE Trans. Geosci. Remote Sens.* **2022**, *60*, 1–14. [CrossRef]
36. Shi, J.; Gao, Y.; Gu, D.; Li, Y.; Chen, K. A novel approach to detect electricity theft based on conv-attentional Transformer Neural Network. *Int. J. Electr. Power Energy Syst.* **2023**, *145*, 108642. [CrossRef]
37. Kim, J.; El-Khamy, M.; Lee, J. T-gsa: Transformer with gaussian-weighted self-attention for speech enhancement. In Proceedings of the ICASSP 2020—2020 IEEE International Conference on Acoustics, Speech and Signal Processing (ICASSP), Barcelona, Spain, 4–8 May 2020; pp. 6649–6653.
38. Nie, Y.; Nguyen, N.H.; Sinthong, P.; Kalagnanam, J. A time series is worth 64 words: Long-term forecasting with transformers. *arXiv* **2022**, arXiv:2211.14730.
39. Finardi, P.; Campiotti, I.; Plensack, G.; de Souza, R.D.; Nogueira, R.; Pinheiro, G.; Lotufo, R. Electricity theft detection with self-attention. *arXiv* **2020**, arXiv:2002.06219.
40. Zhu, Y.; Zhang, Y.; Liu, L.; Liu, Y.; Li, G.; Mao, M.; Lin, L. Hybrid-order representation learning for electricity theft detection. *IEEE Trans. Ind. Inf.* **2022**, *19*, 1248–1259. [CrossRef]
41. Gao, H.X.; Kuenzel, S.; Zhang, X.Y. A hybrid ConvLSTM-based anomaly detection approach for combating energy theft. *IEEE Trans. Instrum. Meas.* **2022**, *71*, 1–10. [CrossRef]
42. Ding, X.; Guo, Y.; Ding, G.; Han, J. ACNet: Strengthening the kernel skeletons for powerful CNN via asymmetric convolution blocks. In Proceedings of the IEEE/CVF International Conference on Computer Vision, Seoul, Republic of Korea, 27 October 2019–2 November 2019; pp. 1911–1920.
43. Ioffe, S.; Szegedy, C. Batch normalization: Accelerating deep network training by reducing internal covariate shift. In Proceedings of the International Conference on Machine Learning, Lille, France, 6–11 July 2015; pp. 448–456.
44. Breiman, L. Random forests. *Mach. Learn.* **2001**, *45*, 5–32. [CrossRef]
45. Dempster, A.; Schmidt, D.F.; Webb, G.I. Minirocket: A very fast (almost) deterministic transform for time series classification. In Proceedings of the 27th ACM SIGKDD Conference on Knowledge Discovery & Data Mining, Virtual, 14–18 August 2021; pp. 248–257.

MDPI
St. Alban-Anlage 66
4052 Basel
Switzerland
www.mdpi.com

Sensors Editorial Office
E-mail: sensors@mdpi.com
www.mdpi.com/journal/sensors

www.ingramcontent.com/pod-product-compliance
Lightning Source LLC
Chambersburg PA
CBHW040821250926
51850CB00007BA/1230
9783036594101